Emerging Sensor Technology in Agriculture

Emerging Sensor Technology in Agriculture

Editors

Sigfredo Fuentes
Carlos Poblete-Echeverria

MDPI • Basel • Beijing • Wuhan • Barcelona • Belgrade • Manchester • Tokyo • Cluj • Tianjin

Editors
Sigfredo Fuentes
University of Melbourne
Australia

Carlos Poblete-Echeverria
Stellenbosch University
South Africa

Editorial Office
MDPI
St. Alban-Anlage 66
4052 Basel, Switzerland

This is a reprint of articles from the Special Issue published online in the open access journal *Sensors* (ISSN 1424-8220) (available at: https://www.mdpi.com/journal/sensors/special_issues/ESTA).

For citation purposes, cite each article independently as indicated on the article page online and as indicated below:

LastName, A.A.; LastName, B.B.; LastName, C.C. Article Title. *Journal Name* **Year**, *Article Number*, Page Range.

ISBN 978-3-03943-613-2 (Hbk)
ISBN 978-3-03943-614-9 (PDF)

Cover image courtesy of Carlos Poblete-Echeverría.

Contents

About the Editors

Sigfredo Fuentes is a professor in Digital Agriculture, Food and Wine Sciences at the University of Melbourne. Previously, he worked at the Universities of Adelaide, Technology, Sydney; Western Sydney (Ph.D.) and Chile. His scientific interests include climate change impacts on agriculture, the development of new computational tools for sensory, animal, plant physiology, food, and wine sciences, new and emerging sensor technology, proximal-, short- and long-range remote sensing using robots and UAVs, machine learning and artificial intelligence. For more information visit: www.vineyardofthefuture.com.

Carlos Poblete-Echeverria is a senior lecturer of Advanced Viticulture and a coordinator of the research group of Digital Viticulture at Stellenbosch University, South Africa. Carlos was working as a lecturer and researcher in Chile for 10 years at the University of Talca and the Catholic University of Valparaiso, in the areas of water management and the use of new technologies. Currently, he is the research leader of the Winetech project "Integrated vineyard monitoring system to improve water management". His research areas are related to estimations of water consumption using models and micrometeorological techniques, climate change, machine vision, and the detection of biotic and abiotic stresses using thermography, spectroscopy, remote sensing, UAVs, and robots. Carlos has published more than 80 scientific and popular articles. For more information, please visit: https://www.researchgate.net/profile/Carlos_Poblete-Echeverria.

 *sensors*

Editorial

Editorial: Special Issue "Emerging Sensor Technology in Agriculture"

Carlos Poblete-Echeverría [1] and Sigfredo Fuentes [2,*]

[1] Department of Viticulture and Oenology, Faculty of AgriSciences, Stellenbosch University, Private Bag X1, Matieland 7602, South Africa; cpe@sun.ac.za

[2] Digital Agriculture, Food and Wine Sciences Group, School of Agriculture and Food, Faculty of Veterinary and Agricultural Sciences, University of Melbourne, Melbourne, VIC 3010, Australia

* Correspondence: sfuentes@unimelb.edu.au; Tel.: +61-4245-04434

Received: 3 July 2020; Accepted: 6 July 2020; Published: 9 July 2020

Research and innovation activities in the area of sensor technology can accelerate the adoption of new and emerging digital tools in the agricultural sector by the implementation of precision farming practices such as remote sensing, operations, and real-time monitoring. The agricultural industry has been greatly affected by climate change; therefore, to be successful in overcoming these effects and remain competitive and sustainable in the market, there is the need to support research and application development of new and emerging sensor technologies and their applications in agriculture. A total of 13 papers were published in this Special Issue entitled: "Emerging Sensor Technology in Agriculture", and the topics addressed include different emerging technologies with applications on ecosystems (grasslands) [1] and several agriculture crops such as peppers [2], apples [2,3], grapevines [2,4–7], cocoa trees [6], citrus [8], legumes [9], wheat and rice [10,11]. Two papers were also related to the use of remote sensing to detect forage quality [9], regions of interest of pigs [12], and pesticide droplet deposition [13] using machine learning.

In Rueda-Ayala et al. [1] an aerial (Unmanned aerial vehicle—UAV) and an on-ground (Kinect sensor—RGB-D (Depth camera)) methods were used to characterize grass ley fields (plant height, biomass, and volume) composed of different species mixtures, using digital grass models. In this study, both methods presented a good performance. From a comparison point of view, the authors took into consideration some basic economic and practical aspects of the methodologies. Hacking et al. [4] used a similar approach to determine yield in grapevines. Another UAV-based study investigated the effect of eddies formed at low altitude in wheat to estimate water status effectively and other physiological parameters in rice [11]. Yield estimation is a key topic in agriculture in general, and it is very relevant in viticulture since winegrowers need such information to manage several logistic aspects at the cellars. In Hacking et al. [4], 2D (RGB images) and 3D (RGB-D) approaches were tested and compared, providing promising results and perspectives in terms of the potential application of these technologies at the vineyard scale (in situ yield estimation). Another interesting use in viticulture was presented in the research of Palacios et al. [5], where they combined computer vision (RGB images) and machine learning for assessing cluster compactness (degree of the aggregation of its berries) under field conditions (system mounted on an all-terrain vehicle). In this study, the bunches were detected and classified to perform the cluster compactness determination using a Gaussian process regression model. The authors highlighted the potential applicability of this method to determine the spatial variability of cluster compactness in commercial vineyards. As was stated by Palacios et al. [5] and Hacking et al. [4], fruit detection is the first mandatory step to perform other calculations. In this regard, Zemmour et al. [2], presented an automatic parameter-tuning procedure for fruit detection. They developed a tuning process to determine the best fit values of the algorithm parameters to enable easy adaption to different kinds of fruits (shapes, colors) and environments (illumination conditions). In this study, the algorithm was tested under challenging conditions in three crops: red apples, green grapes, and sweet yellow peppers. The algorithm presented successfully detected apples and peppers in variable lighting conditions; however, for green grapes, the authors indicated that there is the need to incorporate some additional features such as morphological parameters to improve the detection process. Estimations of the amount of fruit are important for yield predictions, but also for the right moment to harvest them [4]. The study presented by Valente et al. [3] explored the use of a small-sized electrochemical sensor mounted on a UAV for sensing ethylene concentration in an apple orchard. The latter was the first study focused on investigating the feasibility of ethylene-sensitive sensors in a fruit orchard. However, the results are not conclusive for harvest decisions (fruit maturity). This study opens a research area in this field.

As RGB and RGB-D information, temperature is another variable that can be remotely measured to detect some plant conditions, such as water status and stresses (biotic and abiotic). New technologies of infrared sensors/cameras and computational analysis have allowed a faster and accurate characterization of canopy temperatures. Romero-Bravo et al. [10] presented an application of thermography for estimating grain yield and carbon isotope discrimination in wheat genotypes growing under water stress and full irrigation conditions. The results of this study show that the water regime influences the thermal approach, showing better results under water stress conditions. The authors highlighted that more complex models are needed to estimate grain yield and carbon isotopes since the environmental conditions have a strong influence on the temperature profile of the plants.

Bushfires are one of the climatic anomalies that have increased in number, severity, and window of opportunity within agricultural seasons. For grapevines, they present a critical problem due to smoke contamination and smoke taint. Fuentes et al. [7] proposed the first artificial intelligence approach to model smoke contamination in canopies and smoke taint in grapes and wines using non-invasive infrared thermal imagery (IRTI) and near-infrared spectroscopy (NIR), producing highly accurate machine learning models. From the same research group, further applications of remote sensing and machine learning modelling rendered one of the first specific models to assess aroma profiles of cocoa beans for chocolate manufacturing based on canopy architecture profiles at harvest [6]. These two technological developments can assist growers in combatting environmental hazards and predict quality traits of final products.

Mechanization and agricultural management practices can require significant labor and investment that may not necessarily secure efficiency. Mechanical harvesting can be considered a hot topic in agriculture that requires technology to monitor different aspects to increase productivity. A vibration monitoring system for citrus harvesting was proposed and tested to improve fruit detachment frequency with promising results [8]. Other management practices such as pesticide application require accurate monitoring methods to assess efficiency in the distribution of droplets within crop canopies to minimize detrimental effects in the environment and maximize application efficiency. A rapid method to detect spraying deposit was developed based on capacitance sensors [13].

Agriculture involves not only crop production but also animal farming. Digital technologies have been applied in recent years to monitor the quality of animal feed and to detect animals in order to extract information from remote sensing systems that can provide information on physiological stresses and the general welfare of the animals. One paper researched the use of NIR to predict forage quality of warm-season legumes using machine learning modelling with high accuracy [9]. For animals, a different region of interest from pig bodies was successfully detected using convolutional network deep learning techniques, which may allow for more efficient extraction of information from animals to identify biotic or abiotic stress-related problems [12].

The diversity in applications within this Special Issue makes evident the importance of novel research on new and emerging technologies for the agricultural industry.

Acknowledgments: The guest editors would like to extend their gratitude to all the authors who contributed to this Special Issue and to the reviewers who dedicated their valuable time providing the authors with critical and constructive recommendations.

Conflicts of Interest: The guest editors declare no conflict of interest.

References

1. Rueda-Ayala, V.P.; Peña, J.M.; Höglind, M.; Bengochea-Guevara, J.M.; Andújar, D. Comparing UAV-based technologies and RGB-D reconstruction methods for plant height and biomass monitoring on grass ley. *Sensors* **2019**, *19*, 535. [CrossRef] [PubMed]
2. Zemmour, E.; Kurtser, P.; Edan, Y. Automatic parameter tuning for adaptive thresholding in fruit detection. *Sensors* **2019**, *19*, 2130. [CrossRef] [PubMed]
3. Valente, J.; Almeida, R.; Kooistra, L. A Comprehensive Study of the Potential Application of Flying Ethylene-Sensitive Sensors for Ripeness Detection in Apple Orchards. *Sensors* **2019**, *19*, 372. [CrossRef] [PubMed]
4. Hacking, C.; Poona, N.; Manzan, N.; Poblete-Echeverría, C. Investigating 2-d and 3-d proximal remote sensing techniques for vineyard yield estimation. *Sensors* **2019**, *19*, 3652. [CrossRef] [PubMed]
5. Palacios, F.; Diago, M.P.; Tardaguila, J. A Non-Invasive Method Based on Computer Vision for Grapevine Cluster Compactness Assessment Using a Mobile Sensing Platform under Field Conditions. *Sensors* **2019**, *19*, 3799. [CrossRef] [PubMed]
6. Fuentes, S.; Chacon, G.; Torrico, D.D.; Zarate, A.; Gonzalez Viejo, C. Spatial variability of aroma profiles of cocoa trees obtained through computer vision and machine learning modelling: A cover photography and high spatial remote sensing application. *Sensors* **2019**, *19*, 3054. [CrossRef] [PubMed]
7. Fuentes, S.; Tongson, E.J.; De Bei, R.; Gonzalez Viejo, C.; Ristic, R.; Tyerman, S.; Wilkinson, K. Non-invasive tools to detect smoke contamination in grapevine canopies, berries and wine: A remote sensing and machine learning modeling approach. *Sensors* **2019**, *19*, 3335. [CrossRef] [PubMed]
8. Castro-Garcia, S.; Aragon-Rodriguez, F.; Sola-Guirado, R.R.; Serrano, A.J.; Soria-Olivas, E.; Gil-Ribes, J.A. Vibration Monitoring of the Mechanical Harvesting of Citrus to Improve Fruit Detachment Efficiency. *Sensors* **2019**, *19*, 1760. [CrossRef] [PubMed]
9. Baath, G.S.; Baath, H.K.; Gowda, P.H.; Thomas, J.P.; Northup, B.K.; Rao, S.C.; Singh, H. Predicting Forage Quality of Warm-Season Legumes by Near Infrared Spectroscopy Coupled with Machine Learning Techniques. *Sensors* **2020**, *20*, 867. [CrossRef] [PubMed]
10. Romero-Bravo, S.; Méndez-Espinoza, A.M.; Garriga, M.; Estrada, F.; Escobar, A.; González-Martinez, L.; Poblete-Echeverría, C.; Sepulveda, D.; Matus, I.; Castillo, D. Thermal imaging reliability for estimating grain yield and carbon isotope discrimination in wheat genotypes: Importance of the environmental conditions. *Sensors* **2019**, *19*, 2676. [CrossRef] [PubMed]
11. Yao, L.; Wang, Q.; Yang, J.; Zhang, Y.; Zhu, Y.; Cao, W.; Ni, J. UAV-borne dual-band sensor method for monitoring physiological crop status. *Sensors* **2019**, *19*, 816. [CrossRef] [PubMed]
12. Psota, E.T.; Mittek, M.; Pérez, L.C.; Schmidt, T.; Mote, B. Multi-pig part detection and association with a fully-convolutional network. *Sensors* **2019**, *19*, 852. [CrossRef] [PubMed]
13. Wang, P.; Yu, W.; Ou, M.; Gong, C.; Jia, W. Monitoring of the pesticide droplet deposition with a novel capacitance sensor. *Sensors* **2019**, *19*, 537. [CrossRef] [PubMed]

Article

Comparing UAV-Based Technologies and RGB-D Reconstruction Methods for Plant Height and Biomass Monitoring on Grass Ley

Victor P. Rueda-Ayala [1,†], José M. Peña [2], Mats Höglind [1], José M. Bengochea-Guevara [3], Dionisio Andújar [3,*,†]

[1] Department of Grassland and Livestock, Norwegian Institute of Bioeconomy Research, NIBIO Særheim, Postvegen 213, 4353 Klepp Stasjon, Norway; patovicnsf@gmail.com (V.P.R.-A.); mats.hoglind@nibio.no (M.H.)
[2] Institute of Agricultural Sciences, Consejo Superior Investigaciones Científicas (CSIC), Serrano 115b, 28006 Madrid, Spain; jmpena@ica.csic.es
[3] Centre for Automation and Robotics, Consejo Superior Investigaciones Científicas (CSIC), Ctra. de Campo Real km 0.200 La Poveda, 28500 Arganda del Rey (Madrid), Spain; jose.bengochea@car.upm-csic.es
* Correspondence: d.andujar@csic.es
† These authors contributed equally to this work.

Received: 10 December 2018; Accepted: 23 January 2019; Published: 28 January 2019

Abstract: Pastures are botanically diverse and difficult to characterize. Digital modeling of pasture biomass and quality by non-destructive methods can provide highly valuable support for decision-making. This study aimed to evaluate aerial and on-ground methods to characterize grass ley fields, estimating plant height, biomass and volume, using digital grass models. Two fields were sampled, one timothy-dominant and the other ryegrass-dominant. Both sensing systems allowed estimation of biomass, volume and plant height, which were compared with ground truth, also taking into consideration basic economical aspects. To obtain ground-truth data for validation, 10 plots of 1 m^2 were manually and destructively sampled on each field. The studied systems differed in data resolution, thus in estimation capability. There was a reasonably good agreement between the UAV-based, the RGB-D-based estimates and the manual height measurements on both fields. RGB-D-based estimation correlated well with ground truth of plant height ($R^2 > 0.80$) for both fields, and with dry biomass ($R^2 = 0.88$), only for the timothy field. RGB-D-based estimation of plant volume for ryegrass showed a high agreement ($R^2 = 0.87$). The UAV-based system showed a weaker estimation capability for plant height and dry biomass ($R^2 < 0.6$). UAV-systems are more affordable, easier to operate and can cover a larger surface. On-ground techniques with RGB-D cameras can produce highly detailed models, but with more variable results than UAV-based models. On-ground RGB-D data can be effectively analysed with open source software, which is a cost reduction advantage, compared with aerial image analysis. Since the resolution for agricultural operations does not need fine identification the end-details of the grass plants, the use of aerial platforms could result a better option in grasslands.

Keywords: 3D crop modeling; remote sensing; on-ground sensing; depth images; parameter acquisition

---

1. Introduction

Pastures are botanically diverse and difficult to characterize, due to their complex species composition. A good characterization of a forage crop parameters is crucial for successful grassland management [1]. Technological advancement and current sensing technologies are powerful tools for elaborating accurate plant architecture models for phenotyping [2,3]. Digital modeling can be used to detect environmental stress problems, diseases, or the necessity of applying agricultural operations, at the right location and timing. Sensor data could be acquired throughout the whole plant life cycle and be available for model development and validation [4]. Spatial and temporal crop parameter information, for instance biomass and nitrogen content [5], add up value to vegetation models [6], strengthening decision-support systems for site-specific agronomic applications. Because perennial crops have a great biomass building potential, continuous supervision of crop development parameters along their life cycle is recommended [7]. Continuous supervision by means of spatial models could improve decision-making for forage grass production. Moreover, spatial models facilitate estimation of above-ground biomass, canopy height or plant cover in a non-destructive manner, which allow better programming of specific tasks, such as cutting time, fertilization and grasslands renewal [8].

Spatial vegetation models are based on data from digital imaging, spectrometry, fluorescence, thermal or distance measurements, which relate to some plant traits. Spectral reflectance of plant leaves, ranging from ultraviolet (UV), through visible light and near-infrared (NIR) and infrared (IR) wavelengths have been found to be particularly important for calculation of various vegetation indices [9,10]. Vegetation indices often correlate with leaf area index (LAI), biomass or dry matter yield [11]. Biomass estimation using the normalized difference vegetation index (NDVI) has given good results on annual pastures under grazing, although, poor data quality caused large estimation discrepancies on grazed or partially grazed paddoks [12]. Ground-based and aerial visible imaging data acquired at a specific time and the use of algorithms able to segment the RGB spectrum have been proposed for quick and simple description of plant growing dynamics [5,13,14]. However, data assessment from RGB images is limited in some aspects, such as leaf overlapping, that can make important parts of the plant difficult to detect, especially in grass mixtures. Distance sensors can measure distances by different principles (e.g., time-of-flight) and enable estimation of plant height or derive biomass weight by indirect relationships with height [15,16]. Distance sensors, which are normally divided into ultrasonic devices and LiDAR (Light Detection and Ranging), have been widely applied in modern agriculture operations [17–20]. Because they are easy to handle, these sensors can be used to assess big field areas in short time.

The use of 3D technologies from on-ground or aerial platforms open new scenarios for plant modeling. Characterization of plants with the aid of 3D models is available for use in breeding programs and agricultural decision making [8,21]. Various processes are available for capturing the three dimensions, height, width and depth, as 3D point clouds with X-Y-Z coordinates. The most explored, fastest and accurate 3D sensing system is LiDAR combined with sequential displacement of the sensor to acquire the Z coordinate [22]. A drawback of this system is the requirement of calibration and displacement across the sampling space, which increases the associated costs, as the resolution increases [23]. Fortunately, RGB-Depth (RGB-D) cameras and image processing based on Time-of-Flight can compensate those drawbacks by combining depth information with the color scene in a single shoot [17]. RGB-D cameras have been used for several agricultural research and application purposes. The most common is Microsoft Kinect® v2, which allows reconstruction of 3D models, associated with color information. Microsoft launched the fist version of this development in 2010 with the Kinect, and since then, several other devices have appeared in the market: RGB-D cameras, such as the Intel RealSense, Primesense Carmine, Google Tango, or Occiptial's Structure Sensor. These sensors are available at low price, and can capture pixel color and depth images at adequate resolution and at a high rate. Due to the similar output produced by those, the reconstruction method can be easily replicated. This methodology has been successfully applied on many crops, except on grasslands. Wang and Li [24] calculated onion volume with a high accuracy, compared to real measurements. Foliar density of threes was estimated for autonomous spraying [25]. Andújar et al. [17] used a dual methodology separating crops and weeds from soil in maize crops under field conditions. The latter methodology included height selection and RGB segmentation, using a unique model for plant discrimination. Combination of various frames allows reconstruction of big crop surface areas [21,26]. Live use of RGB-D on outdoors scenarios is possible with the current version of Kinect® v2.

UAV's can cover large areas and operate independently of soil conditions [27], which allows more flexibility than ground-based systems, at reduced operational time and costs. Photogrammetry on aerial imagery has shown a high functionality in different studies. High spatial resolution images can be obtained when flying at low altitudes, with large overlapping between images. The data can be processed through Structure-from-Motion reconstruction for building the 3D model. This method has been tested in olive trees to calculate canopy area, tree height and crown volume by generation of digital surface models and OBIA algorithm [28]. Hyperspectral aerial imagery can be used to calculate plant height and values related with dry biomass [29]. Nevertheless, this technology is rapidly improving for application in complex grassland scenarios [30].

Current challenges in both, agricultural research and production rely on sensing devices and technological advancement directed to improve crop quality and increment yield levels. As for other crop producers, forage farmers can immensely benefit from advanced technological support of digital grass modeling, to enhance forage productivity. Digital models could objectively deal with the complexity of grass-mixtures, and assist in the optimization of inputs, leading to better distribution or reduction of fertilizers, pesticides or seeds, e.g., by site-specific fertilization and renewing grass mixtures. In addition, some grassland farming activities in the field depend on biomass estimation to evaluate productivity, normally done via destructive methods (i.e., in this study referring to cutting numerous grass samples). Therefore, this study was carried out with the aim of evaluating aerial and on-ground methods to characterize grass ley fields, composed of different species mixtures. Specifically, it was attempted to objectively estimate plant height, biomass and volume, using digital grass models, and avoiding the unnecessary destruction of the swards.

2. Materials and Methods

2.1. Experiments and Modeling Systems

Two digital characterization systems, a on-ground and a UAV-based system, were used to map pasture architecture on two fields located at NIBIO Særheim research station (Klepp Stasjon, Norway, lat. 58.76 N, long. 5.65 E). The site is characterized by a cold maritime climate with cool summers and cold winters. The precipitation is about 1180 mm annually, especially in autumn and spring. The on-ground system used a RGB-D camera, while the UAV-based sytem used a RGB camera with geo-positioning (geo-tagging) for data acquisition. These systems were tested and compared previously on a small area [31], at the same location. Two fields, each of 0.5 ha in size, were mapped. To obtain ground-truth data for validation, 10 plots of 1 m^2 were sampled on each field (Figure 1a). Each plot was subdivided in four quadrants for measuring the variability within the 1 m^2 area (Figure 1c). Field 1 (ryegrass dominant) was composed of 80% perennial ryegrass (*Lolium perenne* L.), 5% annual ryegrass *Lolium multiflorum* L. and 15% white clover (*Trifolium repens* L.). Field 2 (timothy dominant) was composed of 85% timothy (*Phleum pratense* L.), 10% perennial ryegrass, and 5% annual ryegrass. Both fields were established in 2015. From 2016, they were fertilized annually with 10, 8 and 6 tons of liquid manure at early spring (March–April), after first cut (June–July) and after second cut (August–September), respectively, corresponding to about 260 kg ha^{-1} year^{-1}. Field assessments were conducted during July–August 2017 when the swards were fully developed, at anthesis stage.

Figure 1. Field test conducted at NIBIO Særheim, orthophoto of the ryegrass-dominant field (**a**) with 10 sampling plots and a zoomed in 1 m^2 sampling plot subdivided in four quadrants (**b**); UAV sampling system (**c**) and RGB-D sampling system (**d**).

For the on-ground system the RGB-D Microsoft Kinect® v2 (Microsoft, Redmond, WA, USA) was used, as described by Andújar et al. [17]. Kinect V2 is the most widely used among RGB-D sensors. Although the device is no longer supported by Microsoft, its capabilities are similar to any other option in the market. In addition, readings sensors of this type show a common output, and the processing methodology is similar. The device is equipped with a standard RGB camera of 1080p, a depth camera, an infrared camera and an array of microphones. The RGB camera has a resolution of 1920 × 1080, which can adapt automatically the exposure time of the RBG to obtain brighter images at limiting light conditions. The IR camera can take a clear view into the darkness with a resolution of 512 × 424 pixels. The opening field of view (FOV) is different for every camera. The IR camera has a FOV of 70 degrees horizontally and depth perception is limited to 60 degrees vertically. The range of depth that can be measured with this camera goes from 0.5 to 4.5 m of distance from the sensor, although in outdoors conditions, the maximum range decreases. Studies conducted outdoors under different daytime illumination conditions showed valid depth measurements up to 1.9 m during sunny days, while the distance increases up to 2.8 m under the diffuse illumination of an overcast day [17]. The req uired overlap to fuse the acquired images and create the models is reached by a frame rate than can be set up to 30 fps during data acquisition. The distance is calculated for every pixel in the scene by the method of Time-of-Flight method by phase detection, i.e., the distance is calculated based on the time that a pulse of light takes to travel from the light source to the impacted plant and back to the sensor.

An Intel laptop computer with Windows 8 supported by Kinect SDK (software development kid) was used for data collection. The SDK helps acquiring data by classes, functions and structures, providing the necessary drivers for the sensor, and some sample functions that were implemented for the measurements combined with some OpenCV (The Open Source Computer Vision https://www.opencv.org/). The sensor was hand held pointing out the field samples from top view. The developed method for point cloud generation and reconstruction of large regions, using the fusion of different overlapped depth images was based on a previous development [32]. Storing information only on the voxels closest to the detected object and accessing to the stored information by using a hash table. Following that, for every new input depth image and knowing camera position, the ray-casting technique [33] was applied to project a ray from the camera focus for each pixel of the input depth image to determine the voxels in the 3D world crossing each ray. Then, the voxels related to the depth information are determined. Next step was conducted with a variant of the iterative closest point (ICP) algorithm, which provides a point cloud as output. Thus, the modified algorithm creates a point cloud by detecting the overlapping areas in sequential frames by assessing the relative position of the Kinect sensor for each frame to create a 3D model and removing outliers from the mesh [26]. Outliers could appear isolated in the point cloud. A point was considered an outlier if the average distance to its 64 nearest neighbours is greater than the standard deviation of the distance to the neighbours of all the points (Figure 2). The time to complete the acquisition was lower than 2 s from the top view. The system was supplied with electric power by a field vehicle that allows field measurement and support every device used during the acquisition process.

Figure 2. Section of the 3D reconstruction: before filtering (**a**); removed points are marked in fluorescent (**b**) and after filtering (**c**).

One 3D model was build on the sward before harvesting and one inmmediately after. Once those 3D representations of the sampled plots were available (Figure 3), plant height and volumes could be estimated. For this purpose, both models were overlayed and plant height was estimated by difference between the two models, using cloudcompare. Firstly, an alpha shape [34] or volume that enveloped the set of 3D points was obtained. The alpha parameter specifies how tight the body fits the points. To address this issue, the R package alphashape3d [35] was employed. Figure 4 shows different alpha shapes according to the alpha value selected for the same point cloud. Higher values showed very loose shapes, whereas lower values generated tight bodies. The volume was estimated by applying the same function library that allows calculation of the alpha shapes.

Figure 3. Point clouds created by RGB-D (Microsoft Kinect® v2) system.

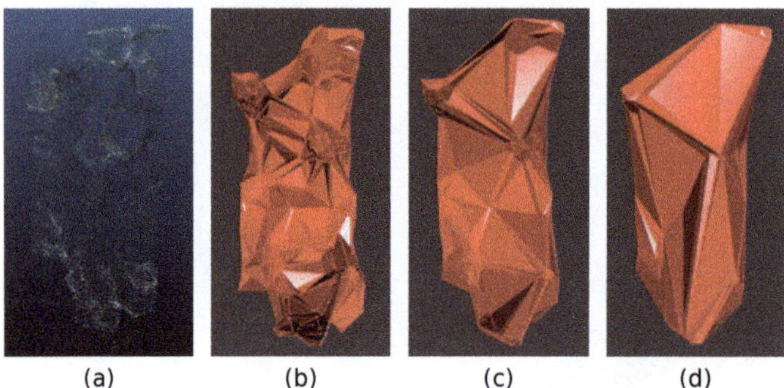

(a)   (b)   (c)   (d)

Figure 4. Alpha shapes for the same point cloud using alpha = 0.1 (**a**) and (**b**), alpha = 0.2 (**c**) and alpha = 0.4 (**d**).

The aerial system consisted of a DJI Mavic Pro quadcopter, combining a 4K digital camera and location information, was used for aerial imaging. The camera mounted on the UAV had a 28 mm lens with a Field of Fiew (FOV) of 78.8 degrees and a resolution of 4000 × 3000, capable of shooting 12.35 megapixel photos; the camera was 3-axis stabilized by its drone's gimbal (https://www.dji.com). The acquired aerial imagery was tested and compared with the RGB-D on-ground system. The UAV flew autonomously following the programed route by an internal GPS receiver using Litchi APP. The route was set up to take images at an interval of 1 s, creating minimum overlaps of 90% forward and 60% sidewards, at 30 m of flying altitude, and ensuring a necessary overlapping between images for photogrammetry post-processing, mosaicking and Digital Surface Model (DSM) generation.

Agisoft PhotoScan Professional Edition (Agisoft LLC, St. Petersburg, Russia) version 1.0.4 was used for 3D model building. This software provides a fully automatic process for image alignment, building field geometry, and orthophoto generation. Quality analysis of all acquired images was done with this software, and images with a value higher that 0.7 were used to reconstruct the DSMs by photogrammetry process. The whole process was fully automatic, except for the manual location of reference points used to correct the model. The model building included several phases: acquisition of very high spatial resolution images with the UAV, and importing them into the software; image alignment; building field geometry by applying close-range photogrammetry methods; dense point cloud generation; application of advanced image analysis to extract the selected geometric features. After that, common points and camera position for each image were located and matched to ensure the refinement of the camera calibration parameters. Then, the software searches for more points in the images to create a dense 3D point cloud, followed by the creation of 3D polygon mesh, from which the final model was generated (Figure 5a).

The DSM and orthomosaics were joined to create a 4-band multi-layer file, i.e., RGB bands and DSM. This file was processed using an OBIA algorithm developed with the eCognition Developer 9 software (Trimble GeoSpatial, Munich, Germany). The software tool for image segmentation and classification applies the multiresolution algorithm Otsu's of automatic thresholding. 3D features (volume) were calculated by integrating the volume of the individual pixels below the top of the crop as a solid generated object (Figure 5b,c, respectively) [36]. This technique has been successfully applied in UAV images both in agriculture and grassland, as well as urban areas and forestry. A desktop computer equipped with an Intel Core i7-4771@3.5 GHz processor, 16 GB of RAM, and NVIDIA GeForce GTX 660 graphic card was used for image processing and 3D modeling.

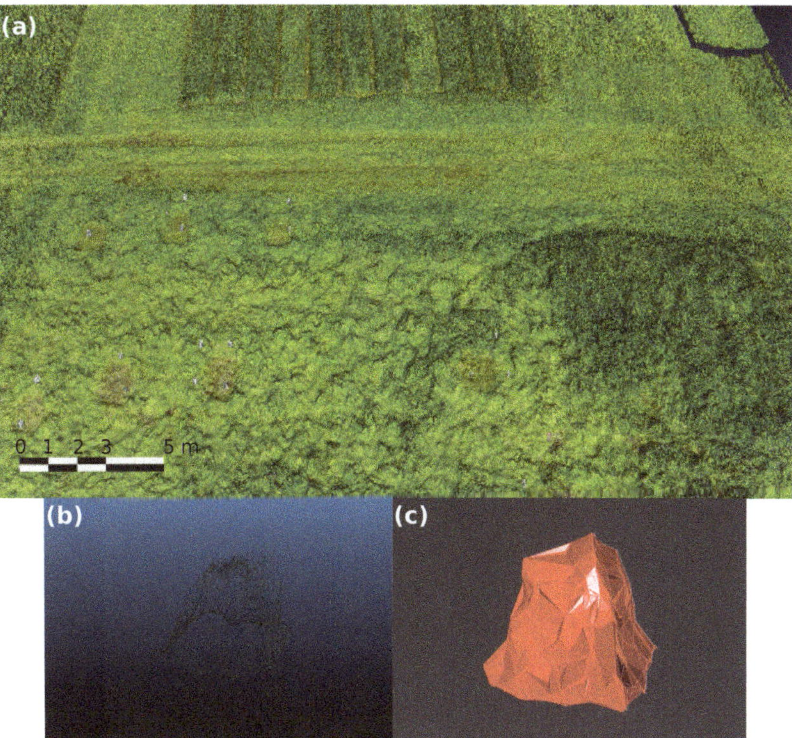

Figure 5. Model constructed by photogrammetry methods (**a**) and processes of point cloud of DSM model (**b**) and solid generation (**c**).

After sensor data acquisition, actual height of every sampled plot was determined with the aid of a measuring tape, on the four quadrants plus the center of each plot. Additionally, the compressed sward height was determined using a rising plate meter, which represented the average height at each sampling plot. The compressed height is used by pasture managers as an indicator of the herbage yield, for decision support. Thereafter, all plants inside in the sampling plot were cut at ground level, then oven-dried at 80 degrees Celsius during 48 h, and finally the dry biomass was measured. The calculated ground-truth data was compared with that extracted from 3D models. From the Kinect-based models, plant volume, maximum height, average height and cover area were extracted.

2.2. Statistical Analysis

Actual field measurements of plant height and dry biomass were compared with the RGB-D-based and UAV-based 3D models assessments, within each field. Simple linear regression were the tested on all relationships, using the Pearson's correlation and R^2 coefficients, with their corresponding standard errors in the evaluation for best fit. Differences between both assessed fields were determined through Anova and subsequent lack-of-fit tests for linear regression models.

3. Results and Discussion

Plant Height, Volume and Biomass

The studied sampling systems differed in data resolution, thus differences were also visible for the estimation capability. Accurate measurement of plant height and volume in pastures is difficult, because single grass plants vary enormously in height, even within areas as small as 0.25 m^2. Measurements of compressed sward height with a raising plate-meter or of undisturbed sward height with a measuring tape disregard such variation, as well. The former bends down the largest leaves to a 'common height' (comparable to an average value) at which all grass plant tips support the plate weight. Similarly, using a measuring tape is based on an 'average height', but determined visually. Despite this difficulty, there was a reasonably good agreement the RGB-D-based estimates and the manual height measurements on both fields. These relationships were stronger for the averaged estimates by 1 m^2 sampling plots, and to a lesser extend for the measurements at the four quadrants. Poor quality of UAV-images resulted due to difficult weather conditions for flying, typicall from the south-wester Norwegian region were this study was carried out. Consequently, UAV-based plant heights could not be reliably estimated, but volume was used instead to evaluate the system. The volume was calculated from the 3D surface, using the alpha shapes (Figure 4). An alpha shape represented the outline surrounding a set of 3D points. Spheres of radius alpha, which did not contain any point inside were generated, and in turn, their surfaces were in contact with more than one point. After connecting those points with the ones of the nearest spheres, the surrounding outline was made, generating the volume. The alpha parameter specified how tight the outline to the points was. Although the height measurement could be done, the exact positioning within the frame in the model was difficult to locate. The height varied significantly as the plot was positioned, resulting in false measurements, consequently, height values were avoided as validation information. The actual plant height (raising plate-meter) averaged for the 20 sampled plots was 49.37 cm while UAV measurements underestimated on average 6.18 cm on the 20 reconstructed models.

RGB-D height assessments by quadrant in sampling plots showed a good linear relation between the measured heights at each quadrant, with $R^2 = 0.88$ for field 1 and $R^2 = 0.81$ for field 2 (Figure 6a). This relationship improved greatly when the assessments were averaged per sampling plot $R^2 = 0.98$ and 0.99, respectively for fields 1 and 2 (Figure 6b). Although end-details of grass plant leaves were difficult to reconstruct in the model, the RGB-D-based system showed its powerful capability to estimate accurate height measurements, which is suitable even in small sampling areas, like those in the present study. Field measurements with a raising plate-meter or using a measuring tape also disregard such end-details. Similarly, UAV missions could not reconstruct the end-details. However, good relationship agreements have been found between UAV-estimated heights and ground-truth in other studies [1,37].

UAV-based plant height estimation generally offers two major advantages over the on-ground technologies. UAV can be properly defined as non-destructive monitoring method, and UAVs can cover huge areas in short time. Technically, the use of on-ground measurements should not be considered fully non-destructive when a whole field is to be scanned. Driving field vehicles to carry out the assessments would lead to a high sward biomass destruction, because of the absence of appropriate sampling pathways across the grass field. On the other hand, on-ground monitoring is more time consuming and of a higher operational cost than aerial inspections with UAVs. These type of results was found using multi-temporal crop surface models derived from UAVs and from a terrestrial laser scanner (TLS) [37]. The crop density was well related with the 3D model reconstructions, but differences between the studied methods were evident. Comparable to our results, UAV-derived plant height was generally lower than TLS estimations at all growth stages. However, the coefficient of variation was expected to be higher for the TLS than in those models created from UAV data [37]. Furthermore, Bareth and Schellberg [1] showed the temporal stability of UAV measurements in grassland fields using Structure from Motion and Multi-view Stereopsis techniques, reaching an overall agreement of $R^2 = 0.86$ between rising plate meter heights and model estimations.

RGB-D-based grass heights correlated poorly with actual dry biomass on the ryegrass field (Figure 7a, left). This results indicate that plant height is a weak proxy for grass biomass on ryegrass dominant pastures. Conversely, the correlation between RGB-D estimated heights and measured biomass on the timothy dominant field showed a much better fit, with an $R^2 = 0.88$ (Figure 7a, right). These differing results may be explained by the different growth habit of the two species. In the studied fields, timothy built biomass primarily by growing tall, whereas ryegrass built biomass only partly by growing tall, but more by tillering and development of biomass close to ground. However, the models created from the UAV system showed more stability regarding the relationship between dry biomass and plant heights. Equivalently to height measurements, aerial models provide a baseline to avoid the noise caused by some leaves or steams above the average coverage.

In general, the RGB-D sward height estimates were slightly lower than assessments with measuring tape and raising plate-meter heights. UAV systems seem to offer highly reliable assessments, closer to the reality. Nevertheless, capturing fine details on grass plants, such as tips of leaves, would require low flying heights, increasing the amount of images to be acquired and enormous computational power needed for the corresponding analysis. These aspects would therefore, increment the risk of disturbing the estimation results. This effect was more common in UAV flights. The use of on-ground methods could improve the models in some breading programs when high fidelity is demanded, keeping in mind that these type of agronomic applications need fast and higher scanning capacity with non-destructive methods.

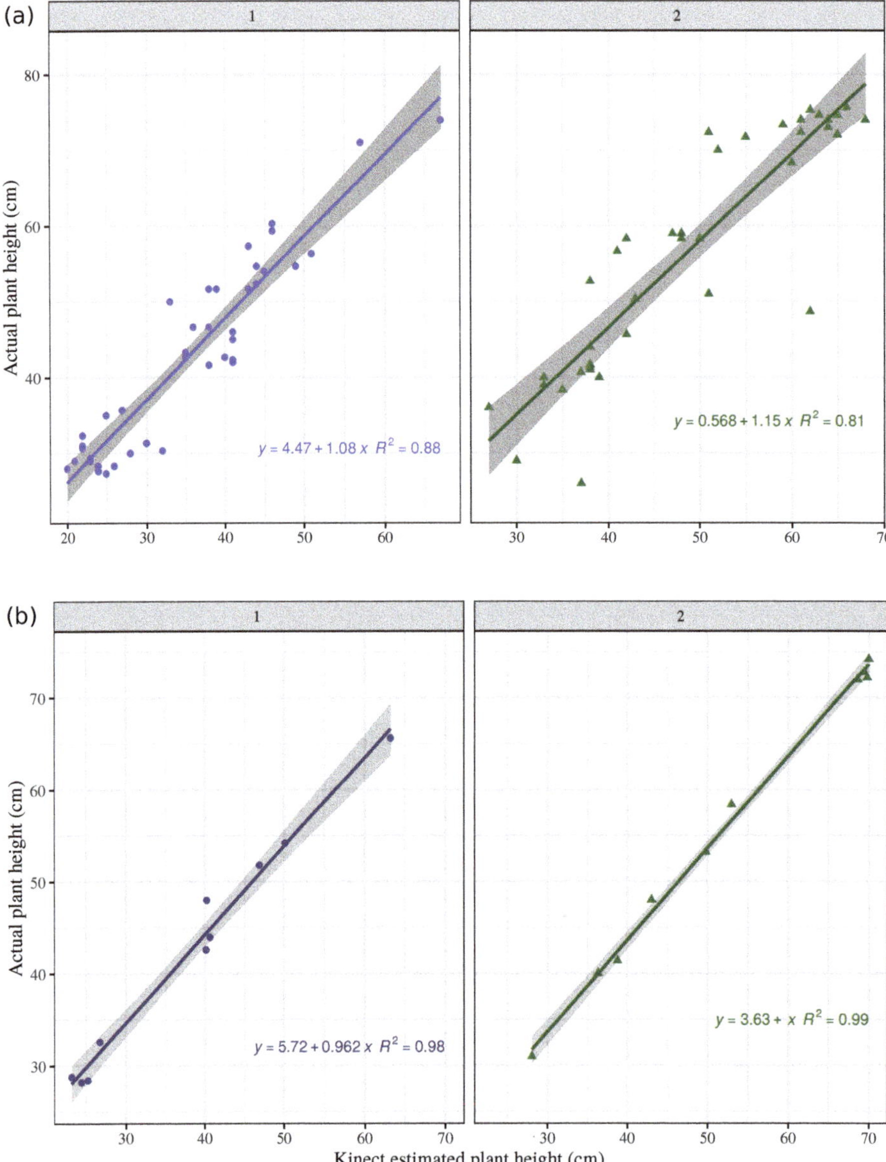

Figure 6. RGB-D estimated grass height compared with field measurements on all four quadrants per sampling plot (**a**), and raising plate-meter height per sampling plot (**b**), on fields 1 and 2. Shadow indicates upper and lower confidence limits.

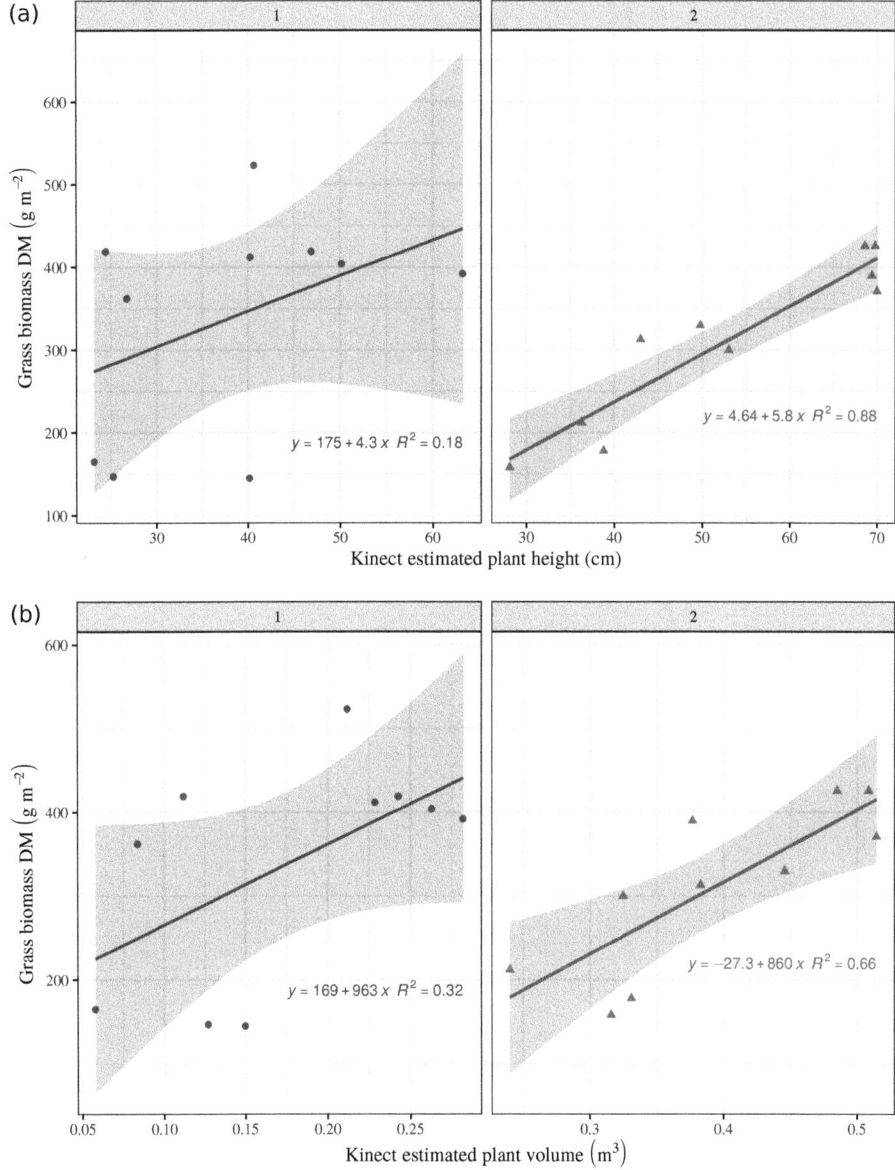

Figure 7. RGB-D estimated grass height (**a**) and volume (**b**) compared with dry biomass per sampling plot, on field 1 and 2. Shadow indicates upper and lower confidence limits.

RGB-D-based volume estimates showed low and intermediate correlation with the assessed grass biomass in fields 1 and 2, with $R^2 = 0.32$ and 0.66, respectively (Figure 7b). Apparently, the higher content of leaves in ryegrass (Figure 7b, left), contributes more to biomass than to the visible and measurable plant volume to which plant height contributes more than plant density. Conversely, biomass on the timothy dominant pasture (Figure 7b, right), corresponded better with the RGB-D-based volume estimates, as it built yield primarily by growing tall. This same trend was observed comparing the actual measured data of plant height and biomass produced, where this relationship was rather poor for ryegrass (Figure 8, left), while it was good for timothy (Figure 8, right).

A different tendency was observed for aerial models. Volume estimated with the UAV-system on the 20 reconstructed plots had a mean value of 0.39 m^3 and a standard deviation on 0.17 m^3 (min = 0.15; max = 0.67). The created models showed an intermediate agreement between the assessed grass biomass and the calculated volume, with an $R^2 = 0.54$. This result shows good capabilities of this method for volume calculation. The developed models showed an irregular shape of the different plots (Figure 5a) and the typical corridors in the experiment. Thus, the accuracy of this method is high and only a few centimeters were underestimated. In the models is also observed that the procedure for 3-D reconstruction was more problematic on areas with a low canopy density. An analogous problem was found in tree reconstruction of orchards, when visible-light images were used. 3-D structure of some of the trees was not properly built, consequently, the mosaicked images showed some blurry areas [28].

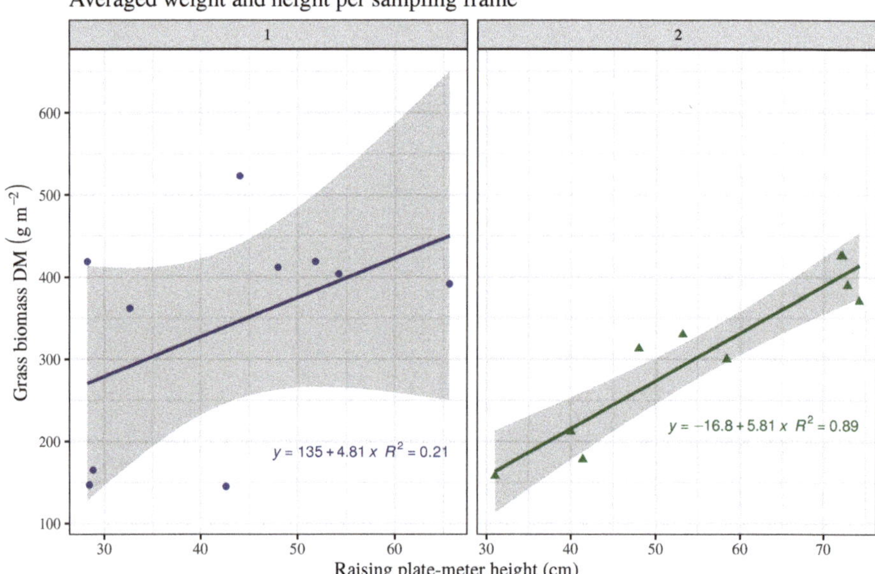

Figure 8. Actual plant height (raising plate-meter) averaged by plot compared with average dry biomass. Shadow indicates upper and lower confidence limits.

The estimated grass volume and the rising plate meter height (average height) per plot showed identical values for both, aerial and on-ground methods. Plant volume estimated with UAV system correlated somewhat low with plant dry biomass ($R^2 = 0.54$, Figure 9a). The aerial model showed a correlation between both values of $R^2 = 0.57$ (Figure Figure 9b). RGB-D estimated plant volume for the ryegrass dominant pasture showed a high correlation ($R^2 = 0.87$) with plant height measured and averaged per plot (Figure 10, left), but just an intermediate correlation ($R^2 = 0.6$) for the timothy pasture (Figure 10, right). The timothy pasture showed much more variability in height among individual tillers within the 0.25 m^2 quadrants (Figure 6), which may explain the low correlation with estimated plant volume for this species. Even though ryegrass plants had a higher number of leaves occupying more volume than timothy per unit area, their leaves bent almost uniformly to a common plant height, which was better measured by the RGB-D system. This fact have been shown in similar studies. A good plant volume estimation using UAV-based image analysis was showed for small weeds [36]. In addition, the combination with multispectral images could improve the results. Estimating above-ground biomass helped monitor crops to predict yield in cereals [29]. The method was proven to be reliable in several scenarios, for instance, relating model biomass estimations to crop nitrogen requirements.

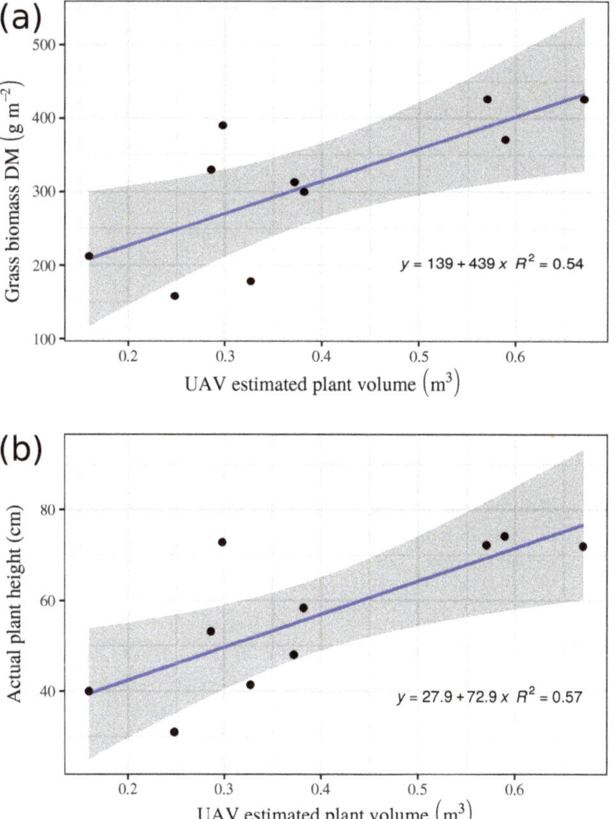

Figure 9. UAV estimated grass volume compared with measured dry biomass (**a**) and with raising plate-meter height (**b**) averaged per sampling plot, on field 1 and 2. Shadow indicates upper and lower confidence limits.

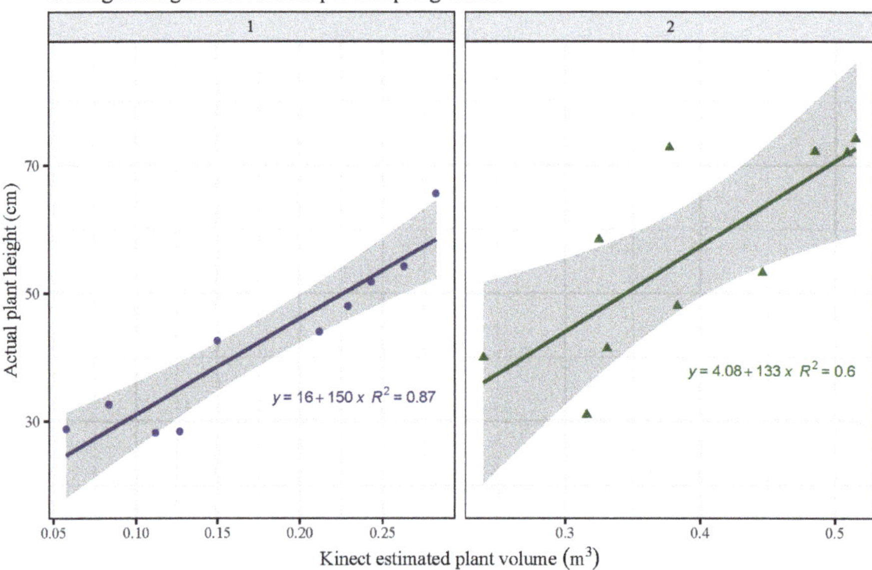

Figure 10. RGB-D estimated grass volume compared with raising plate-meter height averaged per sampling plot, on field 1 and 2. Shadow indicates upper and lower confidence limits.

Comparing costs of UAV-based with RGB-D-based systems, considerable differences exist. It has been argued that cost of aerial imaging is lower and can cover bigger areas [38]. The advantage of using UAV-based sampling was notorious in our study, where whole-field coverage could be achieved in less than 12 min. Contrarily, the RGB-D-based system needed considerably more time for all sampling plots. However, the RGB-D-based system in grassland production could be mounted on a tractor and monitoring can be done simultaneously with other agronomic operations, e.g., fertilization or reseeding, thus diminishing the cost.

4. Conclusions

The use of UAV-based sampling systems offer a higher operative capability, being also affordable from an economic point of view. This system is more affordable, easier to operate and can cover a larger surface than on-ground systems. Since the resolution for agricultural operations does not need fine identification the end-details of the grass plants (i.e., tips of leaves), the use of aerial platforms could result a better option in grasslands. However, resolution of UAV acquired imagery is affected by other conditions external to the camera sensor, such as sunlight, clouds, wind speed and climate, which also affect the imagery resolution and thus the models for parameter estimations. Conversely, on-ground techniques with RGB-D cameras can produce highly detailed models. Nevertheless, far from higher fidelity models, the results showed more variability than UAV models. Increasing speeds for on-ground platforms would improve the performance of these systems, to monitor more area. On-ground RGB-D data can be effectively analysed with open source software, as it was done in this study, which may compensate and challenge the expenses, compared with aerial sampling. However, this technique can be destructive in pasture scenarios. Although not part of this study, the use of on-ground reconstruction method could be more reliable for row-crops or breeding programs. Particularly, the inclusion of depth information in vegetation models, could contribute to improve the results in breading programs.

Author Contributions: Conceptualization, V.P.R.-A. and D.A.; methodology, V.P.R.-A., D.A. and J.M.P.; software, J.M.B.-G.; validation, V.P.R.-A., D.A. and J.P.; formal analysis, D.A., V.P.R.-A., J.M.P. and J.M.B.-G.; investigation, V.P.R.-A., and D.A.; resources, V.P.R.-A., and D.A.; data curation, D.A. and V.P.R.-A.; writing—original draft preparation, D.A. and V.P.R.-A.; writing—review and editing, D.A., V.P.R.-A., J.P. and M.H.; visualization, D.A. and V.P.R.-A.; supervision, D.A. and V.P.R.-A.; project administration, D.A. and V.P.R.-A.; funding acquisition, V.P.R.-A., D.A. and M.H.

Funding: This research was funded by the projects AGL2017-83325-C4-1-R and AGL2017-83325-C4-3-R (Spanish Ministry of Economy and Competition); the RYC-2016-20355 agreement, Spain, as well as, by the Norwegian research funding for agriculture and the food industry (NRF), project 255245 (FOREFF) and the Department of Grassland and Livestock, NIBIO, Norway.

Conflicts of Interest: The authors declare no conflict of interest.

References

1. Bareth, G.; Schellberg, J. Replacing Manual Rising Plate Meter Measurements with Low-cost UAV-Derived Sward Height Data in Grasslands for Spatial Monitoring. *PFG J. Photogramm. Remote Sens. Geoinf. Sci.* **2018**, *86*, 157–168. [CrossRef]

2. Coppens, F.; Wuyts, N.; Inzé, D.; Dhondt, S. Unlocking the potential of plant phenotyping data through integration and data-driven approaches. *Curr. Opin. Syst. Biol.* **2017**, *4*, 58–63. [CrossRef]

3. Fahlgren, N.; Gehan, M.A.; Baxter, I. Lights, camera, action: high-throughput plant phenotyping is ready for a close-up. *Curr. Opin. Plant Biol.* **2015**, *24*, 93–99. [CrossRef] [PubMed]

4. Heege, H.J.; Thiessen, E. Sensing of Crop Properties. In *Precision in Crop Farming: Site Specific Concepts and Sensing Methods: Applications and Results*; Heege, H.J., Ed.; Springer: Dordrecht, The Netherlands, 2013; pp. 103–141.

5. Näsi, R.; Viljanen, N.; Kaivosoja, J.; Alhonoja, K.; Hakala, T.; Markelin, L.; Honkavaara, E. Estimating Biomass and Nitrogen Amount of Barley and Grass Using UAV and Aircraft Based Spectral and Photogrammetric 3D Features. *Remote Sens.* **2018**, *10*, 1082. [CrossRef]

6. Senf, C.; Pflugmacher, D.; Heurich, M.; Krueger, T. A Bayesian hierarchical model for estimating spatial and temporal variation in vegetation phenology from Landsat time series. *Remote Sens. Environ.* **2017**, *194*, 155–160. [CrossRef]

7. Hopkins, A. *Grass: Its Production and Utilization*; British Grassland Society: Kenilworth, UK, 2000.

8. Jimenez-Berni, J.A.; Deery, D.M.; Rozas-Larraondo, P.; Condon, A.T.G.; Rebetzke, G.J.; James, R.A.; Bovill, W.D.; Furbank, R.T.; Sirault, X.R.R. High Throughput Determination of Plant Height, Ground Cover, and Above-Ground Biomass in Wheat with LiDAR. *Front. Plant Sci.* **2018**, *9*, 237. [CrossRef] [PubMed]

9. Glenn, E.P.; Huete, A.R.; Nagler, P.L.; Nelson, S.G. Relationship Between Remotely-sensed Vegetation Indices, Canopy Attributes and Plant Physiological Processes: What Vegetation Indices Can and Cannot Tell Us About the Landscape. *Sensors* **2008**, *8*, 2136–2160. [CrossRef]

10. Fitzgerald, G.J. Characterizing vegetation indices derived from active and passive sensors. *Int. J. Remote Sens.* **2010**, *31*, 4335–4348. [CrossRef]

11. Capolupo, A.; Kooistra, L.; Berendonk, C.; Boccia, L.; Suomalainen, J. Estimating Plant Traits of Grasslands from UAV-Acquired Hyperspectral Images: A Comparison of Statistical Approaches. *ISPRS Int. J. Geoinf.* **2015**, *4*, 2792–2820. [CrossRef]

12. Edirisinghe, A.; Hill, M.J.; Donald, G.E.; Hyder, M. Quantitative mapping of pasture biomass using satellite imagery. *Int. J. Remote Sens.* **2011**, *32*, 2699–2724. [CrossRef]

13. Peteinatos, G.; Weis, M.; Andújar, D.; Rueda-Ayala, V.; Gerhards, R. Potential use of ground-based sensor technologies for weed detection. *Pest Manag. Sci.* **2014**, *70*, 190–199. [CrossRef]

14. Fonseca, R.; Creixell, W.; Maiguashca, J.; Rueda-Ayala, V. Object detection on aerial image using cascaded binary classifier. In Proceedings of the 2016 IEEE Applied Imagery Pattern Recognition Workshop (AIPR), Washington, DC, USA, 18–20 October 2016; pp. 1–6.
15. Andújar, D.; Escolà, A.; Dorado, J.; Fernández-Quintanilla, C. Weed discrimination using ultrasonic sensors. *Weed Res.* **2011**, *51*, 543–547. [CrossRef]
16. Zhang, L.; Grift, T.E. A LIDAR-based crop height measurement system for Miscanthus giganteus. *Comput. Electron. Agric.* **2012**, *85*, 70–76. [CrossRef]
17. Andújar, D.; Dorado, J.; Fernández-Quintanilla, C.; Ribeiro, A. An Approach to the Use of Depth Cameras for Weed Volume Estimation. *Sensors* **2016**, *16*, 972. [CrossRef] [PubMed]
18. Andújar, D.; Escolà, A.; Rosell-Polo, J.R.; Fernández-Quintanilla, C.; Dorado, J. Potential of a terrestrial LiDAR-based system to characterise weed vegetation in maize crops. *Comput. Electron. Agric.* **2013**, *92*, 11–15. [CrossRef]
19. Andújar, D.; Rueda-Ayala, V.; Moreno, H.; Rosell-Polo, J.R.; Escolà, A.; Valero, C.; Gerhards, R.; Fernández-Quintanilla, C.; Dorado, J.; Griepentrog, H.W. Discriminating Crop, Weeds and Soil Surface with a Terrestrial LIDAR Sensor. *Sensors* **2013**, *13*, 14662–14675. [CrossRef] [PubMed]
20. Rueda-Ayala, V.; Peteinatos, G.; Gerhards, R.; Andújar, D. A Non-Chemical System for Online Weed Control. *Sensors* **2015**, *15*, 7691–7707. [CrossRef] [PubMed]
21. Jiang, Y.; Li, C.; Paterson, A.H. High throughput phenotyping of cotton plant height using depth images under field conditions. *Comput. Electron. Agric.* **2016**, *130*, 57–68. [CrossRef]
22. Rosell-Polo, J.R.; Auat-Cheein, F.; Gregorio, E.; Andújar, D.; Puigdomènech, L.; Masip, J.; Escolà, A. Chapter Three—Advances in Structured Light Sensors Applications in Precision Agriculture and Livestock Farming. In *Advances in Agronomy*; Academic Press: Cambridge, MA, USA, 2015; Volume 133, pp. 71–112.
23. Yandún Narváez, F.J.; del Pedregal, J.S.; Prieto, P.A.; Torres-Torriti, M.; Cheein, F.A.A. LiDAR and thermal images fusion for ground-based 3D characterisation of fruit trees. *Biosyst. Eng.* **2016**, *151*, 479–494. [CrossRef]
24. Wang, W.; Li, C. Size estimation of sweet onions using consumer-grade RGB-depth sensor. *J. Food Eng.* **2014**, *142*, 153–162. [CrossRef]
25. Correa, C.; Valero, C.; Barreiro, P.; Ortiz-Cañavate, J.; Gil, J. Usando Kinect como sensor para pulverización inteligente. In *VII Congreso Ibérico de Agroingeniería y Ciencias Hortícolas*; UPM: Madrid, Spain, 2013; pp. 1–6.
26. Bengochea-Guevara, J.M.; Andújar, D.; Sánchez-Sardana, F.L.; Cantuña, K.; Ribeiro, A. A Low-Cost Approach to Automatically Obtain Accurate 3D Models of Woody Crops. *Sensors* **2018**, *18*, 30. [CrossRef] [PubMed]
27. Sankey, T.; Donager, J.; McVay, J.; Sankey, J.B. UAV lidar and hyperspectral fusion for forest monitoring in the southwestern USA. *Remote Sens. Environ.* **2017**, *195*, 30–43. [CrossRef]
28. Torres-Sánchez, J.; López-Granados, F.; Serrano, N.; Arquero, O.; Peña, J.M. High-Throughput 3-D Monitoring of Agricultural-Tree Plantations with Unmanned Aerial Vehicle (UAV) Technology. *PLoS ONE* **2015**, *10*, e0130479. [CrossRef] [PubMed]
29. Bendig, J.; Yu, K.; Aasen, H.; Bolten, A.; Bennertz, S.; Broscheit, J.; Gnyp, M.L.; Bareth, G. Combining UAV-based plant height from crop surface models, visible, and near infrared vegetation indices for biomass monitoring in barley. *Int. J. Appl. Earth Obs. Geoinf.* **2015**, *39*, 79–87. [CrossRef]
30. Lu, B.; He, Y. Species classification using Unmanned Aerial Vehicle (UAV)-acquired high spatial resolution imagery in a heterogeneous grassland. *ISPRS J. Photogramm. Remote Sens.* **2017**, *128*, 73–85. [CrossRef]
31. Rueda-Ayala, V.; Peña, J.; Bengochea-Guevara, J.; Höglind, M.; Rueda-Ayala, C.; Andújar, D. Novel Systems for Pasture Characterization Using RGB-D Cameras and UAV-imagery. In Proceedings of the AgEng conference, Session 14: Robotic Systems in Pastures, Wageningen, The Netherlands, 8–12 July 2018.
32. Curless, B.; Levoy, M. A Volumetric Method for Building Complex Models from Range Images. In Proceedings of the 23rd Annual Conference on Computer Graphics and Interactive Techniques, New York, NY, USA, 4–9 August 1996; pp. 303–312.
33. Roth, S.D. Ray casting for modeling solids. *Comput. Graph. Image Process.* **1982**, *18*, 109–144. [CrossRef]
34. Edelsbrunner, H.; Mücke, E.P. Three-dimensional Alpha Shapes. *ACM Trans. Graph.* **1994**, *13*, 43–72. [CrossRef]

35. Lafarge, T.; Pateiro-Lopez, B. alphashape3d: Implementation of the 3D Alpha-Shape for the Reconstruction of 3D Sets from a Point Cloud. Available online: https://cran.r-project.org/web/packages/alphashape3d/alphashape3d.pdf (accessed on 21 December 2017).
36. de Castro, A.I.; Torres-Sánchez, J.; Peña, J.M.; Jiménez-Brenes, F.M.; Csillik, O.; López-Granados, F. An Automatic Random Forest-OBIA Algorithm for Early Weed Mapping between and within Crop Rows Using UAV Imagery. *Remote Sens.* **2018**, *10*, 285. [CrossRef]
37. Bareth, G.; Bendig, J.; Tilly, N.; Hoffmeister, D.; Aasen, H.; Bolten, A. A Comparison of UAV- and TLS-derived Plant Height for Crop Monitoring: Using Polygon Grids for the Analysis of Crop Surface Models (CSMs). *Photogramm. Fernerkun. Geoinf.* **2016**, *2016*, 85–94. [CrossRef]
38. Anderson, J.E.; Plourde, L.C.; Martin, M.E.; Braswell, B.H.; Smith, M.L.; Dubayah, R.O.; Hofton, M.A.; Blair, J.B. Integrating waveform lidar with hyperspectral imagery for inventory of a northern temperate forest. *Remote Sens. Environ.* **2008**, *112*, 1856–1870. [CrossRef]

Article

Automatic Parameter Tuning for Adaptive Thresholding in Fruit Detection

Elie Zemmour *, Polina Kurtser and Yael Edan

Department of Industrial Engineering and Management, Ben-Gurion University of the Negev,
Beer-Sheva 8410501, Israel; kurtser@post.bgu.ac.il (P.K.); yael@bgu.ac.il (Y.E.)
* Correspondence: eliezem@post.bgu.ac.il

Received: 27 March 2019; Accepted: 2 May 2019; Published: 8 May 2019

Abstract: This paper presents an automatic parameter tuning procedure specially developed for a dynamic adaptive thresholding algorithm for fruit detection. One of the major algorithm strengths is its high detection performances using a small set of training images. The algorithm enables robust detection in highly-variable lighting conditions. The image is dynamically split into variably-sized regions, where each region has approximately homogeneous lighting conditions. Nine thresholds were selected to accommodate three different illumination levels for three different dimensions in four color spaces: RGB, HSI, LAB, and NDI. Each color space uses a different method to represent a pixel in an image: RGB (Red, Green, Blue), HSI (Hue, Saturation, Intensity), LAB (Lightness, Green to Red and Blue to Yellow) and NDI (Normalized Difference Index, which represents the normal difference between the RGB color dimensions). The thresholds were selected by quantifying the required relation between the true positive rate and false positive rate. A tuning process was developed to determine the best fit values of the algorithm parameters to enable easy adaption to different kinds of fruits (shapes, colors) and environments (illumination conditions). Extensive analyses were conducted on three different databases acquired in natural growing conditions: red apples (nine images with 113 apples), green grape clusters (129 images with 1078 grape clusters), and yellow peppers (30 images with 73 peppers). These databases are provided as part of this paper for future developments. The algorithm was evaluated using cross-validation with 70% images for training and 30% images for testing. The algorithm successfully detected apples and peppers in variable lighting conditions resulting with an F-score of 93.17% and 99.31% respectively. Results show the importance of the tuning process for the generalization of the algorithm to different kinds of fruits and environments. In addition, this research revealed the importance of evaluating different color spaces since for each kind of fruit, a different color space might be superior over the others. The LAB color space is most robust to noise. The algorithm is robust to changes in the threshold learned by the training process and to noise effects in images.

Keywords: adaptive thresholding; fruit detection; parameter tuning

---

1. Introduction

Fruit detection is important in many agricultural tasks such as yield monitoring [1–8], phenotyping [9–11], precision agriculture operations (e.g., spraying [12] and thinning [13–15]), and robotic harvesting [16–18]. Despite intensive research conducted in identifying fruits, implementing a real-time vision system remains a complex task [17,18]. Features like shape, texture, and location are subject to high variability in the agricultural domain [18]. Moreover, fruits grow in unstructured environments with highly-variable lighting conditions [19,20] and obstructions [21] that influence detection performance. Color and texture are fundamental characteristics of natural images and play an important role in visual perception [22].

Nevertheless, despite the challenges, several algorithms have been developed with impressive detection rates of over 90–95%. However, these detection rates were achieved only for specific fruit (apples, oranges, and mangoes) [13,23,24]. These crops are known for their high ratio of fruits per image allowing easier acquisition of large quantities of data. Other crops such as sweet peppers and rock melons [23] with a lower fruit-to-image ratio yield lower results of 85–90% [19,23], even with the employment of cutting edge techniques such as deep learning [25]. Additionally, the crops with high detection rates with these results are very distinct from their background in terms of color, a central feature for a color-based-only detection. The only recent research made to detect green crops is on weed detection [26,27], but these green crops are held against a brown background. In grape clusters' detection [28], a rate of about 90% accuracy was achieved. Some work in the development of cucumber harvesters [29] has been done, but the success rate of harvesting was not distinguished from detection success rates and therefore cannot be reported.

In this research, we focus on the detection of three different types of challenging crops: red apples (a high ratio of fruits per image; however, we used a very small dataset http://icvl.cs.bgu.ac.il/lab_projects/agrovision/DB/Sweeper05/#/scene), green grapes ("green-on-green" dataset [28]), and yellow sweet peppers (a low fruit-to-image ratio http://icvl.cs.bgu.ac.il/lab_projects/agrovision/DB/Sweeper06/#/scene).

The adaptive thresholding algorithm presented in this paper is based on previous work [19] that was developed for sweet peppers' detection for a robotic harvester. A set of three thresholds was determined for each region of the image according to its lighting setting. Preliminary results of the same algorithm for an apple detection problem have been previously presented [30].

The current paper advances previous research [30] with several new contributions: (1) a new parameter tuning procedure developed to best-fit the parameters to the specific database; (2) the application and evaluation of the adaptive thresholding algorithm for different color spaces; (3) application of the algorithm to different types of fruits along with intensive evaluation and sensitivity analyses; (4) comparing the contribution of the new developments (Items 1–2) to previous developments.

2. Literature Review

2.1. Detection Algorithms in Agriculture

While this paper does not aim to be a review paper, in addition to the many recent reviews (e.g., [16,18,25]), a summary of previous results helps place the outcomes of this paper into context (Table 1).

As can be seen in the table, most algorithms focus on pixel-based detection (e.g., segmentation). This is indeed a common method in fruit detection (e.g., [31–34]). Many segmentation algorithms have been developed [35] including: K-means [36], mean shift analysis [37], Artificial Neural Networks (ANN) [38], Support Vector Machines (SVM) [39], deep learning [25], and several others.

A common challenge facing agriculture detection research is the lack of data [20], due to the harsh conditions for image acquisition and the tedious related ground truth annotation [40]. Current advanced algorithms (e.g., deep learning) require collecting many data. Therefore, to date, the best detection results are provided for crops with high fruit-to-image ratios (e.g., apples, oranges, and mangoes) and fruits that grow in high density, and hence, each image provides many data. Some research [26,41] aimed to cope with the need for large quantities of highly-variable data by pre-training a network on either non-agricultural open access data [26] or by generating synthetic data [41]. Both methods have shown promising results.

In this paper, we present an alternative direction, focusing on the development of algorithms based on smaller datasets. This research focuses on segmenting objects in the image using an adaptive thresholding method. Observing the histogram of the image color implies that a threshold can be determined to best differentiate between the background and the object distributions [42].

The threshold is computed by finding the histogram minimum (Figure 1) separating two peaks: the object and the background. However, the global minimum between the distributions is very hard to determine in most cases [43].

Currently, most optimal thresholding algorithms determine the threshold only in a one-dimensional space, for example in the RGB space, either R, or G, or B, or a linear combination of their values (e.g., grayscale transformation) [44]. In the transformation from three color dimensions into one, information is lost. In this research, a three-dimensional thresholding algorithm based on [19] was applied and evaluated also for additional color spaces (RGB, HSI, LAB, and NDI color spaces); a threshold is determined for each dimension in the color space.

There are two common adaptive thresholding algorithm concepts: (1) global thresholding, in which for each image, a different threshold is determined according to specific conditions for the entire image that is then transformed into a binary image; (2) local thresholding, in which the image is divided into sections and a different threshold is calculated for each section; the sections are then combined to a binary image. There are several methods that utilize dynamic local thresholding algorithms [45,46]. A common approach is to use multi-resolution windows that apply a bottom-up method, merging pixels while a criterion is met [45,46]. Another approach is the top down method, where the image is divided into subregions according to specific criteria. The top-down approach reduces execution speed and improves generalization [47] and was therefore used in this research.

The previously-developed algorithm by Vitzrabin et al., [19], which this research is based on, dynamically divides the image into several regions, each with approximately the same lighting conditions. The main contribution of the adaptive local 3D thresholding is a very high True Positive Rate (TPR) and a Low False Positive Rate (FPR) in the fruit detection task in an unstructured, highly-variable, and dynamic crop environment. Another contribution is the ability to change in real time the task objective to enable fast adaption to other crops, varieties, or operating conditions requiring small datasets and fast training. The algorithm's adaptation to the desired ratio between TPR and FRP makes it specifically fit for robotic harvesting tasks for which it was originally designed; it contributes to a better success rate in robotic operations in which at first, FPR should be minimum (to reduce cycle times), and when approaching the grasping operation itself [19], TPR should be maximized (to increase grasping accuracy). This can be applicable towards other fruit detection tasks (e.g., same as above for spraying, thinning and in yield detection first maximizing TPR for deciding on harvesting timing and then minimizing TPR for accurate marketing estimation).

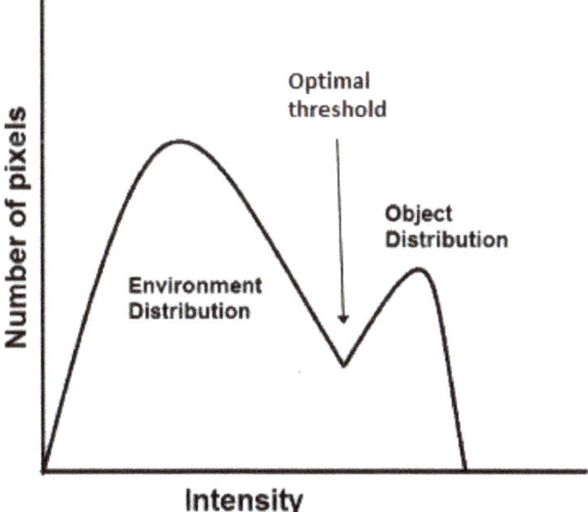

Figure 1. Optimal threshold in the bimodal histogram.

Table 1. Summary of previously-published results.

Paper	Crop	Dataset Size	Detection Level	Algorithm Type	FPR %	TPR %	F %	A %	P %	R %
Arad et al., 2019 [20]	Peppers	156 img	PX W	AD DL	NA	NA	NA	NA	65(95) 84	94(95) NA
Ostovar et al., 2018 [34]	Peppers	170 img	PX	AD	NA	NA	NA	91.5	NA	NA
Chen et al., 2017 [24]	Apples Oranges	1749 (21 img) 7200 (71 img)	PX	DL	5.1 3.3	95.7 96.1	95.3 * 96.4*	NA	NA	NA
McCool et al., 2017 [26]	Weed	Pre-train:10^6 img tuning and testing: 60 img	PX	D-CNN	NA	NA	NA	93.9	NA	NA
Milioto et al., 2017 [27]	Weed	5696 (867 img) 26,163 (1102 img)	PX	CNN	NA	NA	NA	96.8 99.7	97.3 96.1	98.1 96.3
Sa et al., 2016 [23]	Sweet pepper Rock melon Apple Avocado Mango Orange	122 img 135 img 64 img 54 img 170 img 57 img	W	DL	NA	NA	82.8 84.8 93.8 93.2 94.2 91.5	NA	NA	NA
Vitzrabin et al., 2016 [19]	Sweet pepper	479 (221 img)	PX	AD	4.6	90.0	92.6 *	NA	NA	NA
Song et al., 2014 [11]	Pepper plants	1056 img	W	NB+SVM	NA	NA	65	NA	NA	NA
Nuske et al., 2011 [6]	Grapes	2973 img	PX	K-NN	NA	NA	NA	NA	63.7	98
Berenstein et al., 2010 [28]	Grapes	100 img	PX	Morphological	NA	NA	NA	90	NA	NA
Zheng et al., 2009 [37]	Vegetation	20 img 80 img	PX	Mean-Shift	NA	NA	NA	95.4 95.9	NA	NA
Our results	Sweet pepper Apples Grapes	73 (30 img) 113 (9 img) 1078 (129 img)	PX	AD	0.81 2.59 33.35	99.43 89.45 89.48	99.31 93.17 73.52	NA	NA	NA

DL = Deep Learning; PX = Pixels' segmentation; AD = Adaptive threshold; NB = Naive Bias; W = Window detection; F = F-score; A = Accuracy; P = Precision; R = Recall; * Calculated F-score based on reported TPR and FPR according to Equation (4)

2.2. Color Spaces

Images can be represented by different color spaces (e.g., RGB, HSI, LAB, and NDI), each one emphasizing different color features [22]. RGB is the most common color space representing each pixel in the image in three color channels as acquired: red, green, and blue. HSI represents every color with three components: hue (H), saturation (S), and intensity (I), also known as HSV [37]. The LAB color space is an approximation of human vision [36] and presents for each pixel the L* (Lightness) from black to white, a* from green to red, and b* from blue to yellow. An additional color space commonly employed in the agriculture field [19] is the Normalized Difference Index (NDI) space. The NDI is used to differentiate between fruit and background [48] since it helps to overcome changes in illumination and shading due to its normalization technique [49]. Each dimension in the NDI space is the normalized difference index between two colors in the RGB space, resulting in three dimensions (Equation (1)). These operations are applied for all pixel locations in the image, creating a new image with this contrast index. These equations yield NDI values ranging between −1 and +1.

$$NDI_1 = \frac{R-G}{R+G}; NDI_2 = \frac{R-B}{R+B}; NDI_3 = \frac{B-G}{B+G} \tag{1}$$

3. Materials and Methods

3.1. Databases

The algorithm was evaluated on three databases representing three different fruit colors: red (apples), green (grapes), and yellow (peppers) and different types of fruits (high image and low image ratios) for two environmental settings (greenhouse and field) in different illumination conditions. Images were acquired with different cameras. Each image was processed by a human labeler who performed manual segmentation of the image into targets and background (Figures 2 and 3) by visually analyzing the image and marking all the pixels considered as a fruit, in accordance with the common protocols used in the computer vision community [50].

3.1.1. Apples

The orchard apples database included 113 "Royal Gala" apples in 9 images acquired from an orchard in Chile in March 2012 under natural growing conditions with a Prosilica GC2450C (Allied Vision Technologies GmbH, Stadtroda, Germany) camera with 1536 × 2048 resolution; the camera was attached to a pole. The images were captured in daylight; half of the images were acquired under direct sunlight, and half of the images were acquired in the shade.

Figure 2. Apple (**top**) and grape (**bottom**) RGB image (**left**) and ground truth (**right**) examples

3.1.2. Grapes

The images were acquired in a commercial vineyard growing green grapes of the "superior" variety. An RGB camera (Microsoft NX-6000) with 600 × 800 resolution was manually driven, at mid-day, along a commercial vineyard in Lachish, Israel, during the summer season of 2011, one month before harvest time. The images were captured from five different growing rows. A set of 129 images was acquired and included 1078 grape clusters.

3.1.3. Peppers

The dataset included 30 images of 73 yellow peppers acquired in a commercial greenhouse in Ijsselmuiden, Netherlands, using a 6 degree of freedom manipulator (Fanuc LR Mate 200iD/7L), equipped with an iDS Ui-5250RE RGB camera with 600 × 800 resolution. Two different datasets were created by marking the images twice. The first dataset included only peppers with high visibility (denoted as "high visibility peppers"; this was done for 10 images of 25 yellow peppers). In the second dataset, all peppers were marked including peppers in dark areas that were less visible in the image (denoted as "including low-visibility peppers", done for all 30 images) (Figure 3).

Figure 3. Peppers tagging example. **Top**: RGB image (**left**) and ground truth (**right**) example in high visibility. **Bottom**: RGB image (**left**) and labeled image (**Right**). "High-visibility peppers" marked in red and "low-visibility peppers" marked in blue.

3.1.4. Performance Measures

Metrics included the TPR (True Positive Rate, also noted as hit), FPR (False Positive Rate, also noted as false alarms), and the F-score (the harmonic mean of precision and recall [51]. The TPR metric (Equation (2)) states the number of correctly-detected objects relative to the actual number of objects, while the FPR metric calculates the number of false objects detected relative to the actual number of objects (Equation (3)). The F-score (Equation (4)) balances between TPR and FPR equally.

$$TPR = \frac{N_{TDF}}{N_F} \tag{2}$$

where N_{TDF} is the number of pixels detected correctly as part of the fruit and N_F is the actual number of pixels that represent the fruit.

$$FPR = \frac{N_{FDF}}{N_B} \tag{3}$$

where N_{FDF} is the number of pixels falsely classified as fruit and N_B is the number of pixels that represent the background.

$$F(TPR, FPR) = \frac{2 * (TPR * (1 - FPR))}{TPR + (1 - FPR)} \tag{4}$$

3.2. Analyses

The following analyses were conducted for the three databases, apples, grapes, and peppers, using 70% of the data for training and 30% for testing [52]. This rate was chosen to make the algorithm performances more rigid since the number of images in each DB was relativity small. In addition, to ensure robust results, each split into training and testing was randomly performed 5 times, and all detection results reported are an average of the 5 test sets.

- Tuning parameters: Parameters were computed for each database with procedures defined in Section 4.3 and compared to previous predefined parameters
- Color spaces' analyses: Algorithm performances were tested on all databases for four different color spaces: RGB, HSI, LAB, and NDI.
- Sensitivity analysis: Sensitivity analyses were conducted for all the databases and included:

  1. Noise: Noise was created by adding to each pixel in the RGB image a random number from the mean normal distribution for noise values up to 30%. The artificial noise represents the algorithms' robustness toward other cameras with more noise, or when capturing images with different camera settings. Noise values of 5%, 10%, 20%, and 30%, were evaluated.
  2. Thresholds learned in train process: Thresholds were changed by $\pm5\%$, $\pm10\%$, and $\pm15\%$ according to the threshold in each region.
  3. Stop condition: The selected STD value was changed by 5% and 10% to test the robustness of the algorithm to these parameters.
  4. Training vs. testing: The algorithm performances were evaluated while using different percentages of DB images for the training and testing processes.

- Morphological operation (erosion and dilation) contribution: Performances were tested for imaging with and without a morphological operations process.

4. Algorithm

4.1. Algorithm Flow

The overall flow of the algorithm is outlined in Figure 4, and it is as follows. The RGB images were the inputs for the training process. Some areas in the images contained more illumination than others, depending on the position of the light source and shading caused by leaves, branches, and the covering net when it existed. To overcome this issue, the algorithm divided each image into multiple sub-images, with approximately homogeneous illumination conditions (Figure 5). These sub-images were categorized into three illumination conditions: low, medium, and high. The illumination level was obtained by calculating the average on the grayscale sub-images. The grayscale image showed values between zero (completely dark) and 255 (completely white). In the previous algorithm [19], the sub-images were categorized into groups using levels selected empirically as 10, 70, and 130, corresponding to low-, medium-, and high-level images based on manual image analyses. The high value was set as 130 in order to filter overexposed areas in the images. In the current algorithm, a tuning parameter process (detailed in Section 4.3) was developed to determine these three values.

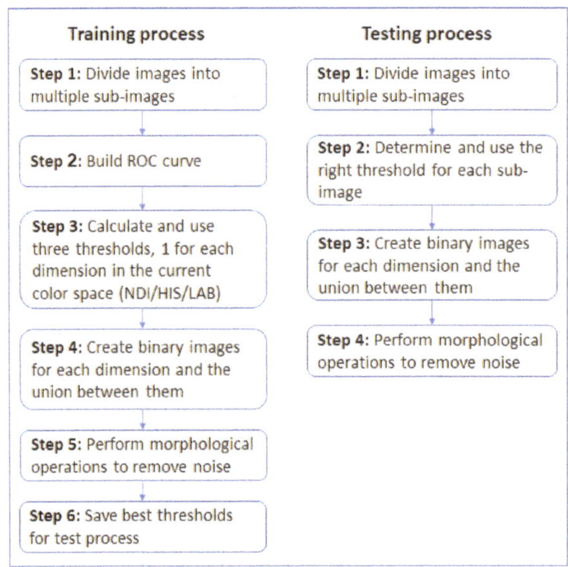

Figure 4. Algorithm flowchart.

Figure 5. Image split into sub-images: visualization.

The algorithm then created a 3D color space image (transformed the RGB image to NDI, HSI, and LAB space or used the RGB space directly). For each color dimension, a binary image (mask) was created, where each pixel that represents the fruit received a value of one and all other pixels received a value of zero. Finally, the algorithm created an ROC (Receiver Operator characteristics Curve) representing TPR as a function of FPR [53] including all nine thresholds learned from the training process. Figure 6 presents an example of nine ROC curves computed for three sub-images with different Light levels (L1, L2, L3) in the NDI color space. In this example, the sub-image with Light Level 2 (L2) in the first NDI dimension obtained the best performances (high TPR and low FPR).

In the test process, the algorithm received RGB images from the camera in real time, transformed the representation to the relevant color space (HSI/LAB/NDI), and created a binary image by applying the thresholds as follows: three thresholds, one for each dimension, were calculated from the nine thresholds learned by linear interpolation between two of the three illumination regions (low, medium, and high) selected as closest to the calculated illumination level for the specific sub-image from the grayscale image and using Equation (5).

$$T = \frac{T(LL[i]) * (currentLL - LL[i])(LL[i+1] - currentLL)}{LL[i+1] - LL[i]} \tag{5}$$

where LL is an array of the light level values for each group: LL = [low,medium,high]), and i is the light level index representing the group that is the closest to the current image light level from below.

For example, if the current light level is 40 and the thresholds in the train process for the low, medium, and high light levels were 10, 70, and 130, the threshold would be calculated in the following way (Equation (6)):

$$T = \frac{T(10)(40 - 10) + T(70)(70 - 40)}{70 - 10} \tag{6}$$

The end of the process results in binary images where white areas (pixels with a value of one) in the binary image represent the fruits and the black areas (pixels with a value of zero) represent the background (see Figures 2 and 3). In total, the algorithm created 7 binary images, 3 images corresponding to the three color space dimensions and 4 binary images corresponding to the intersections between the first 3 binary images. For example, the intersection between Binary Images 1 and 2 resulted in a binary image 1 ∩ 2 that contained white pixels only, where the same pixels were white in both Images 1 and 2 (Figure 7).

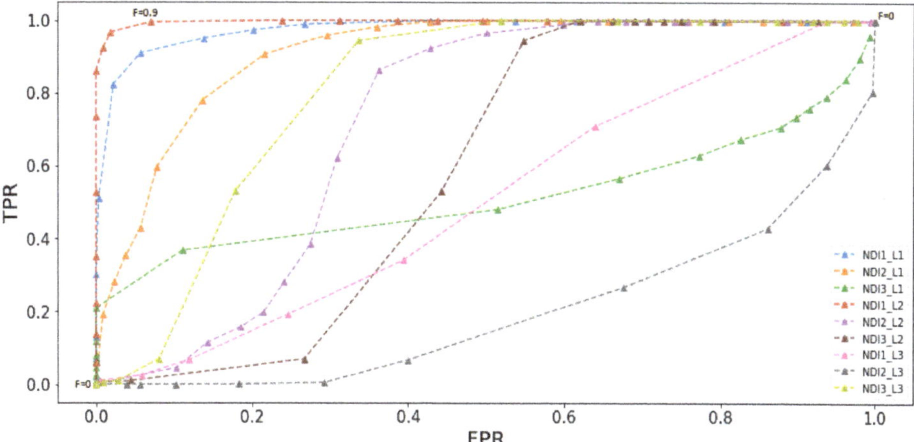

Figure 6. Nine ROC curves: 3 dimensions × 3 light levels $NDI_i - L_j$; i represents the color space dimension; j represents the illumination level.

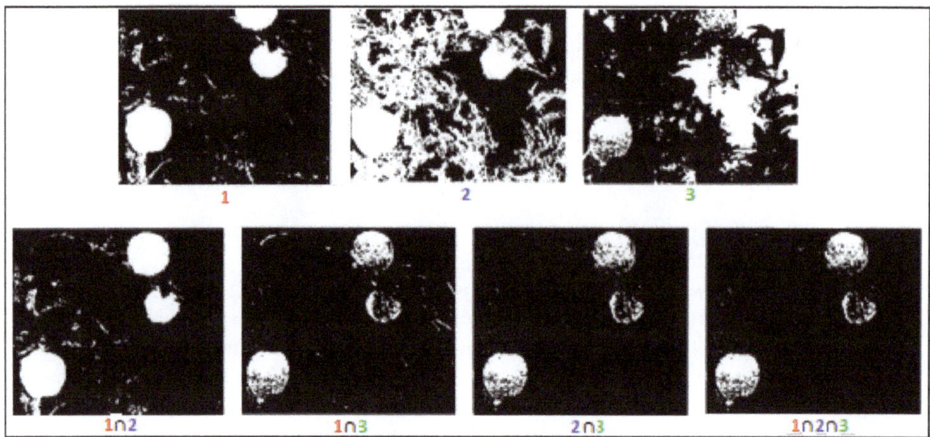

Figure 7. Use of dimension intersection to increase performance.

4.2. Morphological Operations: Erosion and Dilation

The algorithm result is a binary image with major fruit detected and small clusters of pixels that were wrongly classified as fruits (e.g., Figure 8). In addition, some fruits were split between several clusters (e.g., Figure 8). To overcome these problems, several morphological operations were performed based on previous research that indicated their contribution [19]: erosion followed by dilation with a neighborhood of 11 × 11-pixel squares. The square function was used since there was no pre-defined knowledge about the expected fruit orientation. To connect close clusters, the closing morphological operation was then applied by dilation followed by erosion implemented with a 5 × 5-pixel square neighborhood.

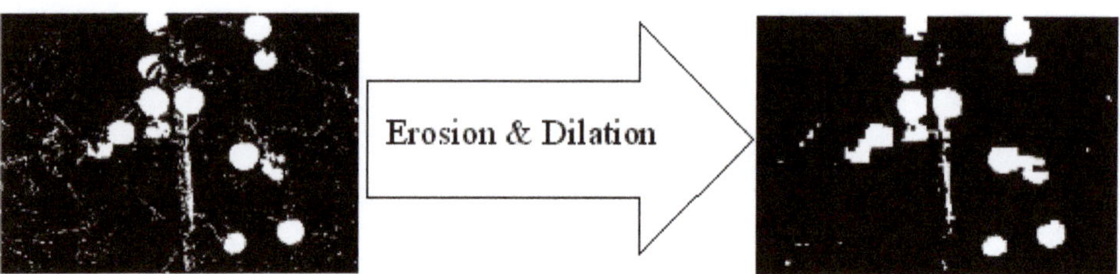

Figure 8. Morphological operation.

4.3. Parameter Tuning

The algorithm used several parameters that influence the algorithm performances: T1, T2, STD, classification rule Direction (D1/D2). The following parameter tuning procedure (Figure 9) was developed and should be performed when exploring images from a new database or when exploring a new color space or new operating conditions (cameras, illumination). The parameters, as detailed below, are: light level thresholds, stop splitting condition, classification rule direction.

4.3.1. Light Level Thresholds (T1, T2)

The algorithm split the images into sub-images set to 1% pixels of the entire image. Then, the algorithm computed the light level of each sub-image by calculating the average pixels values of the grayscale sub-image. Finally, the algorithm grouped the sub-images into three light level categories (see Figure 10) using two thresholds as presented in Equation (7).

$$
i = \begin{cases} Low & 0 < x < T1 \\ Medium & T1 < x < T2 \\ High & x > T2 \end{cases}
\tag{7}
$$

where i is the light level index, as detailed above in Equation (5).

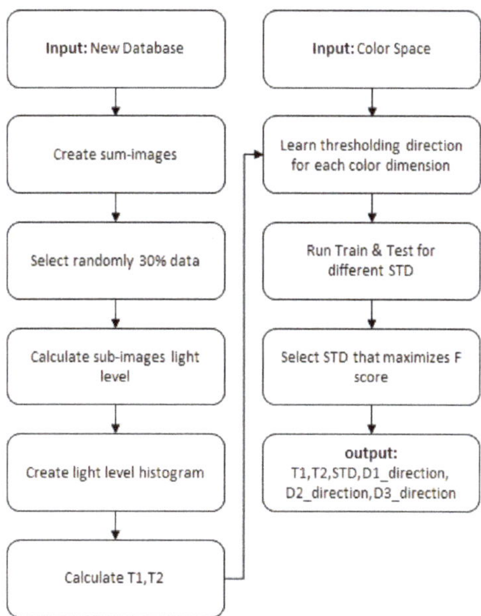

Figure 9. Parameter tuning process.

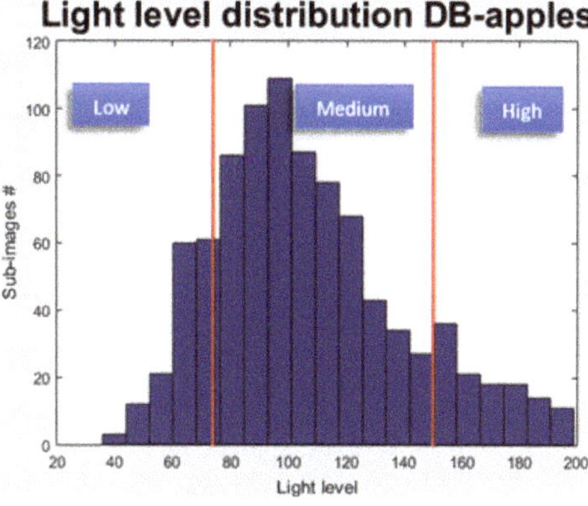

Figure 10. Sub-images level of light distribution.

Research was done to identify the PDF function of the data distributions of each database through a χ^2 goodness of fit test. However, since these tests did not reveal significant results [54], the thresholds were selected as follows: T1 and T2 were chosen so that 15% of the data would be categorized as low, 15% as high, and 70% as medium.

Note that as described in the algorithm flow, the algorithm used a third threshold. Sub-images above that threshold were ignored in the training process since they were almost completely white.

4.3.2. Stop Splitting Condition (STD)

The algorithm split an image into sub-images until the sub-image achieved a predefined Standard Deviation (STD) value. This approach assumes that a larger sub-image contains a higher STD value. To test this assumption, STD was calculated for different sizes of sub-images for the different databases. The stop condition value (STD minimum value) was determined by maximizing the F-score (Equation (4)).

4.3.3. Classification Rule Direction (D1, D2, D3)

As detailed in the Introduction section, as part of the thresholding process, an intensity value was determined to differentiate between objects and background pixels. In order to execute the thresholding process, the algorithm must receive as input the classification rule direction (the algorithm must automatically determine if the intensity of the background distribution is higher or lower than the intensity of the object distribution in each color dimension).

This information was learned as part of the tuning procedure. A simple heuristic rule was used as follows based on the assumption that the images contained more background pixels than objects: (1) execute *image > Threshold*; (2) if the pixels categorized as background represent less than 70% of the image, reverse the thresholding direction *images < Threshold*.

5. Results and Discussion

5.1. Sub-Image Size vs. STD Value

For images with a size of 300×300 or lower, splitting an image into small sub-images (small S) decreases the average STD of the sub-images (in all three databases; Figure 11). Although a direct decrease in very large images is not noted, we still can conclude that splitting a large image to 300×300 or lower will decrease the average STD.

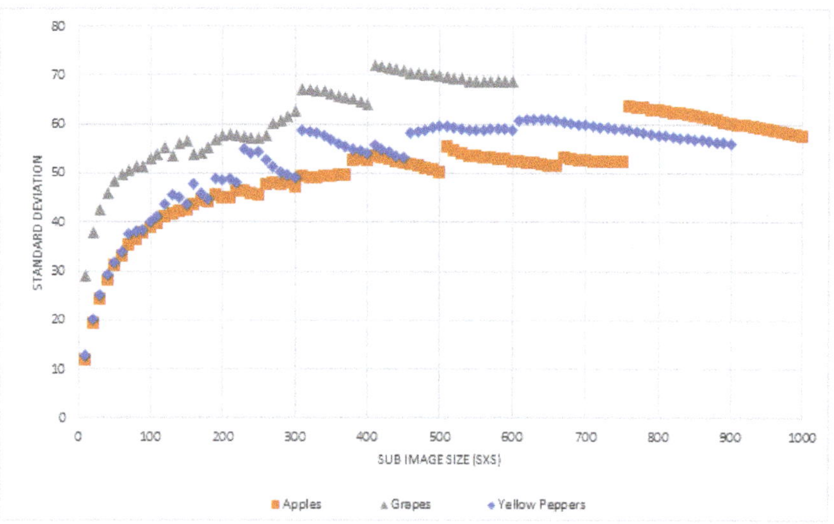

Figure 11. Sub image size vs. average STD.

5.2. Tuning Process

This section presents the tuning procedure results, including thresholds derived to categorize the sub-images into light level groups, as well as the recursive stop condition that achieved the best result for each database.

5.2.1. Light Level Distribution

The light level distribution was computed for each database (Figure 12) along with T1 and T2 (Table 2). The variation in the light distributions between the different databases are described in Table 3. The variance of light in the grape databases was significantly higher than in both the apple and the pepper databases, the pepper database being significantly darker and highly skewed. Therefore, for each database, the selected T1 and T2 were significantly different, implying the importance of the tuning procedure.

Table 2. T1 and T2 values determined for each database.

Measure\DB	Apples	Grapes	Peppers
T1	84	49	18
T2	140	130	47

Table 3. Descriptive statistics of the different light distributions.

DB\Measure	Mean	Std	Skewness	Kurtosis	Median
Apples	118.46	28.04	0.4	−0.17	116.31
Grapes	88	37.9	0.68	−0.13	81.06
Peppers	32.09	18.92	3.16	15.36	26.93

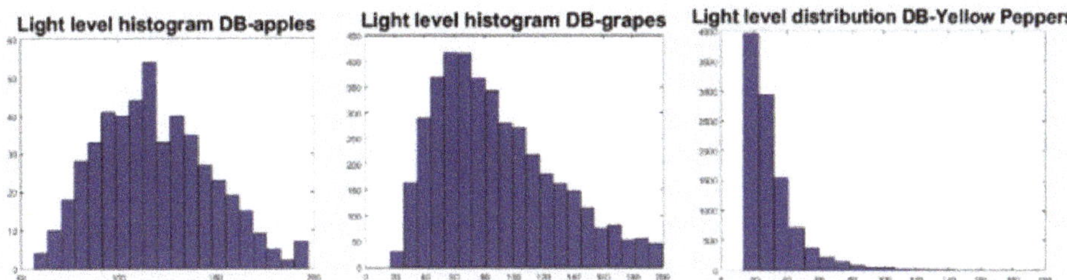

Figure 12. Light level distribution computed for each database.

5.2.2. Stop Splitting Condition

Using a low STD value as a stop condition increased the performance (Figure 13). This happens since smaller sub-images contain less illumination differences. However, small STD values can create also too small sub-images, which may not contain fruit and background pixels in the same frame. In these cases, the algorithm cannot learn a threshold that could differ between them. Additionally, results revealed that when using high STD values, the performances remained constant. This happens since beyond a certain value, the algorithm did not split the image even once.

As part of the parameter tuning process, the STD value was selected by testing the performances of a range of STD [0, 100]. For each STD value, the algorithm ran five iterations where it randomly selected P% of the images, from the selected images; it used 70% for training and 30% testing. The final selected STD values are presented in Table 4 for each database and color space (using P = 30% and 50%).

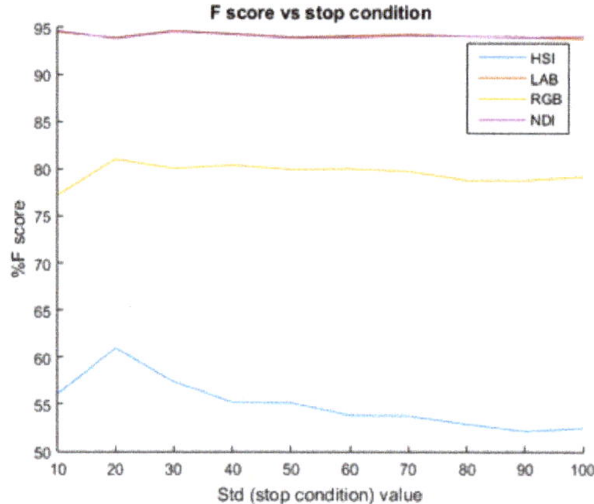

Figure 13. F-score vs. increasing STD value as the stop condition for the recursive function on the apple DB.

5.2.3. Classification Rule Direction

As shown in Table 5, the direction of the classification rule in the thresholding process can be different for each color dimension; therefore, this must be learned as part of the tuning procedure.

Table 4. STD value chosen for each database and color space. D, Direction.

DB		Apples				Grapes				Peppers			
Color space		HSI	LAB	NDI	RGB	HSI	LAB	NDI	RGB	HSI	LAB	NDI	RGB
STD (P = 30%)		20	30	10	20	10	20	60	20	100	10	10	10
STD (P = 50%)		20	10	10	30	20	20	70	20	100	20	10	10
Classification rule direction	D1	>	<	>	>	<	>	>	>	<	>	>	>
	D2	>	>	>	<	>	<	>	>	>	>	>	>
	D3	>	>	>	<	<	>	<	<	>	>	<	<

5.3. Color Space Analyses

In this section, algorithm performance results are presented for each color space followed by a table representing the best color space performances including the performances for all color space dimensions' combinations.

5.3.1. Apples

Results (Figure 14) revealed that NDI and LAB color spaces resulted in similar best performances. In Table 5, the preferences for each dimension in the NDI color space and the performances when using the intersection between them is shown. The NDI first dimension (see Equation (1)) represents the difference between the red and green colors in the image. The objects in this database were red apples, and most of the background was green leaves; therefore, as expected, the first NDI obtained the best F of 93.17%. In the LAB color space, results (Table 5) revealed that the second dimension (A) yielded the best F-score of 93.19.

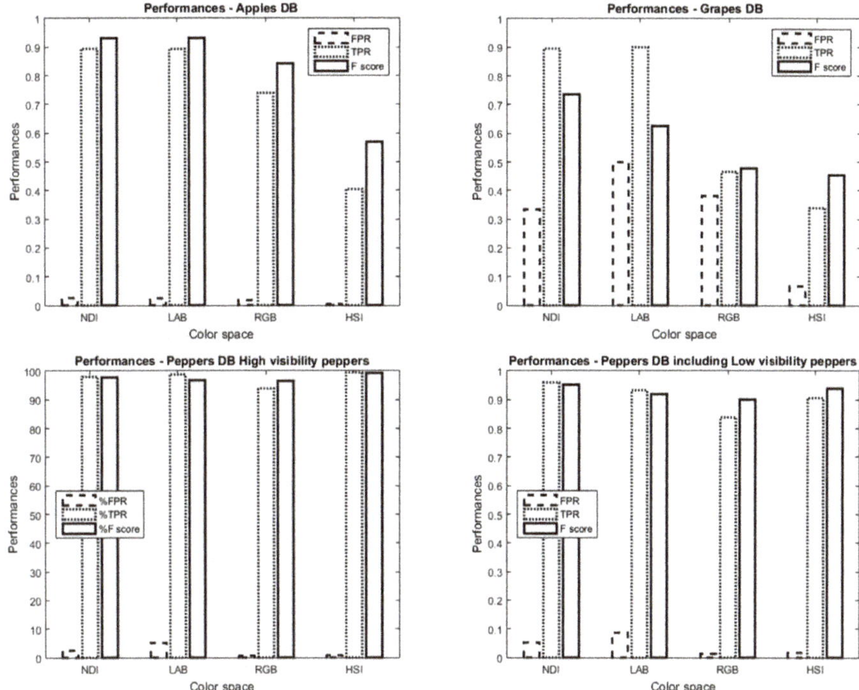

Figure 14. Color space performances for (**top**) apples (**left**), grapes (**right**), peppers (**bottom**) with high visibility (**left**), and peppers with low visibility (**right**).

5.3.2. Grapes

The NDI color space obtained the best result for grapes (Figure 14) with an F-score of 73.52%. The second-best color space was the LAB with an F-score of 62.54%. The best NDI results were obtained using the second dimension (Table 5).

5.3.3. Peppers

High visibility: Figure 14 indicates that the HSI color space obtained the best results with relatively low FPR (0.81%) and very high TPR (99.43%), resulting in a high F-score (99.31%). The second-best color space was NDI with FPR = 2.48% and TPR = 97.96% (F = 97.72%). The best HSI result was obtained using the combination of the first and the second dimensions (Table 5).

Including low visibility: Figure 14 indicates that the NDI color space obtained the best results with relatively low FPR (5.24%) and very high TPR (95.93) resulting in a high F-score (95.19%). Although for the "high visibility" peppers HSI obtained the best performances, when trying to detect peppers in dark areas that were less visible, NDI showed better results. The best NDI result was obtained using the intersection between the first and the second dimensions (Table 5).

Table 5. Performances of each color space for each dimension and intersection for all datasets.

DB	Color Space	Measure	1	2	3	1∩2	1∩3	2∩3	1∩2∩3
Apples	NDI	% FPR	**2.59**	40.91	31.38	1.64	1.32	2.48	0.48
		% TPR	**89.45**	83.53	68.39	78.52	64.82	54.65	54.1
		% F	**93.17**	67.85	67.8	86.75	77.6	69.67	69.69
	LAB	% FPR	33.58	**2.45**	77.08	1.78	28.55	1.55	1.07
		% TPR	61.26	**89.34**	85.26	56.59	52.95	76.27	48.8
		% F	56.79	**93.19**	35.85	69.02	54.61	85.37	62.58
Grapes	NDI	% FPR	35.86	33.35	52.9	4.86	5.5	**32.35**	4.09
		% TPR	44.52	89.48	89.99	38.53	37.27	**87.5**	36.7
		% F	47.19	73.52	58.05	50.12	48.93	**73.2**	48.65
Peppers High Visibility	HSI	% FPR	18.27	1.52	2.16	**0.81**	0.64	0.17	0.14
		% TPR	99.96	99.43	86.42	**99.43**	86.41	86.23	86.23
		% F	89.77	98.95	91.41	**99.31**	92.12	92.24	92.25
Peppers Inc. Low Visibility	NDI	% FPR	66.57	**5.24**	9.23	1.42	1.24	4.57	0.99
		% TPR	85.61	**95.93**	92.49	82.2	78.64	92.33	78.61
		% F	46.96	**95.19**	91.24	88.91	86.59	93.51	86.67

5.4. Sensitivity Analysis

5.4.1. Noise

Analysis showed that the algorithm was robust to noise in the image up to 15% in the apple and pepper databases (Figure 15). The grape images were more sensitive to noise, and performance dropped when noise values of 5% were added. Although better F-score values were obtained for NDI and HSI for grapes and peppers, we can see that the LAB color space yielded more robust performance when adding noise to the images.

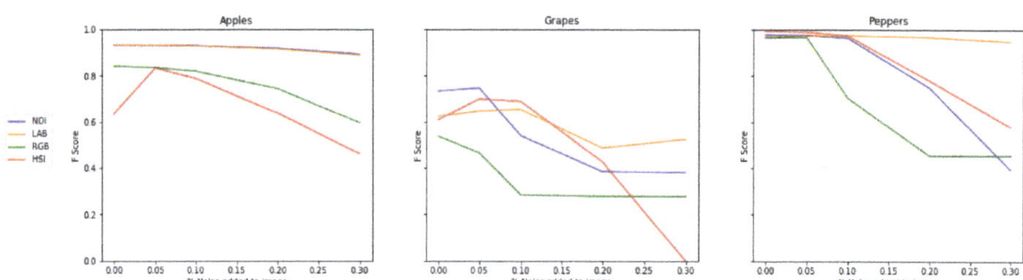

Figure 15. Sensitivity analysis: adding noise to the image.

5.4.2. Thresholds Learned in the Training Process

As expected, TPR decreased when the threshold values changed. The algorithm was relatively robust to the change in the thresholds for apples and peppers. Performance in the grape images was more sensitive to threshold changes, and yielded a significant decrease in TPR when increasing the threshold value (Table 6).

Table 6. Threshold values changed by ±5%, ±10%, and ±15% according to the threshold in each region.

DB	Measure	−15%	−10%	−5%	0%	5%	10%	15%
				Changes in Threshold				
Apples	% FPR	3.58	3.43	3.3	2.59	3.06	2.93	2.81
	% TPR	91.47	91.28	91.07	89.45	90.75	90.57	90.44
Grapes	% FPR	21.59	18.4	15.53	33.35	11.00	9.23	7.72
	% TPR	78.02	72.63	66.42	89.48	50.99	43.63	36.24
Peppers	% FPR	0.98	0.91	0.86	0.81	0.78	0.7	0.65
	% TPR	99.25	99.22	99.2	99.43	99.12	99.07	99.04

5.4.3. Stop Condition

The algorithm showed more robustness to apple and pepper images than grapes (Figure 16).

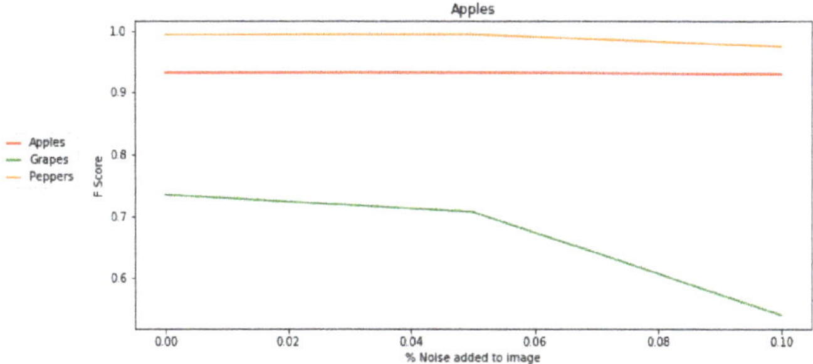

Figure 16. Sensitivity analysis: adding noise to STD stop condition.

5.4.4. Training/Testing

The expectation was that more training images would lead to better performance until over fitting was accommodated. There was a clear increase in TPR; however, FPR increased as well at 80% and 90% training (Table 7).

The tuning process resulted (Table 8) in increased performances for both the grape and pepper databases with a 6.22% and 0.84% increase, respectively. The results for the apple database were similar with only a 0.1% increase, as expected (since this was similar to the database from which the previous parameters were derived).

Table 7. Performances vs. different % images database as the training set.

DB	Measure	10	20	30	40	50	60	70	80	90
				Training %: For a Dataset Size of 129 Images						
Grapes	% FPR	32.81	37.08	28.54	36.16	31.83	29.1	29.63	40.51	40.8
	% TPR	88.79	89.62	87.01	88.44	87.58	82.55	87.14	94.53	95.85
	% F	73.35	70.2	75.19	69.41	72.49	70.24	73.92	72.55	72.54

5.5. Morphological Operations

The morphological operations process increased the F-score by 2.85%, 8.59%, and 2.71% for the apple, grape, and pepper databases respectively, (Figure 17).

Table 8. Parameter tuning contribution to algorithm performances.

DB	Measure	Performances Using Previous Params	Performances Using Tuning Process
Apples	% FPR	2.53	2.59
	% TPR	89.23	89.45
	% F	93.08	93.17
Grapes	% FPR	18.63	33.35
	% TPR	63.7	89.48
	% F	67.3	73.52
Peppers	% FPR	1	0.81
	% TPR	97.97	99.43
	% F	98.47	99.31

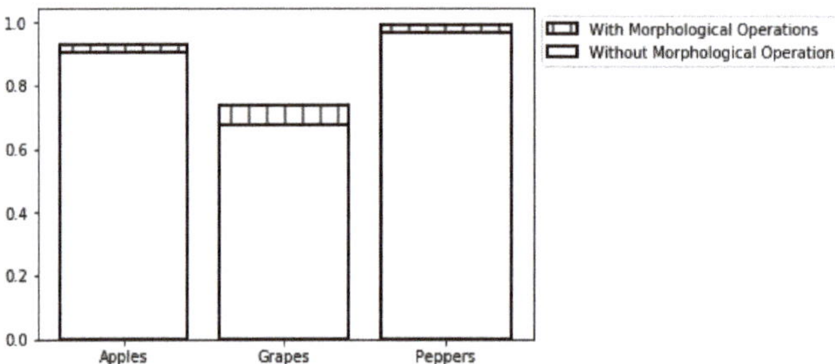

Figure 17. Sensitivity analysis: adding noise to the STD stop condition.

6. Conclusions and Future Work

The algorithm successfully detected apples and peppers (Table 1) in variable lighting conditions resulting in an F-score of 93.17% and 99.31%, respectively, which is one of the best detection rates achieved to date in fruit detection to the best of our knowledge. The average F-score across all datasets was 88.8 (Table 1). Previous research achieved the lowest F-score (65) with the method of [11] for red and green pepper plants, while oranges obtained the highest F-score (96.4) with that of [24]. Previous reported results (Table 1) revealed a 91.5 and 92.6 F-score for peppers in [19,34], respectively, versus our method, which resulted in an F-score of 99.43. For apple images, our method obtained similar F-score performances as in previous work (~93), even though the dataset was much smaller (64 vs. 9 images).

The high F-score was mostly due to low FPR values (except for grapes). In addition, our method achieved high performances using a relatively small dataset.

The algorithm resulted in less impressive results in the grape database of 73.52% due to the difficulties in differentiating between green fruits and green background (leaves). In this case, additional features (e.g., morphological operations fitted for grapes; see [28]) should be used to increase performance. However, this requires the development of specially-tailored features. It is important to note that these results cannot be compared to the weed detection results presented in Table 1, since the background of the green obkects was the ground on which it grew and not the green leaves. Different color spaces yielded the best results for each fruit variety, implying that the color space must be analyzed and fitted to the specific fruit. The LAB color space was more robust to noise in images and hence should be used when images are of low quality. The algorithm was robust to changes in the threshold learned by the training process and to noise effects in images. Morphological operations such as erosion and dilation can improve performance in agriculture images and hence should be utilized. The tuning process developed in this paper enabled the previous algorithm [30] to adapt automatically to changing conditions/objectives (i.e., to detect other fruit with different colors and other outdoor conditions) and, hence, should be used for improved target detection in highly-variable illumination conditions. Finally, this work has presented the feasibility of color-based algorithms solving the challenges that advanced machine learning algorithms face such as small training sets (small number of images and/or small number of fruits per image). This work has shown that for challenging color conditions (e.g., green on green for grapes), additional features should be considered for improved fruit detection.

Author Contributions: E.Z.: Formal algorithm development and analysis, Investigation, Methodology, Software, Writing—Original draft, Writing—review & editing; P.K.: Formal analysis, Methodology, Supervision, Validation; Writing—review & editing; Y.E.: Methodology, Supervision, Validation, Writing—original draft; Writing—review & editing.

Funding: This research was partially supported by the European Commission (SWEEPER GA No. 664313) and by Ben-Gurion University of the Negev through the Helmsley Charitable Trust, the Agricultural, Biological and Cognitive Robotics Initiative, the Marcus Endowment Fund, and the Rabbi W. Gunther Plaut Chair in Manufacturing Engineering.

Conflicts of Interest: The authors declare no conflict of interest.

References

1. Kalantar, A.; Dashuta, A.; Edan, Y.; Gur, A.; Klapp, I. Estimating Melon Yield for Breeding Processes by Machine-Vision Processing of UAV Images. In Proceedings of the Precision Agriculture Conference, Montpellier, France, 8–11 July 2019.
2. Bargoti, S.; Underwood, J.P. Image segmentation for fruit detection and yield estimation in apple orchards. *J. Field Robot.* **2017**, *34*, 1039–1060. [CrossRef]
3. Stein, M.; Bargoti, S.; Underwood, J. Image based mango fruit detection, localisation and yield estimation using multiple view geometry. *Sensors* **2016**, *16*, 1915. [CrossRef] [PubMed]
4. Dorj, U.O.; Lee, M.; Yun, S.S. An yield estimation in citrus orchards via fruit detection and counting using image processing. *Comput. Electron. Agric.* **2017**, *140*, 103–112.
5. Wang, Q.; Nuske, S.; Bergerman, M.; Singh, S. Automated crop yield estimation for apple orchards. In *Experimental Robotics*; Springer: Berlin, Germany, 2013; pp. 745–758.
6. Nuske, S.; Achar, S.; Bates, T.; Narasimhan, S.; Singh, S. Yield estimation in vineyards by visual grape detection. In Proceedings of the IEEE/RSJ International Conference on Intelligent Robots and Systems, San Francisco, CA, USA, 25–30 September 2011; pp. 2352–2358.
7. Zaman, Q.; Schumann, A.; Percival, D.; Gordon, R. Estimation of wild blueberry fruit yield using digital color photography. *Trans. ASABE* **2008**, *51*, 1539–1544. [CrossRef]
8. Oppenheim, D. Object recognition for agricultural applications using deep convolutional neural networks. Master's Thesis, Ben-Gurion University of the Negev, Beer-Sheva, Israel, 2018.
9. Mack, J.; Lenz, C.; Teutrine, J.; Steinhage, V. High-precision 3D detection and reconstruction of grapes from laser range data for efficient phenotyping based on supervised learning. *Comput. Electron. Agric.* **2017**, *135*, 300–311. [CrossRef]
10. Pound, M.P.; Atkinson, J.A.; Wells, D.M.; Pridmore, T.P.; French, A.P. Deep learning for multi-task plant phenotyping. In Proceedings of the IEEE International Conference on Computer Vision, Venice, Italy, 22–29 October 2017; pp. 2055–2063.
11. Song, Y.; Glasbey, C.; Horgan, G.; Polder, G.; Dieleman, J.; Van der Heijden, G. Automatic fruit recognition and counting from multiple images. *Biosyst. Eng.* **2014**, *118*, 203–215. [CrossRef]
12. Oberti, R.; Marchi, M.; Tirelli, P.; Calcante, A.; Iriti, M.; Tona, E.; Hočevar, M.; Baur, J.; Pfaff, J.; Schütz, C.; et al. Selective spraying of grapevines for disease control using a modular agricultural robot. *Biosyst. Eng.* **2016**, *146*, 203–215. [CrossRef]
13. Payne, A.B.; Walsh, K.B.; Subedi, P.; Jarvis, D. Estimation of mango crop yield using image analysis-segmentation method. *Comput. Electron. Agric.* **2013**, *91*, 57–64. [CrossRef]
14. Wang, Z.; Verma, B.; Walsh, K.B.; Subedi, P.; Koirala, A. Automated mango flowering assessment via refinement segmentation. In Proceedings of the International Conference on Image and Vision Computing New Zealand (IVCNZ), Palmerston North, New Zealand, 21–22 November 2016; pp. 1–6.

15. Wouters, N.; De Ketelaere, B.; Deckers, T.; De Baerdemaeker, J.; Saeys, W. Multispectral detection of floral buds for automated thinning of pear. *Comput. Electron. Agric.* **2015**, *113*, 93–103. [CrossRef]

16. Bac, C.W.; van Henten, E.J.; Hemming, J.; Edan, Y. Harvesting robots for high-value crops: State-of-the-art review and challenges ahead. *J. Field Robot.* **2014**, *31*, 888–911. [CrossRef]

17. Kapach, K.; Barnea, E.; Mairon, R.; Edan, Y.; Ben-Shahar, O. Computer vision for fruit harvesting robots–state of the art and challenges ahead. *Int. J. Comput. Vis. Robot.* **2012**, *3*, 4–34. [CrossRef]

18. Gongal, A.; Amatya, S.; Karkee, M.; Zhang, Q.; Lewis, K. Sensors and systems for fruit detection and localization: A review. *Comput. Electron. Agric.* **2015**, *116*, 8–19. [CrossRef]

19. Vitzrabin, E.; Edan, Y. Adaptive thresholding with fusion using a RGBD sensor for red sweet-pepper detection. *Biosyst. Eng.* **2016**, *146*, 45–56. [CrossRef]

20. Arad, B.; Kurtser, P.; Barnea, E.; Harel, B.; Edan, Y.; Ben-Shahar, O. Controlled Lighting and Illumination-Independent Target Detection for Real-Time Cost-Efficient Applications. The Case Study of Sweet Pepper Robotic Harvesting. *Sensors* **2019**, *19*, 1390. [CrossRef] [PubMed]

21. Barth, R.; Hemming, J.; van Henten, E. Design of an eye-in-hand sensing and servo control framework for harvesting robotics in dense vegetation. *Biosyst. Eng.* **2016**, *146*, 71–84. [CrossRef]

22. Arivazhagan, S.; Shebiah, R.N.; Nidhyanandhan, S.S.; Ganesan, L. Fruit recognition using color and texture features. *J. Emerg. Trends Comput. Inf. Sci.* **2010**, *1*, 90–94.

23. Sa, I.; Ge, Z.; Dayoub, F.; Upcroft, B.; Perez, T.; McCool, C. Deepfruits: A fruit detection system using deep neural networks. *Sensors* **2016**, *16*, 1222. [CrossRef] [PubMed]

24. Chen, S.W.; Shivakumar, S.S.; Dcunha, S.; Das, J.; Okon, E.; Qu, C.; Taylor, C.J.; Kumar, V. Counting apples and oranges with deep learning: A data-driven approach. *IEEE Robot. Autom. Lett.* **2017**, *2*, 781–788. [CrossRef]

25. Kamilaris, A.; Prenafeta-Boldú, F.X. Deep learning in agriculture: A survey. *Comput. Electron. Agric.* **2018**, *147*, 70–90. [CrossRef]

26. McCool, C.; Perez, T.; Upcroft, B. Mixtures of lightweight deep convolutional neural networks: Applied to agricultural robotics. *IEEE Robot. Autom. Lett.* **2017**, *2*, 1344–1351. [CrossRef]

27. Milioto, A.; Lottes, P.; Stachniss, C. Real-time blob-wise sugar beets vs weeds classification for monitoring fields using convolutional neural networks. *ISPRS Ann. Photogramm. Remote Sens. Spat. Inf. Sci.* **2017**, *4*, 41. [CrossRef]

28. Berenstein, R.; Shahar, O.B.; Shapiro, A.; Edan, Y. Grape clusters and foliage detection algorithms for autonomous selective vineyard sprayer. *Intell. Serv. Robot.* **2010**, *3*, 233–243. [CrossRef]

29. Van Henten, E.; Van Tuijl, B.V.; Hemming, J.; Kornet, J.; Bontsema, J.; Van Os, E. Field test of an autonomous cucumber picking robot. *Biosyst. Eng.* **2003**, *86*, 305–313. [CrossRef]

30. Zemmour, E.; Kurtser, P.; Edan, Y. Dynamic thresholding algorithm for robotic apple detection. In Proceedings of the IEEE International Conference on the Autonomous Robot Systems and Competitions (ICARSC), Coimbra, Portugal, 26–28 April 2017; pp. 240–246.

31. Wang, J.; He, J.; Han, Y.; Ouyang, C.; Li, D. An adaptive thresholding algorithm of field leaf image. *Comput. Electron. Agric.* **2013**, *96*, 23–39. [CrossRef]

32. Jiang, J.A.; Chang, H.Y.; Wu, K.H.; Ouyang, C.S.; Yang, M.M.; Yang, E.C.; Chen, T.W.; Lin, T.T. An adaptive image segmentation algorithm for X-ray quarantine inspection of selected fruits. *Comput. Electron. Agric.* **2008**, *60*, 190–200. [CrossRef]

33. Arroyo, J.; Guijarro, M.; Pajares, G. An instance-based learning approach for thresholding in crop images under different outdoor conditions. *Comput. Electron. Agric.* **2016**, *127*, 669–679. [CrossRef]

34. Ostovar, A.; Ringdahl, O.; Hellström, T. Adaptive Image Thresholding of Yellow Peppers for a Harvesting Robot. *Robotics* **2018**, *7*, 11. [CrossRef]

35. Zhang, Y.J. A survey on evaluation methods for image segmentation. *Pattern Recognit.* **1996**, *29*, 1335–1346. [CrossRef]

36. Shmmala, F.A.; Ashour, W. Color based image segmentation using different versions of k-means in two spaces. *Glob. Adv. Res. J. Eng. Technol. Innov.* **2013**, *1*, 30–41.

37. Zheng, L.; Zhang, J.; Wang, Q. Mean-shift-based color segmentation of images containing green vegetation. *Comput. Electron. Agric.* **2009**, *65*, 93–98. [CrossRef]

38. Al-Allaf, O.N. Review of face detection systems based artificial neural networks algorithms. *Int. J. Multimed. Its Appl.* **2014**, *6*, 1. [CrossRef]

39. Sakthivel, K.; Nallusamy, R.; Kavitha, C. Color Image Segmentation Using SVM Pixel Classification Image. *World Acad. Sci. Eng. Technol. Int. J. Comput. Electr. Autom. Control Inf. Eng.* **2015**, *8*, 1919–1925.

40. Kurtser, P.; Edan, Y. Statistical models for fruit detectability: Spatial and temporal analyses of sweet peppers. *Biosyst. Eng.* **2018**, *171*, 272–289. [CrossRef]

41. Barth, R.; IJsselmuiden, J.; Hemming, J.; Van Henten, E.J. Data synthesis methods for semantic segmentation in agriculture: A Capsicum annuum dataset. *Comput. Electron. Agric.* **2018**, *144*, 284–296. [CrossRef]

42. Park, J.; Lee, G.; Cho, W.; Toan, N.; Kim, S.; Park, S. Moving object detection based on clausius entropy. In Proceedings of the IEEE 10th International Conference on Computer and Information Technology (CIT), Bradford, West Yorkshire, UK, 29 June–1 July 2010; pp. 517–521.

43. Hannan, M.; Burks, T.; Bulanon, D. A real-time machine vision algorithm for robotic citrus harvesting. In Proceedings of the 2007 ASAE Annual Meeting, American Society of Agricultural and Biological Engineers, Minneapolis, MN, USA, 17–20 June 2007; p. 1.

44. Bulanon, D.M.; Burks, T.F.; Alchanatis, V. Fruit visibility analysis for robotic citrus harvesting. *Trans. ASABE* **2009**, *52*, 277–283. [CrossRef]

45. Gunatilaka, A.H.; Baertlein, B.A. Feature-level and decision-level fusion of noncoincidently sampled sensors for land mine detection. *IEEE Trans. Pattern Anal. Mach. Intell.* **2001**, *23*, 577–589. [CrossRef]

46. Kanungo, P.; Nanda, P.K.; Ghosh, A. Parallel genetic algorithm based adaptive thresholding for image segmentation under uneven lighting conditions. In Proceedings of the IEEE International Conference on Systems Man and Cybernetics (SMC), Istanbul, Turkey, 10–13 October 2010; pp. 1904–1911.

47. Hall, D.L.; McMullen, S.A. *Mathematical Techniques in Multisensor Data Fusion*; Artech House: Norwood, MA, USA, 2004.

48. Woebbecke, D.M.; Meyer, G.E.; Von Bargen, K.; Mortensen, D.A. Plant species identification, size, and enumeration using machine vision techniques on near-binary images. In *Optics in Agriculture and Forestry*; SPIE-International Society for Optics and Photonics: Bellingham, WA, USA, 1993; Volume 1836, pp. 208–220.

49. Shrestha, D.; Steward, B.; Bartlett, E. Segmentation of plant from background using neural network approach. In Proceedings of the Intelligent Engineering Systems through Artificial Neural Networks: Proceedings Artificial Neural Networks in Engineering (ANNIE) International Conference, St. Louis, MO, USA, 4–7 November 2001; Volume 11, pp. 903–908.

50. Deng, J.; Dong, W.; Socher, R.; Li, L.J.; Li, K.; Fei-Fei, L. Imagenet: A large-scale hierarchical image database. In Proceedings of the IEEE Computer Vision and Pattern Recognition, Miami Beach, FL, USA, 20–25 June 2009.

51. Goutte, C.; Gaussier, E. A probabilistic interpretation of precision, recall and F-score, with implication for evaluation. In *European Conference on Information Retrieval*; Springer: Berlin, Germany, 2005; pp. 345–359.

52. Guyon, I. *A Scaling Law for the Validation-Set Training-Set Size Ratio*; AT&T Bell Laboratories: Murray Hill, NJ, USA, 1997; pp. 1–11.

53. Siegel, M.; Wu, H. Objective evaluation of subjective decisions. In Proceedings of the IEEE International Workshop on Soft Computing Techniques in Instrumentation, Measurement and Related Applications, Provo, UT, USA, 17 May 2003.

54. Zemmour, E. Adaptive thresholding algorithm for robotic fruit detection. Master's Thesis, Ben-Gurion University of the Negev, Beer-Sheva, Israel, 2018.

Article

A Comprehensive Study of the Potential Application of Flying Ethylene-Sensitive Sensors for Ripeness Detection in Apple Orchards

João Valente *[†], Rodrigo Almeida[†] and Lammert Kooistra

Laboratory of Geo-information Science and Remote Sensing, Wageningen University & Research,
6708 PB Wageningen, The Netherlands; rodrigo.almeida@wur.nl (R.A.); lammert.kooistra@wur.nl (L.K.)
*   Correspondence: joao.valente@wur.nl; Tel.: +31-628-398-164
†   These authors contributed equally to this work.

Received: 25 November 2018; Accepted: 14 January 2019; Published: 17 January 2019

Abstract: The right moment to harvest apples in fruit orchards is still decided after persistent monitoring of the fruit orchards via local inspection and using manual instrumentation. However, this task is tedious, time consuming, and requires costly human effort because of the manual work that is necessary to sample large orchard parcels. The sensor miniaturization and the advances in gas detection technology have increased the usage of gas sensors and detectors in many industrial applications. This work explores the combination of small-sized sensors under Unmanned Aerial Vehicles (UAV) to understand its suitability for ethylene sensing in an apple orchard. To accomplish this goal, a simulated environment built from field data was used to understand the spatial distribution of ethylene when subject to the orchard environment and the wind of the UAV rotors. The simulation results indicate the main driving variables of the ethylene emission. Additionally, preliminary field tests are also reported. It was demonstrated that the minimum sensing wind speed cut-off is 2 ms^{-1} and that a small commercial UAV (like Phantom 3 Professional) can sense volatile ethylene at less than six meters from the ground with a detection probability of a maximum of 10%. This work is a step forward in the usage of aerial remote sensing technology to detect the optimal harvest time.

Keywords: apple orchards; modeling and simulation; unmanned aerial vehicles; fruit ripeness; ethylene gas detection

---

1. Introduction

Sustainable agriculture is a top priority for all the governments and nations worldwide. Our population is growing fast, and our resources are getting more scarce each day. By 2050, our population will reach nine billion, requiring crop production to double in order to meet food demands [1].

An efficient way to increase the upcoming demands is to avoid fruit spoiling during the harvesting. Immature fruits result in poor quality and are subject to mechanical damage, and overripe fruit results in a soft and flavorless quality, with a very short shelf-life. In general, if the harvesting is done too early or too late, physiological disorders in the fruits will be provoked with the consequence of a shorter shelf-life [2]. These issues becomes more relevant with international trade of fruit and vegetables is increasing, making the shelf-life become an important marketing tool [3]. Therefore, the Optimal Harvest Date (OHD) will dictate the resulting fruit yield.

The OHD is usually obtained from maturity indices that take into account fruit chemical composition, like total soluble solids or total acidity, fruit physical properties, like firmness or color, fruit physiological changes, like aroma and ethylene emission rate, and finally, chronological features, like the number of days after planting or blooming [4].

Fruits' and vegetables' lifespan can be broken down into three steps: maturation (i.e., increase in fruit size), ripening (i.e., increase in flavor), and senescence (i.e., tissue death) [3]. Fruits that ripen after harvesting are denoted as climacteric fruits [4]. For climacteric fruits, like apples, the optimal harvest date occurs when the pre-climacteric minimum happens, equivalent to the end of the maturation process or the beginning of the ripening process, as illustrated in Figure 1.

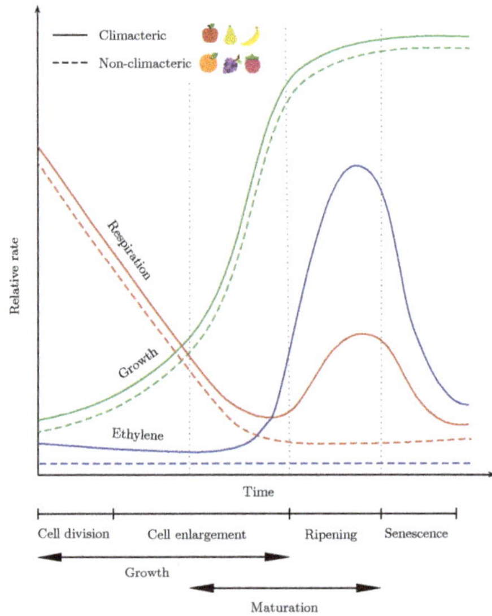

Figure 1. Relative rate of respiration, ethylene production, and growth in climacteric and non-climacteric fruits. Adapted from [5].

The fruits' distinctive aromas are characterized by a wide variety of Volatile Organic Compounds (VOCs) that are released during their maturation process [6]. The VOCs can be detected using a single-gas sensor or an array of gas sensors (also known as an electronic nose) [7]. An important VOC that is associated with fruit ripening is ethylene [8].

Ethylene (C_2H_4) is a gaseous phytohormone that regulates several growth and development processes in plants. In climacteric fruits, ethylene production regulates processes like flesh softening, color changes, and aroma emissions during ripening [9]. Ethylene can be measured via gas chromatography techniques, electrochemical sensors, and optical sensors [10].

Most current destructive and non-destructive methods of assessing fruit maturity require the sampling of individual fruit in the field and in some cases a further assessment in the lab [8,11,12]. That process is both labor intensive, since it requires an operator to physically go to the field and sample fruits, and dependent on the individual fruits that are sampled. Using the electronic noses and gas measurements with the fruit in concentration chambers provides less noise and augments the ethylene signal substantially, but it requires time and manpower to harvest and analyze the fruit, and at the same time, it is a method that is highly reliable on the sampling scheme used for the fruit [13,14].

The increasing availability of UAVs is a potential solution to acquire remotely and quickly data on a plot of land without the manual labor that would be required traditionally. The land manager/owner does not have to survey the plot manually, but can deploy a UAV. There are several aerial remote sensing applications in agriculture that were successful reported as an important contribution and step forward in Precision Agriculture (PA) practices [15].

Using the combination of airborne and electronic nose technology to map ethylene concentration in the orchard might give important information regarding fruit maturity in a fast and more representative way, without the need for additional labor. To the author's knowledge, no studies have been made so far regarding the potential limitations of this mapping application, but one could hypothesize that the sensitivity of the sensor and the atmospheric conditions during the measurements (i.e., wind speed and direction) are decisive.

Although plenty of research has been developed linking ethylene emission or VOC emission in apples to their maturity [8,16–18] and, in some literature, there are indications towards measuring ethylene in the field [11,19], to the authors' knowledge, no work of this sort has been carried out. This work should, because of this, be considered as a first attempt at understanding the potential and the limitations of such measurements, creating with it a theoretical framework from which further work can be developed.

On the other hand, in air quality monitoring systems, some development has occurred considering mobile measurement platforms such as a UAV, especially when it comes to gas source localization and adaptive path planning for gas plume estimations [20,21]. Additionally, the optimal position of a gas sensor in the UAV has been studied using a simulation approach by [22]. Although some successful gas sensing experiments have been reported with the sensor pointing down [23,24], no literature was found regarding the challenges of measuring in an orchard environment, especially when it comes to the dispersion dynamics and its effect on the measurement process. Additionally, most of these works were performed using artificial gas sources that are easily modeled and do not take into account the complexities of a natural emission source such as apples.

The main goal of this work is to evaluate if ethylene produced by apple orchards can be sensed using an electrochemical sensor mounted on a UAV. The evaluation is made using a model-based approach to identify the most influential factors for detection, after which the model results are compared to measurements from a UAV-mounted electrochemical sensor flown over an experimental apple orchard.

2. Materials and Methods

2.1. Study Area

The study area in which this research is based on is located in the Wageningen Plant Research for Flower bulbs, Nursery stock and Fruits in Randwijk, The Netherlands (see Figure 2a). A test plot of 0.17 ha of apple trees was selected (Study Area A). It had a length of 5 m in between tree rows and 1.1 m between trees in the row (5 × 1.1), which results in 14 lines of about 300 trees in the plot. Two apple (*Malus domestica*) cultivars are shown in Figure 2c: Junami and Golden Delicious (on the headers of each line for pollination purposes). Additionally, one other test plot was selected: Study Area B, which is a traditional apple orchard with 5 m between rows and 1 m between apple trees. The variety in this plot is Natyra. Only two lines were selected in Study Area B shown in Figure 2d.

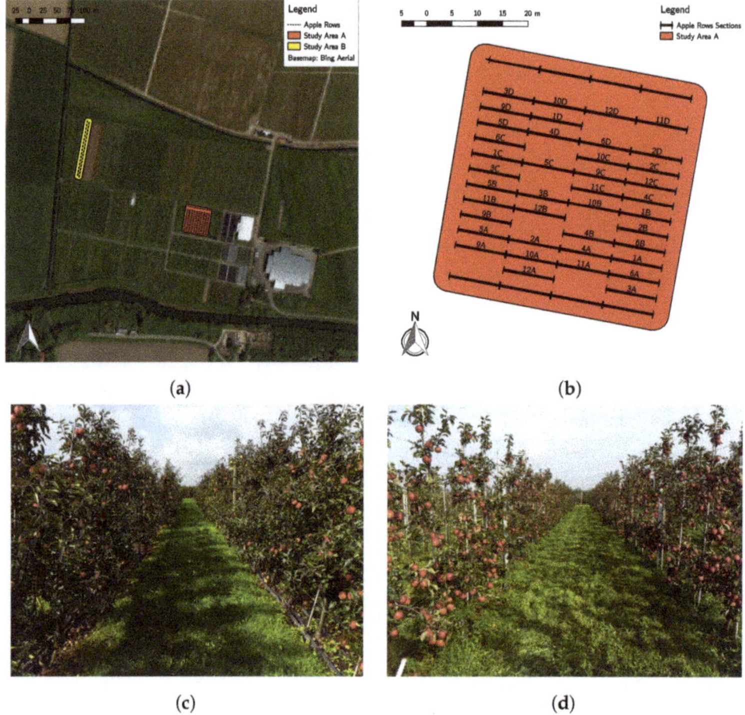

Figure 2. General and detailed map of the study area. (**a**) Map of the selected study areas in Randwijk (A and B). (**b**) Sections of apple lines used for fruit load assessment in Study Area A. Some lines show discontinuities since trees were removed in that section. (**c**) Junami and Golden Delicious cultivar. (**d**) Natyra cultivar.

2.2. Ethylene Flying Detector

The selected ethylene sensor was the Winsen ME4-C2H4, an electrochemical gas sensor, also referred to as a Taguchi gas sensor (TGS). According to the sensor specification sheet, it has a sensing range of 0–100 ppm of C_2H_4 and a response and recovery time of 100 s. Furthermore, the manufacturer indicates that the sensor has less than 10% of error.

In order to test both the ethylene sensor and the entire prototype, several preliminary experiments were conducted. With this, response times (amount of time it takes the sensor to detect the presence of ethylene) and recovery times (amount of time until the sensor signal returns to null after the ethylene source is removed) were tested. Figure 3 illustrates the results from the experiment in a controlled environment during four hours. The ethylene-sensitive sensor was placed inside a sealed plastic container of 40 cm × 50 cm × 40 cm with four *Junami* apples. After 3.5 h, the box was open, and after that, the sensor was placed outside the box.

The UAV-based measurements were conducted with the Phantom 3 Professional. This is a quadcopter drone weighing 1280 g with approximately 23 min of maximum flight time designed primarily for photo and video capture applications. The default payload (an HD camera) was removed and replaced with the ethylene sensor, as illustrated in Figure 4. The maximum payload of the UAV is 300 g, and the total payload was 218 g (very similar to the default payload).

Additionally, the sensor prototype is equipped with a memory card when the device is on, with a configurable measurement frequency, were measurements are recorded with the respective time-stamp and output signal from the sensor. In these experiments, one measurement per second (1 Hz) was determined as the measurement frequency.

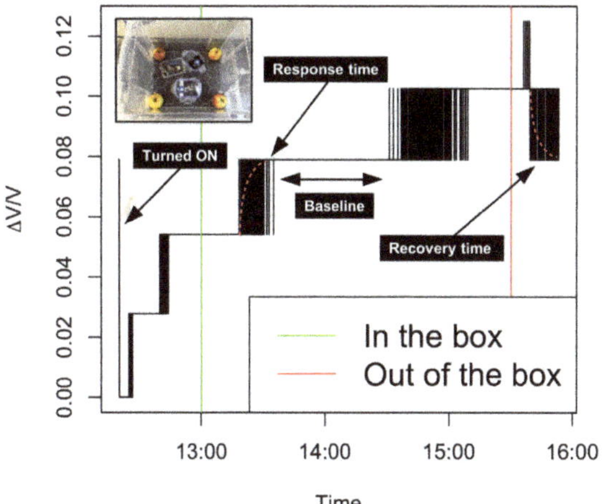

Figure 3. Tests conducted indoors in a sealed environment with an ethylene emission source (apples) that was placed in the box at the green line and removed at the red line.

The complete remote sensing system design for detecting and measuring ethylene is illustrated in Figure 4, and it has three main components: electrochemical sensor, Arduino board, and battery. The system was composed of commercially-available materials and open source tools. Finally, it can be easily acquired with a cost of less than 1000 Euros.

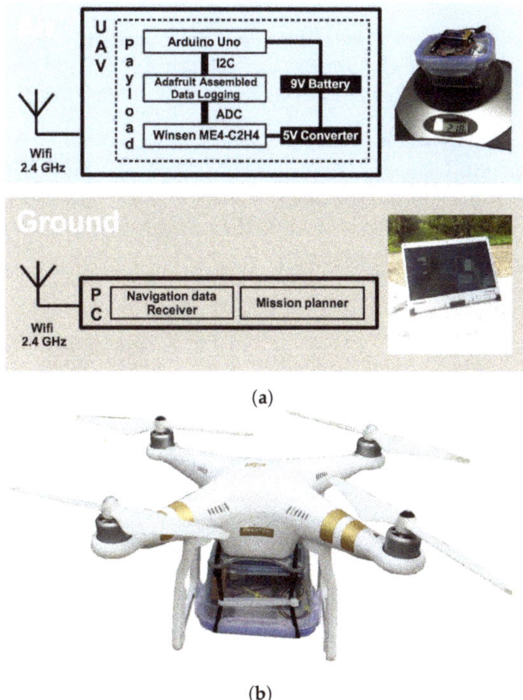

Figure 4. The ethylene flying-detector system: (**a**) air-ground system architecture and (**b**) Phantom 3 Professional (UAV) with the sensor prototype attached.

3. Determining the UAV Hovering Height

Understanding how the ethylene emission distributes above the orchard canopy is very important in order to define a starting sampling strategy. Determining the height above the orchard canopy where ethylene presents a higher concentration is not a trivial task mainly because there are several biophysical parameters such as the wind speed, temperature, and humidity. In this study, the wind speed (environment) and wind flow (UAV rotors) effect on the ethylene distribution was observed, while the temperature and humidity were omitted.

In order to determine the ideal sensing position, a modeling and simulation framework was developed to decrease the system deployment and testing times. Moreover, it allows a more reliable data acquisition by restricting the aerial sampling to areas within the orchard where a minimum ethylene concentration is expected. The modeling and simulation framework used GADEN, a gas dispersion simulation framework developed by [25], which is compatible with the ROS (Robot Operating System) [26].

3.1. Environment Wind Speed Modeling

Several parameters had to be obtained from the orchard field manager and from the research center where the experimental field is located in order to build the simulation workspace in GADEN. The parameters taken into account to simulate the ethylene distribution within the orchard when subject to wind were:

1. Ethylene emission (μ Lh^{-1} kg^{-1}). Each apple in the tree can in principle be at a different maturation stage and, therefore, have a different ethylene emission rate corresponding with three maturity stages.
2. Fruit position (height (m), direction ($°$)). Each apple in the tree can be at a different position in the canopy.
3. Fruit load (kg). Each apple tree can have a different amount of fruit.
4. Wind speed (ms^{-1}). In any given moment, the local wind speed and direction might vary. For this initial evaluation, two wind speed directions were considered.

These parameters were used to define an ethylene emission source for each tree represented in the simulation environment. Only one sample was taken per tree. Ideally, it would be possible to simulate each individual fruit on the tree canopy, but in this case, a simplification was performed using an artificial center for the total emissions of ethylene from a single tree. The distribution of the samples used in the simulations is illustrated in Figure 5.

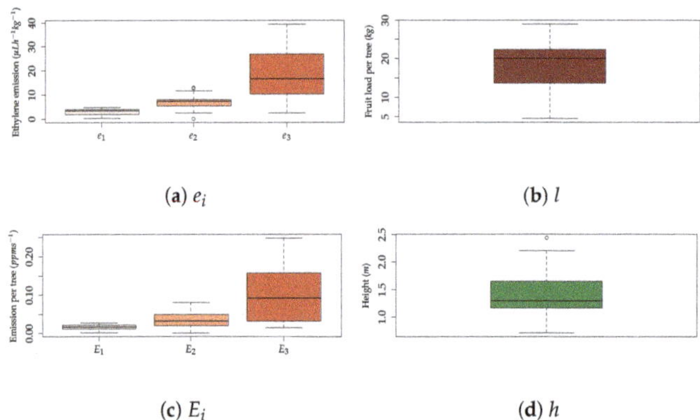

Figure 5. Distribution of the parameters used in the simulations: $\{e, E\}_1$, $\{e, E\}_2$, and $\{e, E\}_3$ stand for pre-climacteric, entering climacteric, and climacteric stages, respectively. Moreover, *l* stands for fruit load per tree and h for height.

This artificial center (P) can be described as the average position of the emission sources of the tree and is defined by a height (h) and direction in relation to the main stem (dir). This dir parameter in relation to the stem is defined in order to make the distribution of this parameter uniform, and therefore, the number of directions must be divisible by the amount of trees. In this case, six directions were defined, and each one was the sixth part of a circle, equivalent to 60°.

This simplification was applied mainly due to computational constraints. The simulator creates for each emission source a separate process, and for each process and time-step, the output is a simulation file of 90 MB. One can imagine that if each apple were simulated individually, the local memory of a standard computer would be very quickly surpassed. At the same time, according to our observations in the field, apples are usually clumped together in a branch, which means that the average distance between apples is in general small. Several branches can be further apart, but usually occupy one zone of the canopy. Figure 6 shows the orchard CAD model and the assumptions previously explained and used in the simulation process.

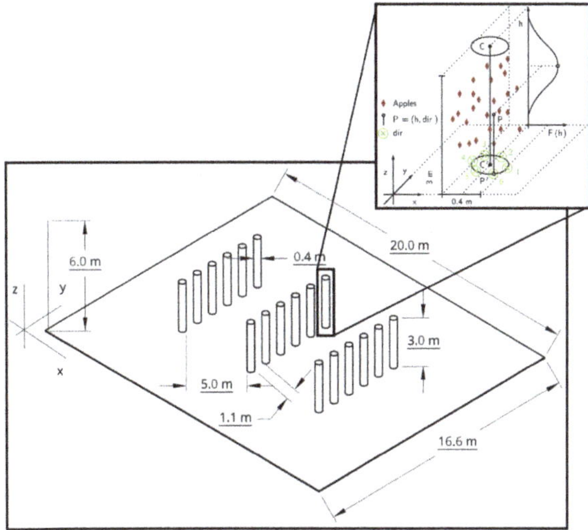

Figure 6. CAD model and respective parameters set in GADEN.

In the designed workspace, two environment inlets were set, the x-plane = 0 and the y-plane = 0. One of these inlets was chosen in order to simulate wind flow in a given direction: x for $\vec{x}$ and y for $\vec{y}$ (see Figure 7). The corresponding wind speed was assigned to this inlet, while the exact opposite plane (at the end of the environment) was set as a pressure outlet. All the other boundaries in the environment were set as walls with a slip setting. The computational fluid dynamics simulations were developed in SimScale, an online CFD software, with the recommended settings given in [25].

The number of ethylene-occupied cells in the environment is another important metric since it provides information on the probability of randomly finding an ethylene-filled cell. To get an understanding about which height is the most suitable to fly above the orchard, we must first look into the percentage of occupied cells with ethylene concentration above the canopy, as represented in Figure 8.

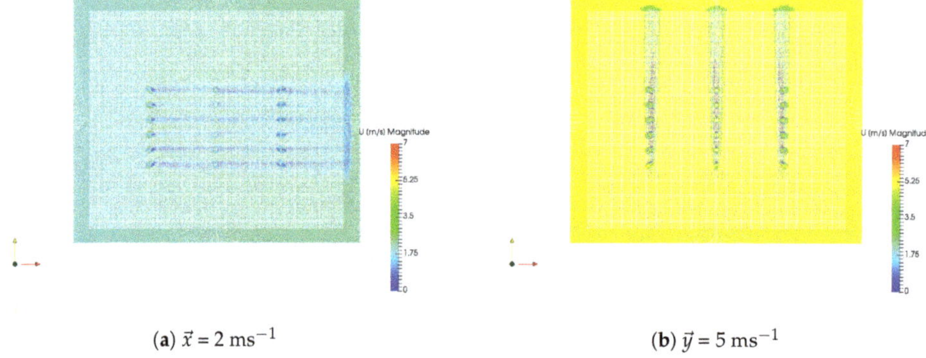

(**a**) $\vec{x} = 2 \text{ ms}^{-1}$

(**b**) $\vec{y} = 5 \text{ ms}^{-1}$

Figure 7. Wind flow simulations used as input for GADEN without considering the rotors' airflow.

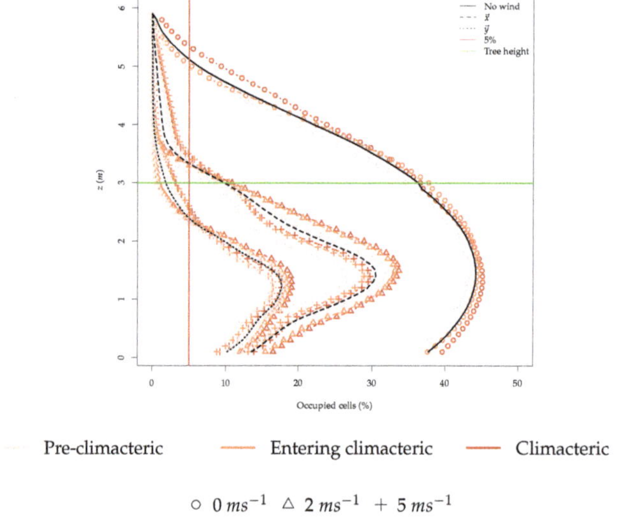

Pre-climacteric   —— Entering climacteric   —— Climacteric

$\circ \; 0 \, ms^{-1} \quad \triangle \; 2 \, ms^{-1} \quad + \; 5 \, ms^{-1}$

Figure 8. Percentage of occupied cells (cells with ethylene concentration higher than zero) in the environment across all time steps and simulations for the z-plane.

In almost all the simulations, less than 5% of the cells above the tree height were filled with ethylene. When wind speed was zero, there were more ethylene-filled cells above the tree height, but the majority of ethylene-filled cells can still be found under the tree height. It is also clear that the ethylene filled cells above the tree height had much lower ethylene concentration than the cells lower than the tree height. Therefore, the more likely place in the z axis to find ethylene-filled cells is between 1 and 2 m, where all simulations showed the biggest percentage of occupied cells.

To evaluate the impact of wind speed on the average ethylene concentration, Figure 9 was constructed. When looking at the environment, one can conclude that on average, a 1-ms^{-1} increase in wind speed results in a 30% decrease in average ethylene concentration. In the rows, the zone with higher average ethylene concentration, this decrease was 440%, while for in-between rows, this was only 110%. This difference is also accompanied by a very large difference in absolute ethylene concentration. This gives us an indication that choosing to sample in the rows might yield a higher concentration, but this measurement is very sensitive to the wind conditions.

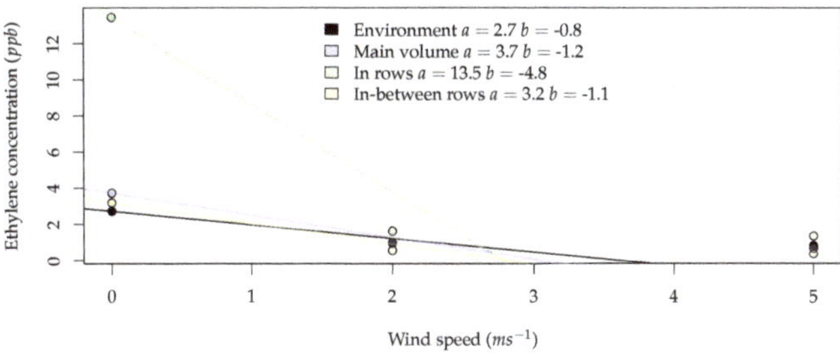

Figure 9. Relation between wind speed and average ethylene concentration in the four different zones. The colored lines represent the trend line for each zone, as given by the equation $y = a + bx$, where b is the decrease in average ethylene concentration (ppb) per additional unit of wind speed (ms^{-1}).

From Figures 8 and 9, it can be inferred that higher concentration levels of ethylene can be found below the trees and that the wind speed cut-off for the best practice is 2 ms^{-1}. In the next section, the rotors' airflow affect will be added to the environment to corroborate the results previously obtained omitting the rotors' airflow.

3.2. Rotors' Airflow Modeling

In order to simulate the effect of a UAV flying in the orchard, two different drone positions were considered: over the row (Position 1) and in between rows (Position 2). The drone over the row was positioned at 4 m, while the drone in between rows was positioned at 2 m. Only one wind scenario was considered for these simulations, $\vec{x} = 2$ ms^{-1}, in line with the results obtained in the previous section. This results in a total of six drone simulations, as exemplified in Table 1.

Table 1. Summary of drone simulator runs. The simulation number (#) will be used as a reference for naming each of these scenarios in the following sections.

#	Ethylene Emission	Wind Direction	Wind Speed (ms^{-1})	Drone Position
1.1	Pre-climacteric	$\vec{x}$	2	$(10, 7.8, 4)$
1.2	Entering climacteric			
1.3	Climacteric			
2.1	Pre-climacteric	$\vec{x}$	2	$(12.5, 7.8, 2)$
2.2	Entering climacteric			
2.3	Climacteric			

The wind flow caused by the rotors of the drone was modeled as four square air inlets with a given wind speed in the negative $\vec{z}$ direction. The squares had a width of 0.1 m, which is approximately the diameter of a single rotor in the Phantom 3 Professional. A relationship exists between the rotation speed of the propeller and the resulting wind speed generated, or thrust [27]. Taking the example of the Phantom 3 Professional in hovering flight in normal conditions, the rotors spin at around 8000 rpm [28], which results in an airflow of about 18 ms^{-1}. This wind speed was assigned to the velocity inlets mentioned above. The resulting wind flow simulations used as input for the GADEN simulations are displayed in Figure 10.

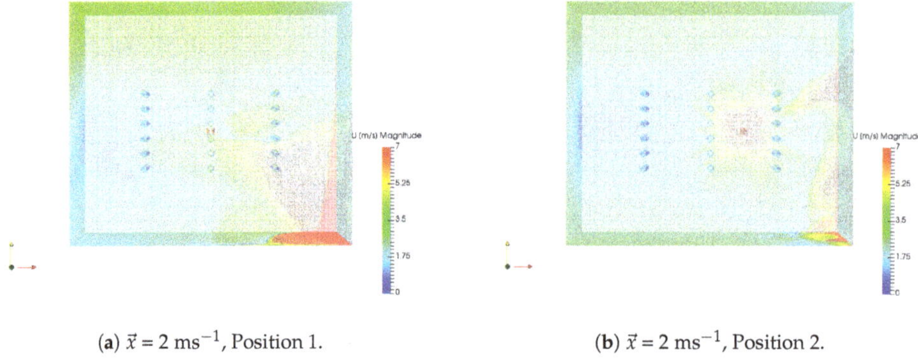

(**a**) $\vec{x}$ = 2 ms^{-1}, Position 1. (**b**) $\vec{x}$ = 2 ms^{-1}, Position 2.

Figure 10. Drone wind flow simulations used as input for GADEN.

The biggest difference between the ethylene concentration distribution with and without a drone appears to be the range of values that are present. When looking at the climacteric simulations with the drone in both positions, the maximum concentration was about 250 ppb, while without the drone, the same conditions yielded a maximum of 300 ppb. This range also decreased substantially with the height of the drone (Position 2 to 1), from 250 to 150 ppb, as Figure 11 clearly shows. There was also a gas concentration effect right under the drone position where it appeared that the wind displacement of the gas decreased.

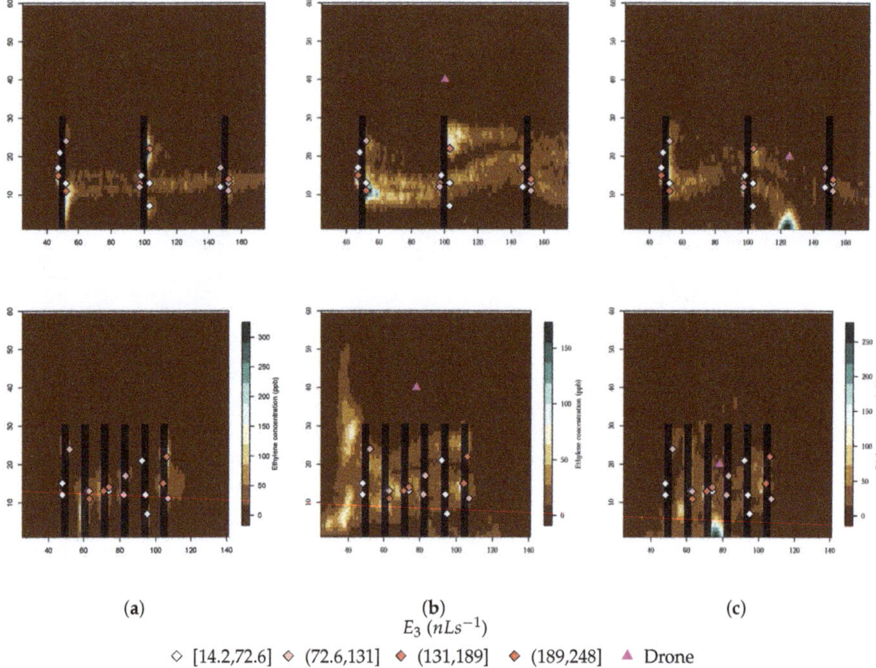

(**a**) (**b**) (**c**)

$E_3\ (nLs^{-1})$

◇ [14.2,72.6] ◇ (72.6,131] ◆ (131,189] ◆ (189,248] ▲ Drone

Figure 11. Maximum ethylene concentration in the *xz*-plane (top plots) and *yz*-plane (bottom plots) for the drone simulations in the climacteric stage: (**a**) omitting rotor wind flow; (**b**) Drone Position 1; and (**c**) Drone Position 2. The ethylene sources' position and emission rate are also provided at the bottom.

When looking at the immediate vicinity of the position of the drone, a clear difference was detected between Position 1 and 2, as Figure 12 illustrates. While no ethylene was detected around Position 1, at Position 2, in every simulated time step, ethylene was present. This is a consequence of the concentration effect mentioned above.

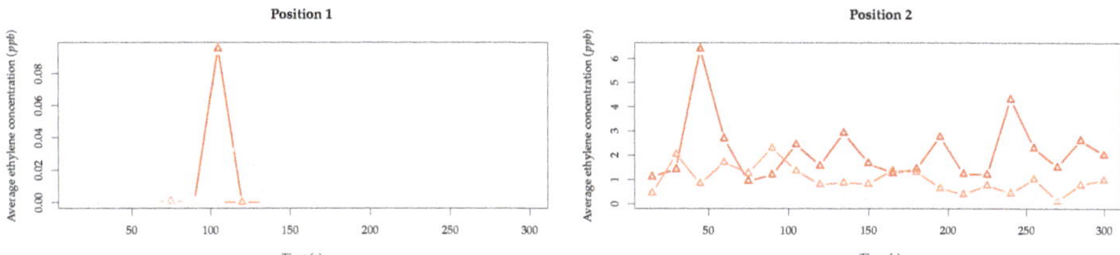

Figure 12. Average ethylene concentration across time in the vicinity of the drone position (± 0.2 m in *xyz*) for Positions 1 and 2.

The distribution of the occupied cells in the environment was also very different, as Figure 13 illustrates. The percentage of occupied cells was in general lower due to the increase in average wind speed in the environment, and especially on the z axis, a compression of the occupied cells closer to $z = 0$ was visible, depending on the height of the drone, which further confirms the concentration of ethylene effect described previously. This compression results in a higher percentage of occupied cells closer to the ground.

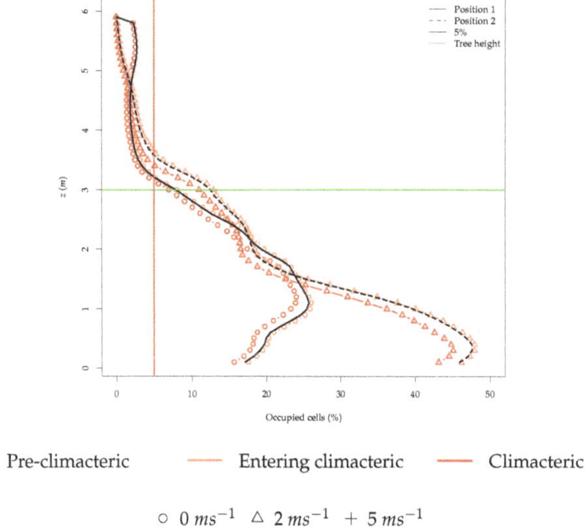

Figure 13. Percentage of occupied cells (cells with ethylene concentration higher than zero) in the environment across all time steps and simulations for the z-plane.

In general, we can say that the drone flying overhead had two main effects: a decrease in average ethylene concentration in the orchard, directly correlated with the height of the drone (4 m caused more gas dispersion than 2 m) and a concentration of gas directly under the drone, close to the ground (an effect that was more discernible at a 2-m height). In general, the drone flying overhead at 4 m caused a decrease in average ethylene concentration of 95%, while at 2 m, a decrease in 90%.

4. Field Tests on the Orchard

The results obtained in Figure 3 provide evidence that both the wind speed and rotor wind flow played an important role in ethylene sensing. The behavior observed in the simulation is used now as the reference to define boundaries in the field tests, observe other measuring heights, and analyze the feasibility of this practice in the real orchard environment from an aerial mission perspective.

4.1. Sampling Scheme

In order to analyze the sensor functioning and detect ethylene in the selected plot on the ground and using a UAV, both a spatial and temporal sampling scheme was defined. The measurements with the UAV were conducted also per measurement point: at 6 m and 12 m during 120 s. This difference in sensing time has to do with practical constraints related to the battery life of the UAV. Please refer to Figure 14.

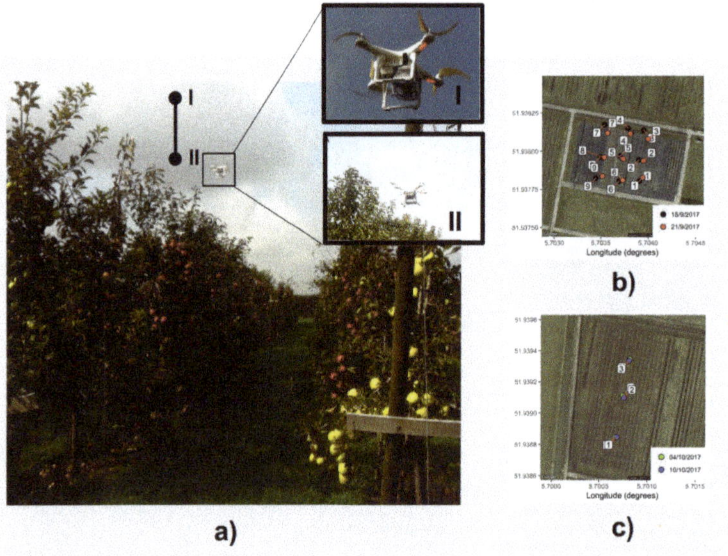

Figure 14. Hovering and sampling at two different position per experiment: (**a**) low and high heights within the orchard; (**b**) samples on Study Field A over the two days; and (**c**) samples on Study Field B over the two days.

For Study Area A (Figure 2a,c), nine measurement points were selected for UAV-based measurements on two different days: 15 and 21 September 2017. These points were selected using a three by three grid in the plot and placing the point roughly in the center of each grid. Additionally, two UAV-based measurements were conducted also per measurement point: at 6 m and 12 m above the ground (3 m and 9 m above the canopy), with a sensing time of less than 3 min. This difference in sensing time has to do with practical constraints related to the battery life of the UAV employed. With a battery charge from the Phantom 3 Professional, Study Area A was sampled a maximum of nine times.

For Study Area B (Figure 2a,d), the same spatial sampling approach as Study Area A was used, but in this case, only two lines were taken into account. Three points were selected in the middle of these two apple lines, where measurements were taken as described before, with a different height in the UAV measurements: 4 and 6 m. These measurements were also performed on two different dates: 4 and 10 October 2017.

4.2. System Deployment and Sensitivity Tests

A weather station adjacent to the plot was used to provide real-time atmospheric measurements of wind speed and direction, air temperature, and air moisture during the sampling period on the different dates. The average wind speed on 15 September was about 3 ms^{-1}, while on 21 September, it was about 3.2 ms^{-1}. The average wind speed on 4 October was about 5 ms^{-1}, while on 10 October, it was about 3.9 ms^{-1}. Furthermore, the first harvesting day for Study Areas A and B was, respectively, 28 September and 10 October.

The average measurements obtained per day are shown in Figure 15. Looking at the UAV-based measurements, no output was measured above 10% of the reference signal, thus achieving a maximum of 0.5 ppm (500 ppb). We can notice that there was no variation in the first two days of measurements (Study Field A). Nevertheless, there was an increase from the 4–10 October, which in this case suggested that there was more ethylene concentration on the second date (which was expected). The decrease in wind speed might also explain some of this variation from one date to the other.

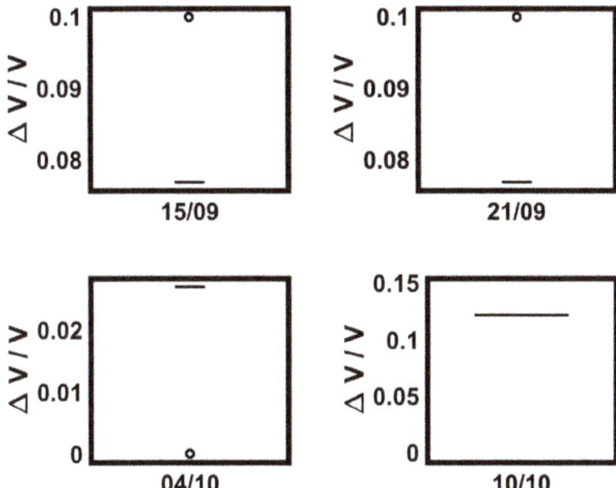

Figure 15. Sensor voltage output for aerial measurements on different days.

When flying at different heights (see Figure 16), an interesting behavior was observed: the measurements performed at a higher altitude (12 m) showed no variation from the baseline; at 6 m, there were more outliers that indicated more detection peaks, and at only 4 m, some variation was actually detected.

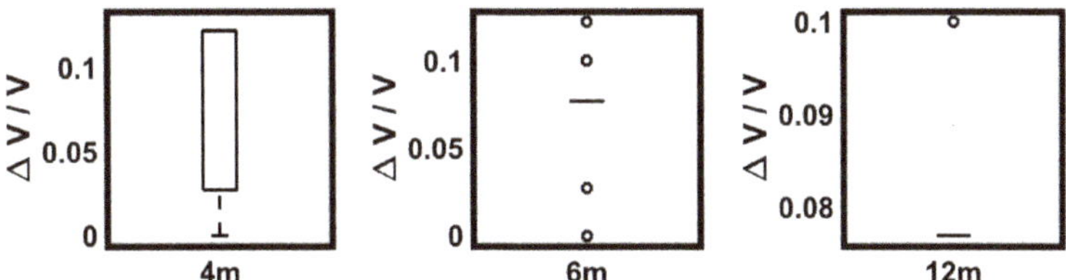

Figure 16. Sensor voltage output for aerial measurements at different heights.

5. Discussion

Although plenty of research has been developed linking ethylene emission or VOC emission in apples to their maturity [8,16–18] and in some literature there are indications towards measuring ethylene in the field [11,19], to the authors' knowledge, no work of this sort has been carried out, and there is to date no investigation addressing flying ethylene-sensitive sensor systems.

The modeling part of this study has shown that several factors influence the ethylene emission from the apple trees: these will be discussed in more detail below. We included the main influencing factors in the model, but additional ones, like time of the day, might be important, as well. These should be evaluated in follow-up studies.

In this section, the lessons learned from this study and major outcomes will be summed up.

5.1. Wind Speed and Rotors' Effect

The hypothesis that wind speed decreases the probability of ethylene detection was verified through simulation. Although, this supposition was expected, in the literature, there was not any discussion of the the maximum wind speed to ensure the minimum sensing. It was shown through simulation that the wind speed cut-off was $2\,\mathrm{ms}^{-1}$ (Figure 9).

Moreover, several authors discussed the rotor effect when using this sensor technology mounted in multi-rotor UAVs. It was stated that this is true, and there is a small margin left for detection that may be improved at a determinate hovering height (Figure 8).

5.2. Theoretical versus Practical Optimal Sampling Height

Simulation and practical results agreed that for wind speeds higher than $2\,\mathrm{ms}^{-1}$, there would be very few or almost no detections. The minimum wind speed recorded during the field campaign was $3\,\mathrm{ms}^{-1}$, and indeed, the sensor variation was very low.

The most important analytical outcome reinforced by the field tests was that flying lower would increase ethylene detection. Furthermore, flying close to or under the tree canopy gave a better result, although the margin for detection was limited, as shown in Figure 13. For the sake of the security and safety of the platform and the flying crew, the drone did not fly under the tree canopy, but it was stated that as the height decreased, more variation was observed (see Figure 16).

This study suggests that the UAV overflight should be performed at the lowest possible height to decrease the impact of the wind flow generated by the rotors on the ethylene distribution. However, further flying maneuvers should be explored when flying within or close to the orchard.

5.3. Discrete versus Continuous Sampling

It is important to note that the UAV measurements will only be considered during hovering flights over the determined measurement point. With that, the data acquired during the path of the UAV in the study area was not determinant, but it is an important point for further research considering moving and continuous measurements. However, the increased response time from the sensors (>90 s) should be taken into consideration in the sampling strategy to adopt.

5.4. Ethylene Detection over the Season and Inferring the OHD

The simulations showed that the range of ethylene concentration in an orchard was in the ppb range, and wind speed had a very big impact on this ethylene concentration. According to [29], the usual measuring range for VOCs starts at 100 ppb, and not very many sensors offer a sub-ppb range. If the state-of-the art of the technology does not provide a sensor with such characteristics, this might decrease the feasibility of this remote sensing strategy.

The observations when entering the climacteric stage were expected to be stronger than the ones observed in Figure 12. In order to determine the OHD, an ethylene increase during this stage was expected. Therefore, this strengths even more the idea that a sensor with more sensibility must be considered in future experiments.

5.5. Feasibility of Using Flying Ethylene-Sensitive Sensors

The challenges of measuring ethylene with a UAV in an orchard environment especially when it comes to the dispersion dynamics and its effect on the measurement process was not found in the literature.

On the other hand, in air quality monitoring systems, some development has happened considering mobile measurement platforms such as a UAV, especially when it comes to gas source localization and adaptive path planning for gas plume tracking [20,21]. The benefits of using the mobile platform for air quality monitoring and also for the purpose of this research are similar: they can offer high resolution sampling both at a spatial and temporal level at a low cost [23]. However, most of these works are performed using artificial gas sources that are easily modeled, and none takes into account the complexities of a natural emission source such as apples trees.

The optimal position of the gas sensor also has been discussed [22,23] and could have been a valuable reference in this study. However, in one work, the authors performed experiments indoors, inside a garage, and in the other work, the outcomes provided were very limited. Both authors suggested different sensors placements: pointing down separated from the main frame and on the top of the platform. In this study, the sensor was used pointing down because the frame of the UAV did not allow other configurations. Further, studies are needed to explore the position suggestion from the previous authors. Nevertheless, the simulations provided in this study reveal that higher concentrations values will be found mostly below the platform (see Figure 13).

5.6. UAV vs. UGV

The discussion of which mobile vehicle will perform better in a determinate agricultural management task is not new. In general, UAVs have more of a sensing role, like aerial surveying, where there is the need to increase spatial resolution; while Unmanned Ground Vehicles (UGV) have more of an actuation role, where there an action should be performed, such as mechanical weeding [30].

In this study, we were interested in a versatile platform that could carry different instrumentation and sample the orchard on different 2D and 3D positions. Moreover, while this could be achieved at different heights with the UAV, it would be limited to a static height with the UGV. Moreover, UAVs are considerable better than UGVs, as regards the price, maintenance, and portability. Summing up, they can offer high resolution sampling both at a spatial and temporal level at a low cost [23].

6. Conclusions

This is the first study to investigate the feasibility of using a flying ethylene-sensitive sensor systems in a fruit orchard only some days before being harvested. A simulated environment built from field data was used to understand the spatial distribution of ethylene within the apple orchard, to define the field sampling boundaries, and to evaluate how this influences the detection from a miniaturized sensor on a UAV. Finally, some preliminary tests in the orchard field were carry out to elucidate the sensor;s sensitivity and to contrast with the theoretical study.

The drone flight effect on the ethylene distribution was tested, and we concluded that flying at a higher altitude will cause more disturbance and lower the average ethylene concentration than flying lower. At the same time, at higher altitude, almost no ethylene is present in the vicinity of the drone. In general, the drone flying overhead at 4 m causes a decrease in average ethylene concentration of 95%, while at 2 m, a decrease of 90%. The detection margin is short and not sufficient to infer the fruit maturity, where increased variability over the season is expected. With these results, the issue of the measurement system sensitivity is further confirmed: a requirement for a sub-ppb ethylene sensor is clearly supported.

The use of a UAV to perform ethylene measurements in an uncontrolled environment such as an apple orchard still needs to be further explored, but it is suggested that future practices using this system are imminent with further research. The effect different UAV propeller spans on the intensity of dispersion of the gas and also detailed response models of different sensor models are, among others, pressing issues to be considered in the future.

Author Contributions: Conceptualization, J.V. and R.A.; methodology, J.V. and R.A.; software, R.A.; validation, R.A.; formal analysis, J.V. and R.A.; investigation, J.V. and R.A.; resources, J.V. and L.K.; writing, original draft preparation, all; writing, review and editing, all; supervision, J.V. and L.K.; project administration, J.V.; funding acquisition, L.K.

Funding: This work was supported by the SPECTORSproject (143081), which is funded by the European cooperation program INTERREGDeutschland-Nederland.

Acknowledgments: The authors would also like to thank Pieter van Dalfsen, whom is with Wageningen Plant Research, for providing the field test and data that helped us to understand more about the apple orchard lifecycle.

Conflicts of Interest: The authors declare no conflict of interest.

References

1. Alexandratos, N.; Bruinsma, J. *World Agriculture Towards 2030/2050: The 2012 Revision*; ESA Working Paper No. 12-03; FAO, Agricultural Development Economics Division: Roma, Italy, 2012.

2. Kader, A.A. Fruit maturity, ripening, and quality relationships. *Int. Symp. Effect Pre- Postharvest Factors Fruit Storage* **1997**, *485*, 203–208. [CrossRef]

3. Barbosa-Cánovas, G.V. *Handling and Preservation of Fruits and Vegetables by Combined Methods for Rural Areas: Technical Manual*; Number 149 in 1; Food & Agriculture Organization: Roma, Italy, 2003.

4. Knee, M. *Fruit Quality and Its Biological Basis*; CRC Press: Boca Raton, FL, USA, 2002; Volume 9.

5. Paul, V.; Pandey, R.; Srivastava, G.C. The fading distinctions between classical patterns of ripening in climacteric and non-climacteric fruit and the ubiquity of ethylene—An overview. *J. Food Sci. Technol.* **2011**, *49*, 1–21. [CrossRef] [PubMed]

6. Baietto, M.; Wilson, A. Electronic-Nose Applications for Fruit Identification, Ripeness and Quality Grading. *Sensors* **2015**, *15*, 899–931. [CrossRef] [PubMed]

7. Arshak, K.; Moore, E.; Lyons, G.; Harris, J.; Clifford, S. A review of gas sensors employed in electronic nose applications. *Sens. Rev.* **2004**, *24*, 181–198. [CrossRef]

8. Ma, L.; Wang, L.; Chen, R.; Chang, K.; Wang, S.; Hu, X.; Sun, X.; Lu, Z.; Sun, H.; Guo, Q.; et al. A Low Cost Compact Measurement System Constructed Using a Smart Electrochemical Sensor for the Real-Time Discrimination of Fruit Ripening. *Sensors* **2016**, *16*, 501. [CrossRef] [PubMed]

9. Génard, M.; Gouble, B. ETHY. A theory of fruit climacteric ethylene emission. *Plant Physiol.* **2005**, *139*, 531–545. [CrossRef] [PubMed]

10. Cristescu, S.M.; Mandon, J.; Arslanov, D.; Pessemier, J.D.; Hermans, C.; Harren, F.J.M. Current methods for detecting ethylene in plants. *Ann. Bot.* **2012**, *111*, 347–360. [CrossRef] [PubMed]

11. Łysiak, G. Measurement of ethylene production as a method for determining the optimum harvest date of "Jonagored" apples. *Folia Hortic.* **2014**, *26*. [CrossRef]

12. Gómez, A.H.; Wang, J.; Hu, G.; Pereira, A.G. Electronic nose technique potential monitoring mandarin maturity. *Sens. Actuators B Chem.* **2006**, *113*, 347–353. [CrossRef]

13. Kathirvelan, J.; Vijayaraghavan, R. An infrared based sensor system for the detection of ethylene for the discrimination of fruit ripening. *Infrared Phys. Technol.* **2017**, *85*, 403–409. [CrossRef]

14. Lihuan, S.; Liu, W.; Xiaohong, Z.; Guohua, H.; Zhidong, Z. Fabrication of electronic nose system and exploration on its applications in mango fruit (M. indica cv. Datainong) quality rapid determination. *J. Food Meas. Charact.* **2017**, *11*, 1969–1977. [CrossRef]

15. Zhang, C.; Kovacs, J.M. The application of small unmanned aerial systems for precision agriculture: A review. *Precis. Agric.* **2012**, *13*, 693–712. [CrossRef]

16. Brezmes, J.; Llobet, E.; Vilanova, X.; Saiz, G.; Correig, X. Fruit ripeness monitoring using an electronic nose. *Sens. Actuators B Chem.* **2000**, *69*, 223–229. [CrossRef]

17. Saevels, S.; Lammertyn, J.; Berna, A.Z.; Veraverbeke, E.A.; Di Natale, C.; Nicolai, B.M. Electronic nose as a non-destructive tool to evaluate the optimal harvest date of apples. *Postharvest. Biol. Technol.* **2003**, *30*, 3–14. [CrossRef]

18. Young, H.; Rossiter, K.; Wang, M.; Miller, M. Characterization of Royal Gala Apple Aroma Using Electronic Nose TechnologyPotential Maturity Indicator. *J. Agric. Food Chem.* **1999**, *47*, 5173–5177. [CrossRef] [PubMed]

19. Pathange, L.P.; Mallikarjunan, P.; Marini, R.P.; O'Keefe, S.; Vaughan, D. Non-destructive evaluation of apple maturity using an electronic nose system. *J. Food Eng.* **2006**, *77*, 1018–1023. [CrossRef]

20. Neumann, P.P.; Bennetts, V.H.; Lilienthal, A.J.; Bartholmai, M.; Schiller, J.H. Gas source localization with a micro-drone using bio-inspired and particle filter-based algorithms. *Adv. Robot.* **2013**, *27*, 725–738. [CrossRef]

21. Rossi, M.; Brunelli, D.; Adami, A.; Lorenzelli, L.; Menna, F.; Remondino, F. Gas-Drone: Portable gas sensing system on UAVs for gas leakage localization. In Proceedings of the IEEE SENSORS 2014 Proceedings, Valencia, Spain, 2–5 November 2014. [CrossRef]

22. Roldán, J.; Joossen, G.; Sanz, D.; del Cerro, J.; Barrientos, A. Mini-UAV Based Sensory System for Measuring Environmental Variables in Greenhouses. *Sensors* **2015**, *15*, 3334–3350. [CrossRef] [PubMed]

23. Villa, T.F.; Gonzalez, F.; Miljievic, B.; Ristovski, Z.D.; Morawska, L. An overview of small unmanned aerial vehicles for air quality measurements: Present applications and future prospectives. *Sensors* **2016**, *16*, 1072. [CrossRef] [PubMed]

24. Neumann, P.P. Gas Source Localization and Gas Distribution Mapping with a Micro-Drone. Ph.D. Thesis, Bundesanstalt für Materialforschung und-prüfung (BAM), Berlin, Germany, 2013.

25. Monroy, J.; Hernandez-Bennets, V.; Fan, H.; Lilienthal, A.; Gonzalez-Jimenez, J. GADEN: A 3D Gas Dispersion Simulator for Mobile Robot Olfaction in Realistic Environments. *Sensors* **2017**, *17*, 1479. [CrossRef] [PubMed]

26. Quigley, M.; Conley, K.; Gerkey, B.P.; Faust, J.; Foote, T.; Leibs, J.; Wheeler, R.; Ng, A.Y. ROS: An open-source Robot Operating System. *ICRA Workshop Open Source Softw.* **2009**, *3*, 5.

27. Allain, R. Modeling the Thrust from a Quadcopter. Available online: https://www.wired.com/2014/05/modeling-the-thrust-from-a-quadcopter/ (accessed on 1 April 2018).

28. DJI. *Phantom 3 Professional User Manual*; DJI: Shenzhen, China, 2017.

29. Spinelle, L.; Gerboles, M.; Kok, G.; Persijn, S.; Sauerwald, T. Review of Portable and Low-Cost Sensors for the Ambient Air Monitoring of Benzene and Other Volatile Organic Compounds. *Sensors* **2017**, *17*, 1520. [CrossRef] [PubMed]

30. Conesa-Muñoz, J.; Valente, J.; del Cerro, J.; Barrientos, A.; Ribeiro, A. A Multi-Robot Sense-Act Approach to Lead to a Proper Acting in Environmental Incidents. *Sensors* **2016**, *16*, 1269. [CrossRef] [PubMed]

Article

Investigating 2-D and 3-D Proximal Remote Sensing Techniques for Vineyard Yield Estimation

Chris Hacking [1], Nitesh Poona [1], Nicola Manzan [2] and Carlos Poblete-Echeverría [3,*]

[1] Department of Geography and Environmental Studies, Stellenbosch University, Private Bag X1, Matieland 7602, South Africa

[2] Dipartimento di Scienze AgroAlimentari, Ambientali e Animali, University of Udine, Via delle Scienze 208, 33100 Udine, Italy

[3] Department of Viticulture and Oenology, Faculty of AgriSciences, Stellenbosch University, Private Bag X1, Matieland 7602, South Africa

* Correspondence: cpe@sun.ac.za; Tel.: +27-21-808-2747

Received: 27 June 2019; Accepted: 20 August 2019; Published: 22 August 2019

Abstract: Vineyard yield estimation provides the winegrower with insightful information regarding the expected yield, facilitating managerial decisions to achieve maximum quantity and quality and assisting the winery with logistics. The use of proximal remote sensing technology and techniques for yield estimation has produced limited success within viticulture. In this study, 2-D RGB and 3-D RGB-D (Kinect sensor) imagery were investigated for yield estimation in a vertical shoot positioned (VSP) vineyard. Three experiments were implemented, including two measurement levels and two canopy treatments. The RGB imagery (bunch- and plant-level) underwent image segmentation before the fruit area was estimated using a calibrated pixel area. RGB-D imagery captured at bunch-level (mesh) and plant-level (point cloud) was reconstructed for fruit volume estimation. The RGB and RGB-D measurements utilised cross-validation to determine fruit mass, which was subsequently used for yield estimation. Experiment one's (laboratory conditions) bunch-level results achieved a high yield estimation agreement with RGB-D imagery (r^2 = 0.950), which outperformed RGB imagery (r^2 = 0.889). Both RGB and RGB-D performed similarly in experiment two (bunch-level), while RGB outperformed RGB-D in experiment three (plant-level). The RGB-D sensor (Kinect) is suited to ideal laboratory conditions, while the robust RGB methodology is suitable for both laboratory and in-situ yield estimation.

Keywords: Kinect sensor; RGB; RGB-D; image segmentation; colour thresholding; bunch area; bunch volume; point cloud; mesh; surface reconstruction

---

1. Introduction

Modern-day viticulture has seen an increase in the use of robust scientific methods combined with new technologies to improve overall production [1]. Precision farming, a direct result of the modernisation in agriculture, can be discipline-specific—i.e., specific to horticulture or viticulture. Precision viticulture aims to effectively manage production inputs to improve yield and grape quality while reducing the environmental impact of farming [2]. The use of remote sensing technology and techniques in precision viticulture allows variability to be monitored at vineyard level, per individual block or on a vine basis. Aspects such as vine shape, size and vigour can be observed, providing more accurate yield and fruit quality information [3].

Yield estimation provides information to the winegrower that can be used to manage the vineyard, optimising quality and yield [4]. Awareness of the estimated yield allows the vineyard manager to manipulate the vines to obtain the desired grape characteristics, and provides an effective plan for use during the winemaking process [5]. Accurate yield forecasting assists with logistical planning, both during and after the harvest; for example, what volume will be harvested, where the grapes will be stored, and an expected market price [5].

Wolpert and Vilas [6] outlined a two-step method for vineyard yield estimation. The start of the process determines the number of bunches situated on individual vines early in the season. Subsequent determination of bunch weights occurs at vèraison. Unfortunately, the two-step method is labour-intensive, error-prone and destructive in the estimation process. Additionally, De la Fuente et al. [7] presented yield prediction models using destructive, manually collected data between fruit-set and vèraison, aligning with the more 'classical' two-step method. To overcome the limitations of the manual methods, modern techniques have employed sensors attached to automatic harvesters to monitor yield during the harvesting process [3]. Yield estimation before harvest is becoming possible, with increasing accuracies when making use of non-invasive proximal remote sensing (PRS) technology and techniques [8–12].

A common PRS approach employs 2-D (2-dimensional) RGB (Red, Green, Blue) imagery, captured with a digital camera for yield estimation; for example [8,9,13]. The 2-D approach can be categorised into two steps: (i) image segmentation (to differentiate the bunch from the background); and (ii) yield estimation, using a suitable bunch metric (i.e., the pixel count of the bunch area in an image).

Diago et al. [13] used image classification (for segmentation) and a bunch metric to classify 'background noise' and 'grape' classes. The authors achieved a testing r^2 of 0.73. An alternative segmentation approach to image classification uses colour thresholding to differentiate grapes from the background. Dunn and Martin [8] presented an RGB colour thresholding approach that applied specific thresholds to the colour properties of an RGB image, generating a binary image—background and grapes. The authors achieved an r^2 of 0.72. A similar thresholding approach was adopted by Liu, Marden and Whitty [9] and Font et al. [14]. Liu, Marden and Whitty [9] introduced a bunch-level experiment under laboratory conditions with manual colour thresholding for image segmentation, which resulted in a yield estimation r^2 of 0.77. Additionally, these authors [9] presented a more complex automatic process for image segmentation on the same dataset presented by Dunn and Martin [8], resulting in an improved r^2 of 0.86. Automatic segmentation stemming from colour thresholding removes the human factor of manual thresholding, resulting in a more robust methodology [15,16]. However, these techniques are more sophisticated than manual thresholding, and therefore require adequate knowledge of computer vision techniques.

The second step, yield estimation, depends on the selected bunch metric. A simple pixel count of the segmented bunches is a favoured metric [8,13], with adaptions to the pixel count presented by Liu, Marden and Whitty [9] and Font et al. [14]. Liu, Marden and Whitty [9] tested five metrics: (i) volume, (ii) pixel count, (iii) perimeter, (iv) berry number and (v) berry size. Yield estimation using the pixel count produced superior results over the remaining metrics [9]. Unlike Liu, Marden and Whitty [9], Nuske et al. [11] avoided image segmentation and used a berry detection algorithm to determine a berry count. The use of the berry count as a yield estimation metric provided an r^2 of 0.74. Subsequent work on multiple multi-temporal datasets produced r^2 values between 0.6–0.73 [4].

A less common PRS approach utilises an RGB-D (RGB-Depth) camera to capture a 3-D (3-dimensional) model in either mesh or point-cloud format, representing the bunch or vine in a 3-D coordinate system [17]. The use of the Microsoft Kinect™ constitutes an ideal low-cost RGB-D sensor for in-situ imaging of vines [17]. The resulting 3-D models can be used to extract volumetric measurements for yield estimation. A limited number of studies have investigated the utility of the Kinect sensor for volume estimation. For example, Wang and Li [18] and Andújar et al. [19] employed the Kinect sensor under laboratory conditions for the volume estimation of sweet onions and cauliflowers, respectively. To date, the only use of an RGB-D Kinect sensor for yield estimation within viticulture was presented by Marinello et al. [10]. The authors assessed the sensor position for volume estimation of table grape bunches by testing two viewing angles, side-on and bottom-up. Multiple sensor-target distances were tested for the side-on viewing angle. Marinello et al. [10] concluded that a side-on viewing angle with a sensor-target distance of 0.8–1.0 m generated the best results.

An alternative PRS technique combines 2-D imagery with computer vision techniques, whereby 3-D models are created from the 2-D RGB images. Volume estimations are extracted from the subsequent 3-D models, enabling yield estimation calculations. Advanced computer vision techniques allow substantial automation in the process [20–22].

RGB and RGB-D technology and techniques incorporated into a suitable methodology present a viable solution for vineyard yield estimation. The established use of RGB imagery is evident, while the novel use of RGB-D imagery shows promise for future yield estimation. However, to date no study has investigated 2-D and 3-D PRS techniques side-by-side for vineyard yield estimation. Examining these two techniques in a commercial vineyard could provide insight into their capabilities and operational potential.

A key aspect for consideration when implementing these methodologies within a vineyard is the canopy coverage. The combination of essential canopy management practices and the vineyard's trellis system—particularly a vertical shoot positioned (VSP) system—directly influences canopy coverage, and inevitably the success of PRS techniques for yield estimation. High canopy coverage in the bunch zone results in bunch occlusion from the sensor. The incorporation of a canopy treatment was therefore proposed for this study.

This study aimed to investigate 2-D RGB and 3-D RGB-D PRS techniques for yield estimation in a VSP vineyard, using bunch area/volume estimation. The study was undertaken as three experiments, occurring under laboratory and field conditions. Field conditions were conducted at both bunch- and plant-level. Two canopy treatments were implemented from direct canopy manipulation and were defined as full canopy (FC) and leaf removal (LR). Hypothetically, the LR treatment will produce better estimation results. Furthermore, to achieve the aim of the study, two objectives were determined: to develop independent 2-D and 3-D yield estimation methodologies and to analyse and compare the success of the two PRS techniques for yield estimation.

2. Materials and Methods

2.1. Study Site

The study was carried out at the end of the 2016/17 growing season in a drip-irrigated Shiraz vineyard at the Welgevallen Experimental Farm located in Stellenbosch, South Africa (33°56′26″ S; 18°51′56″ E). The vineyard was planted in the year 2000, with a grapevine spacing of 2.7 × 1.5 m in a North-South orientation, and lies approximately 157 m above sea level. A seven-wire hedge VSP trellis system with three sets of moveable canopy wires is used in the vineyard. The Stellenbosch area falls within the coastal wine grape region of the Western Cape, which is characterised by a Mediterranean climate with long, dry summers [23]. Thirty-one individual vines were selected across three rows and sampled for this study (Figure 1).

2.2. Data Acquisition

Data was acquired between 28 February and 3 March 2017 (harvest), where data collected in situ was implemented with the two canopy treatments. The purpose behind the canopy treatments was to gain a direct line of sight to the bunches, which were generally hidden by the vine's canopy. No manipulation of the canopy, essentially the normal canopy condition, was classified as the FC treatment. The alternative LR treatment occurs after manual manipulation of the canopy, resulting in complete leaf removal in the bunch zone, effectively displaying the bunches.

Figure 2 illustrates the data-acquisition process that resulted in a total of ten datasets, as outlined by each step:

1. RGB and RGB-D imagery acquired for the FC treatment taken at bunch-level (n = 21; randomly selected and labelled bunches from the 31 vines) and plant-level (n = 31; individual vines). The resulting four datasets included: (i) RGB: bunch; (ii) RGB-D: bunch; (iii) RGB: vine; and (iv) RGB-D: vine.

2. Manual manipulation of the canopy resulted in the environment for the four LR treatment datasets. These datasets were identical to the FC datasets and concluded the in-situ datasets (datasets = 8). At this point, the 31 vines were harvested.

3. RGB and RGB-D imagery of the harvested bunches (n = 21) captured under laboratory conditions resulted in the final two datasets. At this point, all datasets (datasets = 10) have been captured.

4. Reference measurements, captured under laboratory conditions, included mass (g) and displacement (mL) for the individual bunches (n = 21) and individual vines (n = 31).

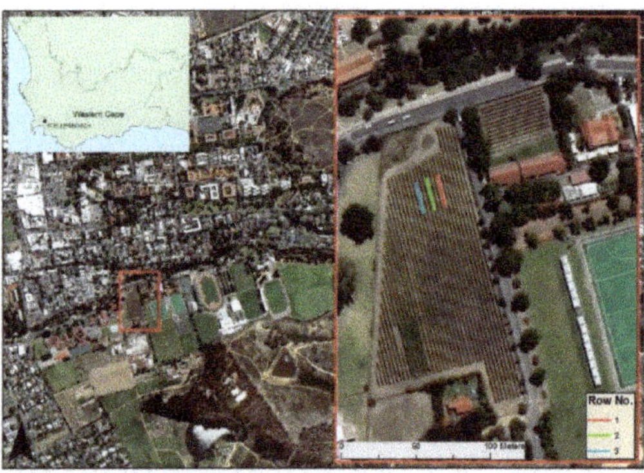

Figure 1. Location of the Shiraz vineyard in Stellenbosch, South Africa. Inset map (red rectangle) shows the three rows used for data collection.

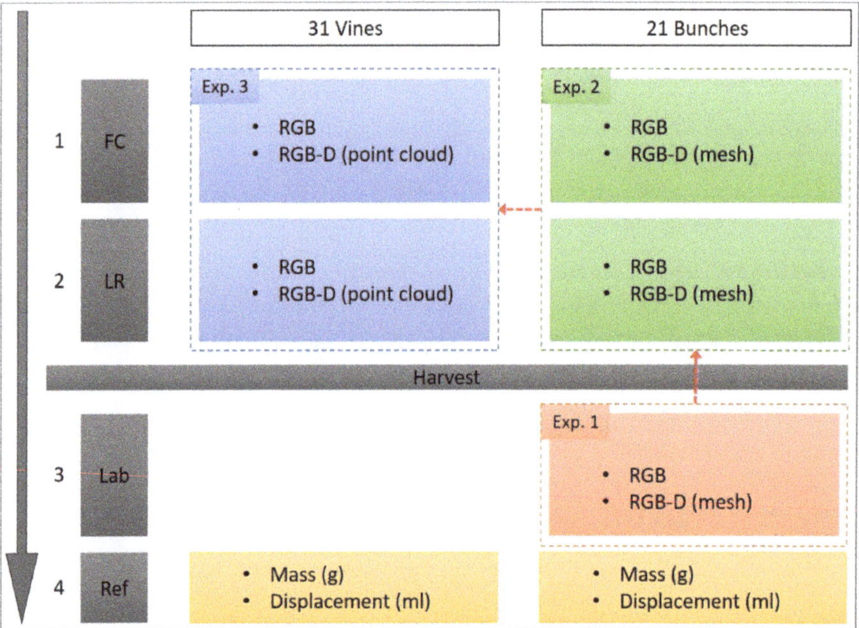

Figure 2. Data-acquisition protocol used in this study. Order of acquisition indicated by the grey arrow. {Key: FC = full canopy; LR = leaf removal; Lab = laboratory; Ref = reference measurements; Exp = experiment}.

RGB and RGB-D images were captured by two PRS sensors. A D3200 digital single-lens reflex camera (Nikon, Tokyo, Japan) was used for capturing 24.2-megapixel RGB images. The camera captured images in *auto* mode, with the flash disabled. The second sensor, a Kinect™ V1 (Microsoft, Redmond, WA, USA), was used to capture RGB-D imagery as either a mesh (bunch-level) or a point cloud (plant-level). The following data-acquisition subsections provide experiment-specific details.

2.2.1. Reference Measurements

Reference measurements were collected under laboratory conditions for the 21 individual bunches and the 31 individual vines. Individual bunch mass (g) was recorded with a Mentor scale (Ohaus, Parsippany, NJ, USA), and individual vine mass (g) was recorded with a Viper SW scale (Mettler Toledo, Columbus, OH, USA). Bunch/vine volume measurements were recorded as the displacement (mL) of water when bunches were submerged in a container of water [24]. The mass and volume measurements were used as reference measurements for the estimated measurements derived from the two PRS techniques.

2.2.2. Experiment One: Individual Bunches under Laboratory Conditions

The 21 individual bunches were imaged using the RGB camera and Kinect sensor under laboratory conditions; i.e., the laboratory was illuminated with white fluorescent lights and natural light entered through the windows. Each bunch was suspended from a tripod against a green background to maximise image contrast (Figure 3a):

(a) *RGB imagery:* The camera was placed parallel to the suspended bunch at a distance of 60 cm. A single image per bunch was captured for image processing. A ruler was included in each image for length reference (Figure 3b).

(b) *RGB-D (Kinect) imagery:* The Kinect sensor was placed 60 cm proximal to the target and captured individual meshes per bunch, resulting in a solid 3-D model of the bunch. The Kinect sensor, coupled with the Kinect Fusion software (part of Microsoft's Software Development Kit 1.8 for Windows [25]) running on a laptop, can capture individual meshes. Meshes were captured in *.stl* format with a resolution of 640 voxels/m, and a voxel resolution of 256 × 256 × 256 voxels. A white surface was positioned directly behind the bunch to improve contrast and depth determination, as illustrated by the mesh seen in Figure 3c. During mesh capture, the entire bunch system was rotated, providing different angles of view.

  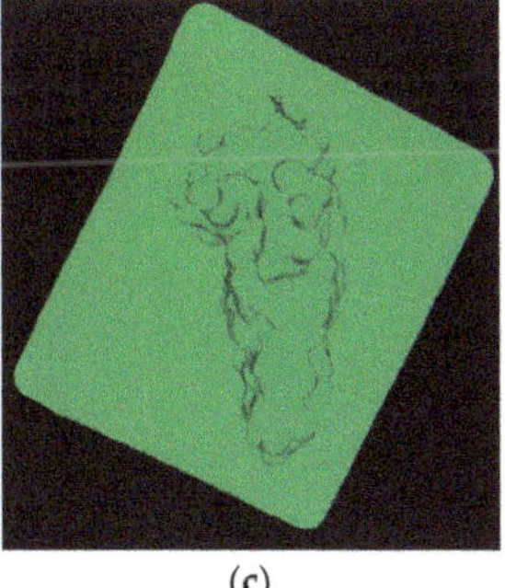

(a)   (b)   (c)

Figure 3. Data acquisition under laboratory conditions. (a) Experimental setup for image capture; (b) RGB image of an individual bunch with a ruler for reference length; and (c) RGB-D (Kinect mesh) of an individual bunch.

2.2.3. Experiment Two: Individual Bunches in Field Conditions

Images of the same 21 individual bunches were captured in in-situ conditions under both canopy treatments with the same RGB and RGB-D proximal sensors that were used in experiment one. Here, individual bunches were still attached to the respective vines:

(a) *RGB imagery:* The camera captured images of the individual bunches for both FC (Figure 4a) and LR (Figure 4b) treatments. The camera was positioned approximately 40 cm from the bunch being imaged, maintaining the reference length (ruler) in each image. Image acquisition occurred between 12H00 and 13H00, under natural solar illumination.

(b) *RGB-D (Kinect) imagery:* The same software setup from experiment one was used, which allowed the Kinect to capture an individual mesh per bunch for FC (Figure 4c) and LR (Figure 4d) treatments. The imagery was captured after sunset (approximately 20H00) with artificial illumination. The Kinect was held approximately 60 cm from the bunch and was moved around the bunch axis by hand. For the LR treatment, a board was placed behind the bunches, thereby creating an artificial background to improve volume extraction (Figure 4d).

(a)   (b)   (c)   (d)

Figure 4. Data acquisition of individual bunches in the field. (**a**) RGB image with full canopy (FC); (**b**) RGB image with leaf removal (LR); (**c**) RGB-D (Kinect mesh) with FC; and (**d**) RGB-D (Kinect mesh) with LR.

2.2.4. Experiment Three: Individual Vines in Field Conditions

The same RGB and RGB-D sensors were used to capture in-situ images at plant-level (31 individual vines imaged for both FC and LR treatments):

(a) *RGB imagery:* Imagery was captured at 12H00, under natural solar illumination. The camera was positioned 2 m from the vine, capturing an image for each side of the canopy. This protocol was repeated for both canopy treatments (FC: Figure 5a; LR: Figure 5b).

(b) *RGB-D (Kinect) imagery:* The Kinect sensor captured point clouds—instead of meshes—of the 31 vines, due to the scale difference. Point clouds consisted of thousands of individual points to create 3-D models; for both the FC and LR treatments (Figure 5c,d). The Real-Time Appearance-Based Mapping (RTAB-Map) software [26] was used for point-cloud modelling and regeneration. In RTAB-Map [26], the default filtering parameter and a 3-D cloud decimation ('thinning' of the point cloud) value of '2' were used during exportation of the point clouds in *.ply* format. The point cloud was captured per individual row, repeated for both sides of the canopy. The Kinect sensor was hand-held approximately 2 m from the vines at a perpendicular angle and was moved along the row in a north to south direction. The imagery was collected at 19H00, immediately before sunset, using the last natural illumination of the day.

Figure 5. Experiment three data examples at plant-level. RGB imagery of FC (**a**) and LR (**b**) treatments. RGB-D (Kinect point cloud) of FC (**c**) and LR (**d**) treatments.

2.3. Data Analysis

Data pre-processing and analysis occurred sequentially from experiment one (indicated via red arrows in Figure 2). The canopy treatments existing in experiment two and three had no effect on how the datasets were analysed for yield estimation. The proposed RGB and RGB-D methodologies were created on the LR datasets, before being directly applied to the FC datasets.

2.3.1. RGB Imagery

RGB images were processed using a custom script in MATLAB® [27], as follows:

1. The reference length (obtained from the ruler) in each image was used to scale the image pixels, creating a calibration value in cm^2.
2. Manual selection of the region of interest (ROI) containing the relevant bunch/bunches was undertaken. ROIs were strategically digitised so as to capture minimal background.
3. The masked RGB images were then converted to the HSV (hue, saturation, value) colour space and segmented using MATLAB's [27] *Colour Thresholder* app, part of the *Image Processing Toolbox™*. It was visually evident when selecting the threshold values that the lighting conditions influenced the values. Separate threshold values were therefore determined for the respective experiments. Threshold values were computed using a random training sample from the specific dataset; equivalent to 25% of the experiment's dataset.
4. After the image segmentation process, the number of segmented pixels was determined (adaption of the pixel count metric [9]) and converted into a pixel area (cm^2) using the calibration value.

Figure 6 illustrates the image segmentation process for experiment one at bunch-level (Figure 6a,b), and experiment three at plant-level (Figure 6c,d).

For experiment three, segmentation produced a segmented bunch area image (Figure 6d) for a single side of the vine, with a similar image for the reverse side of the vine. To obtain a single area per vine, the *Total Bunch Area of Vine (TBAV)* was calculated, as follows:

$$TBAV = (Ae + Aw)/2 \tag{1}$$

where *Ae* was the area of the east-facing side of the vine, and *Aw* represented the vine's west-facing area. The *TBAV* was in cm^2. Segmentation success was assessed on the testing subset (25% of the total dataset). A confusion matrix was computed for each experiment, using the predicted values from the segmented binary image versus the actual values of the original RGB image. F1-score and accuracy metrics calculated from the confusion matrix evaluated the segmentation accuracy.

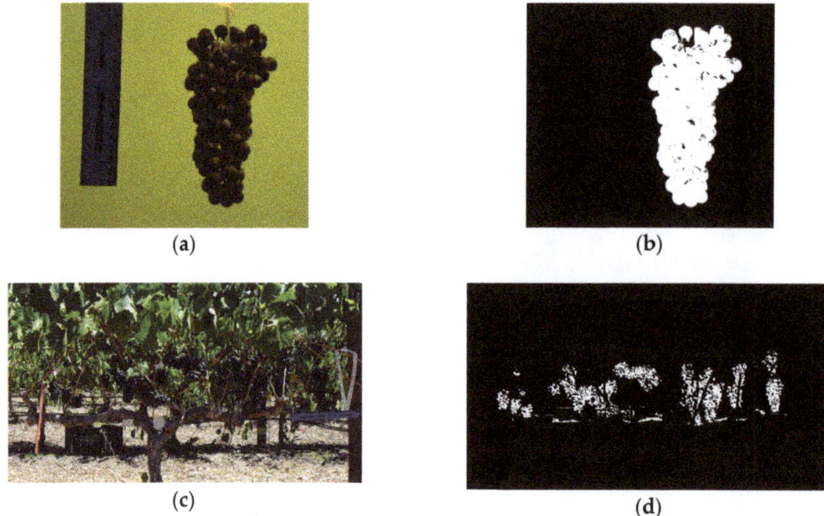

Figure 6. (**a**) Represents the original RGB image, with (**b**) illustrating the segmented binary image at bunch-level. (**c**) An RGB image of an east-facing vine, with (**d**) the segmented binary image at plant-level.

2.3.2. RGB-D (Kinect) Imagery

Due to the two image data types captured—mesh vs. point cloud—RGB-D imagery was processed differently for experiment one and two (mesh: bunch-level), and experiment three (point cloud: plant-level). Data processing for experiments one and two progressed as follows:

1. Each mesh was manually cleaned and sectioned to remove any background, using the *Cross-Section* tool in CloudCompare [28].
2. The cleaned mesh was reconstructed in MeshLab [29], using the *Screened Poisson Surface Reconstruction* [30] method—see Figure 7. Additional cleaning of the mesh was completed ad hoc in MeshLab [29].
3. Volume (cm^3) of the mesh was calculated in CloudCompare [28].

Screened Poisson Surface Reconstruction [30] allows the back of the mesh, which was 'open' (Figure 7a), to be reconstructed. It becomes 'watertight', as seen in Figure 7b.

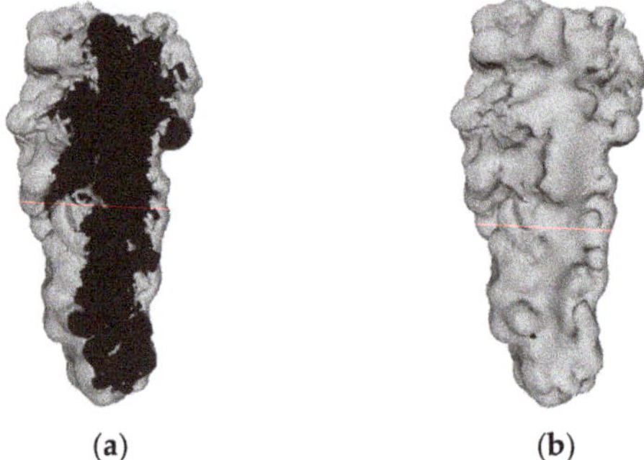

Figure 7. (**a**) Example of mesh prior to reconstruction, and (**b**) the same mesh after Poisson reconstruction.

The nature of the point-cloud data requires different processing steps to that of the mesh data. This effect was attributed mainly to the necessity of closing the points of the point cloud, thereby producing a 'watertight mesh' for volume extraction.

The point cloud datasets from experiment three were processed as follows:

1. Point clouds were imported into CloudCompare [28], and subsequently cleaned and sectioned to individual vines, focusing on the bunch zone.
2. Bunches were segmented from the point cloud using their colour properties. CloudCompare's [28] *Filter Points by Value* tool incorporates user-defined thresholds, manipulating the RGB colour space of the point cloud. Threshold values were determined on a random sample (25%) of the LR dataset.
3. A custom-built script in R statistical software [31] was used for calculating the segmented point cloud's volume, representing the vine's bunches.

Figure 8 illustrates the raw point cloud (Figure 8a) captured by the Kinect sensor, with the segmented point cloud (Figure 8b) displaying the bunches.

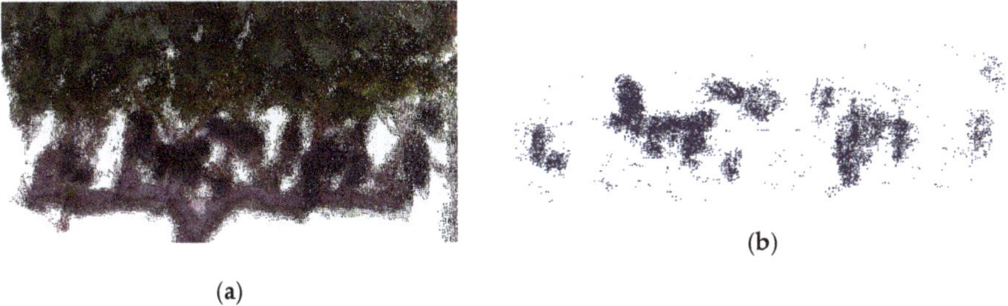

(a)

(b)

Figure 8. (a) The Kinect point cloud for an LR-treated vine (1E—east side) and (b) the segmented point cloud of the same vine.

The custom script in R statistical software [31] incorporated the *alphashape3d* package v1.3 [32] to compute the 3-D shape for volume calculation purposes. The *alphashape3d* [32] package includes the *α-shape* algorithm [33] to recover the geometric structure of the 3-D point cloud for volume calculation. The *α-shape* algorithm [33] requires a specific alpha value for computation; hence, the alpha value directly influences the total volume calculated. To adjust an alpha value for our experimental conditions, various levels of alpha were tested on 25% of the dataset and linearly regressed against the reference values. Similar investigative experiments were conducted by Rueda-Ayala et al. [34] and Ribeiro et al. [35] to determine experiment-specific alpha values.

The process described above for experiment three was repeated for both sides of the vine, resulting in two volume measurements per vine. A single volume value per vine was obtained via the *Total Bunch Volume per Vine (TBVV)* calculation:

$$TBVV = (Ve + Vw)/2 \tag{2}$$

where *Ve* was the volume of the vine's east-facing side, and *Vw* was the volume of the west-facing side. The resultant *TBVV* was in cm^3 per vine.

2.3.3. Cross-Validation

Five-fold cross-validation was used to develop the yield estimation model for each dataset. Cross-validation was implemented using the *Caret* package [36] in R statistical software [31], repeated ten times for model robustness. The model produced 'fitted values', which represented the estimated yield (in grams). Following this, the estimated yield (g) values were linearly regressed against the actual mass (g) to produce a final r^2 value (coefficient of determination), indicating the potential for yield estimation. The Root Mean Square Error (RMSE) was computed from the linear regression, indicating the yield estimation error (in grams).

3. Results

3.1. Reference Measurements

Results of the reference measurements indicated a strong relationship between mass and volume at bunch-level (r^2 = 0.971) and plant-level (r^2 = 0.996). The established relationships between mass and volume served as the basis for the subsequent experiments, which were used to evaluate the 2-D and 3-D techniques.

3.2. Pre-Processing

The complexity of the RGB-D (Kinect) datasets required two additional pre-processing steps. The first step was the determination of an alpha value required for volume calculation, using the *alphashape3d* package [32]. The second step was volume correction for all Kinect datasets, due to volume estimation errors in the datasets. The segmentation accuracy for the RGB pre-processing has also been included in this section.

3.2.1. RGB Segmentation Accuracy

Experiment one's segmentation results yielded an F1-score of 0.976, with an accuracy of 0.971. Experiment two resulted in a lower F1-score of 0.842, with an accuracy of 0.781. Lastly, experiment three's F1-score and accuracy were 0.833 and 0.932, respectively. Solar illumination could contribute to the lowered results presented in the in-situ measurements.

3.2.2. alphashape3d's Adjusted Alpha Value

Figure 9 represents the r^2, and RMSE curves for the various alpha values tested. Values tested ranged from 0.001 to 0.050, in increments of 0.001. It is evident from this figure that the alpha value selected (alpha = 0.010) satisfies a high coefficient of determination (r^2 = 0.605) combined with the lowest RMSE (703.301 cm^3). Figure 10 provides a visual interpretation of the results from Figure 9. A low alpha value, such as 0.005 (Figure 10b), produced an r^2 of 0.520 and an RMSE of 2052.751 cm^3 (sample dataset mean volume = 1961.875 cm^3). Conversely, a high alpha value of 0.050 (Figure 10d) produced an r^2 of 0.506 and an RMSE of 13 936.31 cm^3. The selected alpha value of 0.010 (Figure 10c) was subsequently used for all further analyses.

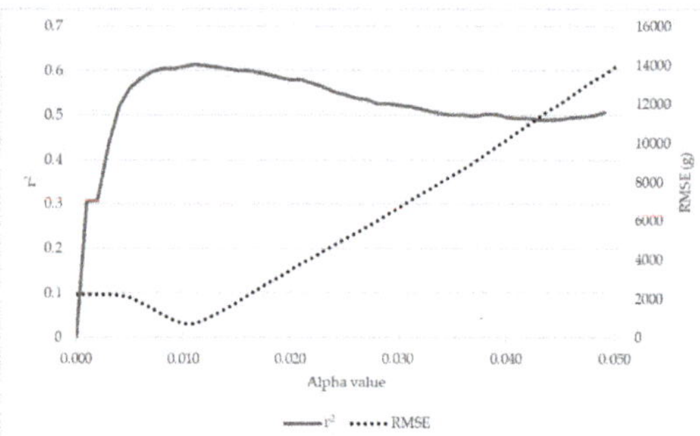

Figure 9. The relevant r^2 and RMSE values for each alpha value tested for the *alphashape3d* package in the custom R script. Alpha values incremented in 0.001, ranging from 0.001–0.050.

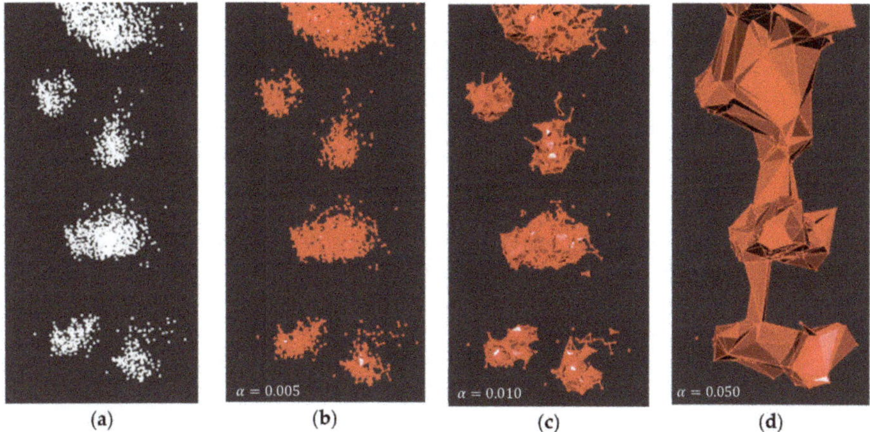

Figure 10. Point-cloud reconstruction testing the alpha value for the *alphashade3d* package. The original point cloud before reconstruction—represented as white points for a visual purpose (**a**), and after reconstruction (**b–d**).

3.2.3. Kinect Volume Correction

Figure 11 shows a volume estimation error present in experiment one's data, which aligns with a subsequent review of the literature [10,18,19]. The 21 bunches have a mean actual volume (the reference volume) of 144.952 cm^3 and a mean estimated Kinect volume of 175.672 cm^3. Overestimation by the Kinect V1 sensor was evident from the results. Volume correction via cross-validation was therefore subsequently incorporated into the methodology, where a mean corrected volume of 144.952 cm^3 was achieved. Thereafter, the correction was applied to the remaining Kinect datasets.

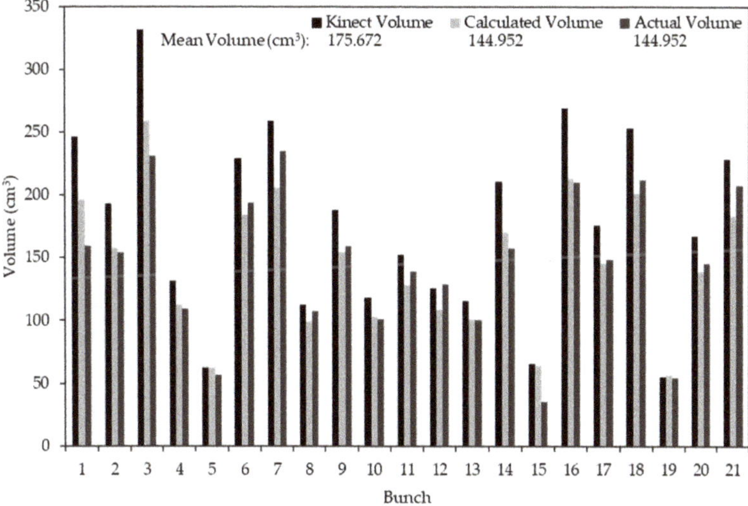

Figure 11. Experiment one's results, for the 21 individual bunches, illustrating the volume estimation error by the Kinect sensor.

3.3. RGB Results

Figure 12 shows the results for the three experiments that used 2-D RGB digital imagery. The best results were obtained in experiment one (Figure 12a), which produced an r^2 of 0.889 and an RMSE of

17.978 g. The level of accuracy achieved in experiment one can be attributed to the controlled laboratory conditions and supports the proposed methodology for yield estimation. Applying this methodology to in-situ bunches produced less accurate results, as seen in experiment two. Experiment two's FC treatment (Figure 12b) produced the lowest yield estimation results for 2-D RGB imagery, with an r^2 of 0.625 and an RMSE of 27.738 g. The LR treatment's (Figure 12c) r^2 and RMSE values were 0.742 and 25.066 g, respectively. The lesser FC values were directly attributed to the canopy coverage present, as the bunches were partially occluded from the sensor's view. The effect of the canopy treatment was evident when comparing the results. At plant-level (experiment three), the same pattern was present between the canopy treatments. The FC (Figure 12d) treatment of experiment three produced an r^2 of 0.779, while the LR (Figure 12e) treatment produced an even higher r^2 of 0.877. The respective RMSE values were 559.357 g and 443.235 g. The success of yield estimation in experiment three, specifically the LR treatment, supports the methodology's capability for 2-D RGB yield estimation. The in-situ yield estimation was preferable at plant-level, which may be attributed to the lighting conditions and the success of the colour thresholding for bunch segmentation at plant-level.

3.4. RGB-D Results

Figure 13 illustrates the RGB-D results obtained for the three experiments. The unrivalled results obtained in experiment one (Figure 13a) produced an r^2 of 0.950 and an RMSE of 12.458 g—the best-performing results presented in this study. The Kinect sensor favoured the controlled conditions of the laboratory, specifically the artificial illumination as a source of light. Applying the same methodology to experiment two (in-situ bunches) resulted in a lower yield estimation performance for both canopy treatments. The FC treatment (Figure 13b) produced an abnormally low r^2 of 0.020, with an RMSE of 8.081 g. A statistical outlier in the data was evident when the results were analysed. With the removal of this outlier, the modified FC results (n = 20) improved drastically, with a new r^2 of 0.609 and an RMSE of 26.790 g (Figure 13c). In contrast, the LR treatment (Figure 13d) resulted in an r^2 of 0.756 and an RMSE of 24.601 g, which aligned with the LR results for bunch-level obtained with RGB imagery (Figure 12c). At plant-level, the unfavourable results of the FC treatment (Figure 13e) generated an r^2 of 0.487 and an RMSE of 673.535 g. However, the LR treatment (Figure 13f) provided some promise for the Kinect sensor at plant-level, achieving an r^2 of 0.594 and an RMSE of 661.739 g. The same effect of the canopy treatment was evident in the RGB-D results as in the RGB results, with the LR treatment producing a better yield estimation agreement. The results of experiment three indicated a limitation within the proposed methodology for RGB-D yield estimation at plant-level. Such limitation could be from the data-acquisition process, or the image segmentation within the data analysis.

The poor results obtained in experiment two's FC dataset can be attributed to the mesh reconstruction step in the data-analysis process. Exaggerated reconstruction of bunches presents a potential limitation of the *Screened Poisson Surface Reconstruction* algorithm [30], as depicted in Figure 14. The statistically-outlying bunch circled in red (Figure 13b) produced a Kinect volume of 856 cm^3, as illustrated in Figure 14a, when the actual volume was only 35 cm^3. An example of an accurately reconstructed mesh (Figure 14b) was included for visual comparison. This anomaly was unavoidable during data processing.

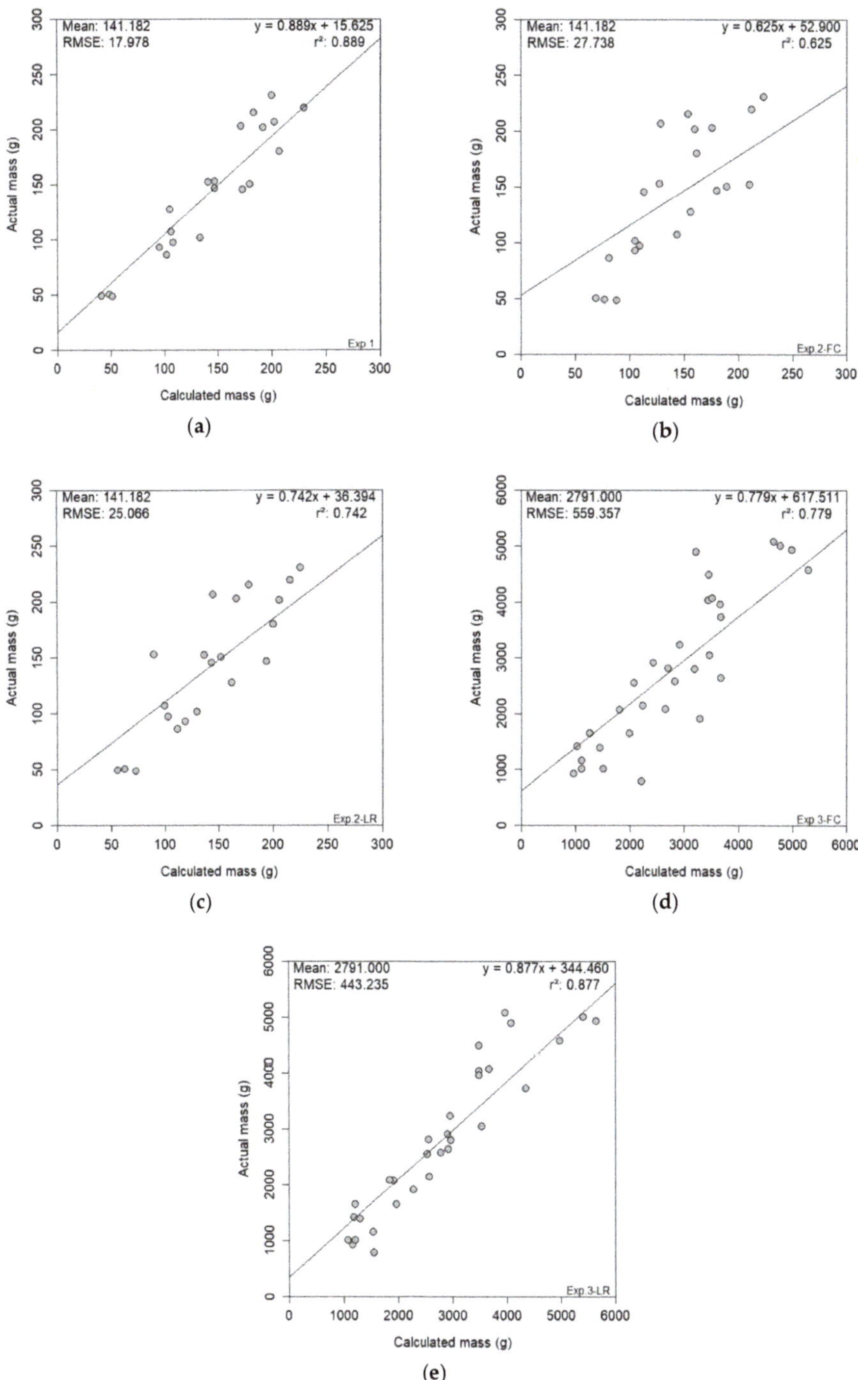

Figure 12. RGB results presented for the three experiments; experiment one (**a**), experiment two FC (**b**) & LR (**c**) and experiment three FC (**d**) & LR (**e**). *{Key: Exp. = experiment; FC = full canopy; LR = leaf removal}.*

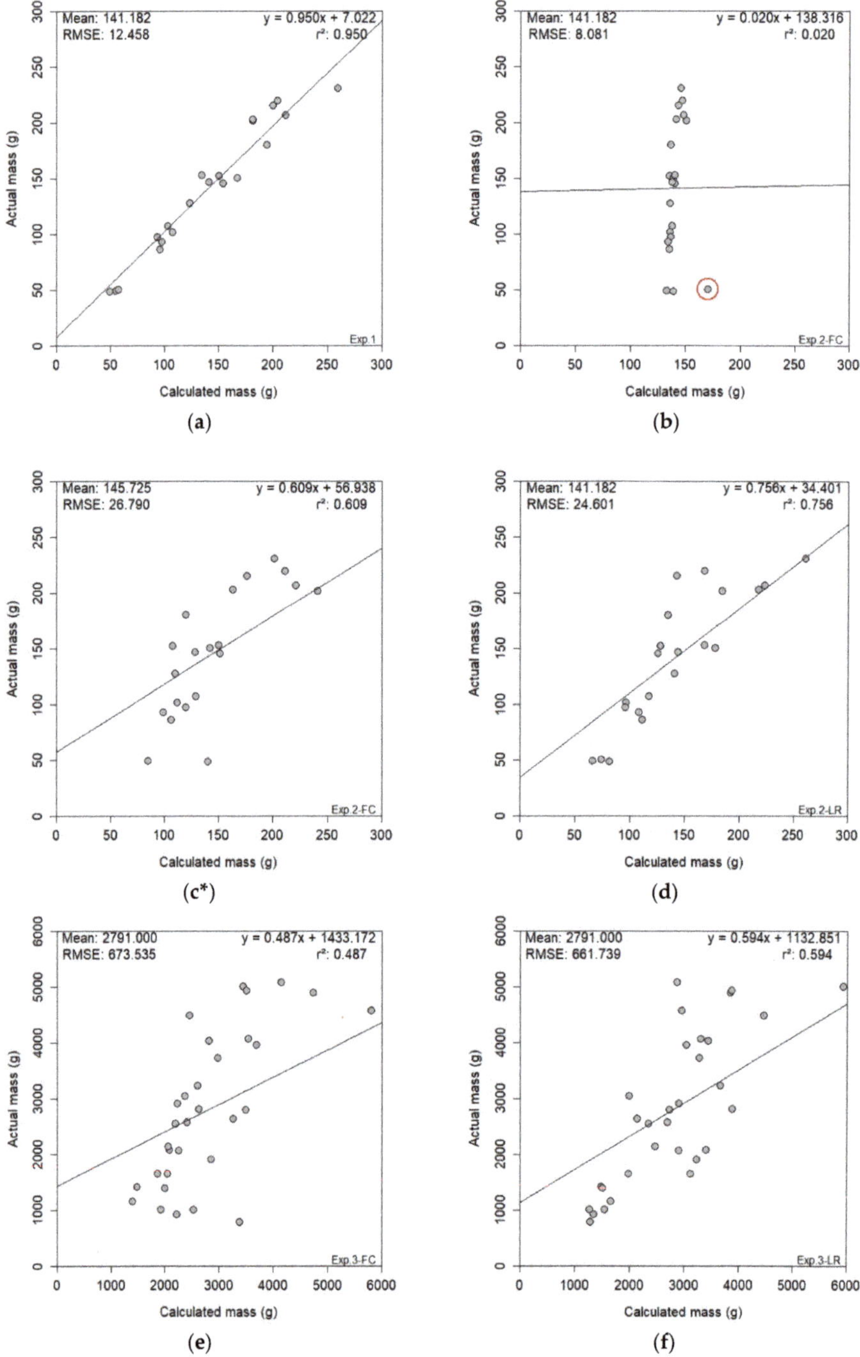

Figure 13. Presented RGB-D results of the three experiments; experiment one (**a**), experiment two FC (n = 21) (**b**), experiment two FC* (n = 20) (**c**), experiment two LR (**d**) and experiment three FC (**e**) & LR (**f**). {Key: Exp. = experiment; FC = full canopy; LR = leaf removal; *statistical outlier removed, resulting in 20 bunches}.

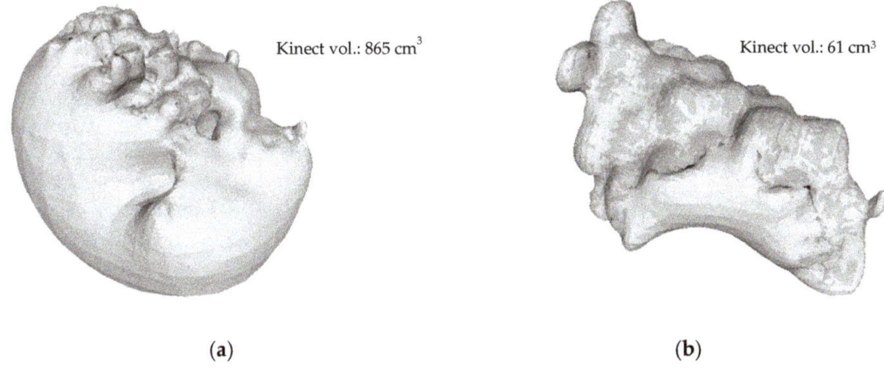

Figure 14. Illustration of *Screened Poisson Surface Reconstruction*. The reconstructed bunch (circled in red—Figure 13b) with the incorrect volume (**a**), and an example of a reconstructed bunch of the correct volume (**b**).

4. Discussion

To date, most studies have employed 2-D RGB imagery for vineyard yield estimation at both bunch- and plant-level. However, only Marinello et al. [9] have investigated the use of 3-D RGB-D (specifically making use of a Kinect sensor) imaging for vineyard yield estimation. The research we have presented investigated 2-D and 3-D PRS techniques for vineyard yield estimation. The study was undertaken as three experiments, consisting of bunch-level and plant-level datasets, with in-situ measurements captured for the two canopy treatments (FC and LR). The following subsections discuss the results of the two PRS techniques in further detail.

4.1. Using 2-D RGB Imagery for Yield Estimation

The presented results using RGB imagery for yield estimation are robust and support the 2-D PRS technique. Experiment one ($r^2 = 0.889$; RMSE = 17.978 g) illustrated the success of RGB imagery in a controlled environment for yield estimation at bunch-level. At plant-level, similar results were presented for the LR treatment in experiment three ($r^2 = 0.877$; RMSE = 443.235 g). The success of the methodology under both laboratory and field conditions supports the use of colour thresholding for image segmentation, and the adapted pixel area metric for yield estimation.

Colour thresholding for image segmentation was favoured by several studies [8,9,14]. In our study, the use of the HSV colour space for thresholding has proven fruitful. The HSV colour space is supported by Font et al. [14], who achieved favourable results (estimation error of 13.55%) when working with the II layer for segmentation purposes. At bunch-level, experiment one's result ($r^2 = 0.889$) shows an improvement to the result ($r^2 = 0.77$) presented by Liu, Marden and Whitty [9]; similarly conducted under laboratory conditions. This aspect was a noteworthy improvement, especially considering the manual nature of the colour thresholding. At plant-level, the manually produced results of experiment three (FC: $r^2 = 0.779$; LR: $r^2 = 0.877$) align with the automated classification results ($r^2 = 0.865$) of Liu, Marden and Whitty [9]. Here, Liu, Marden and Whitty [9] use the same 1×1 m image dataset as Dunn and Martin [8], as opposed to the plant-level imagery used in our study. Additionally, the colour thresholding approach of our study outperformed the image classification approach (test $r^2 = 0.73$) [13] to segmentation.

We presented an adaption of the pixel count metric for yield estimation, which expands on current literature [8,9,13,14]. Of the five metrics tested by Liu, Marden and Whitty [9], the pixel count produced the best results. We found the incorporation of a calibration length (the ruler) for pixel count resulted in an improved quantitative pixel area (cm^2) for yield estimation. Again, our results at bunch-level ($r^2 = 0.889$) improved the bunch-level results ($r^2 = 0.77$) conducted in laboratory conditions—as presented by Liu, Marden and Whitty [9], who were specifically testing the various yield estimation metrics. At plant-level, our LR treatment ($r^2 = 0.877$) outperformed all current literature to date, with our FC treatment ($r^2 = 0.779$) representing a slight improvement for in-situ measurements. The presented pixel area (cm^2) metric also improved on the berry count results [4,11], with the highest berry count r^2 (0.74) presented still being lower than our plant-level pixel area r^2 (0.779) for the FC treatment. However, the potential of the berry count is applicable across all cultivars, as it does not depend on the colour of the berry [11].

The limitation of slight distance variations between the camera and bunches within each image was resolved by the incorporation of the reference length (the ruler). The necessity of determining the calibration length for each image outweighs the added processing requirement of this step. Overall, this allows improved yield estimation success, as represented in our results. Future work could attempt to automate this process. However, a more restrictive limitation was the human involvement in determining the appropriate threshold values; this could explain the lowered in-situ estimation performance at bunch-level. Future work could investigate a more automated methodology, thus alleviating this limitation.

4.2. Using 3-D RGB-D Imagery for Yield Estimation

To date, the work presented by Marinello et al. [10] is the only literature supporting the use of the Kinect V1 sensor for vineyard yield estimation. Marinello et al. [10] used table grape bunches to determine the optimal viewing angle and distance for the sensor. The authors concluded that a side-on view of the bunch, with a distance of between 0.8–1.0 m, produced the least variability in mass estimations. The current findings of our study, which incorporated RGB-D imagery for yield estimation across the three experiments, are therefore the most comprehensive findings to date. The presented study exemplifies the potential of 3-D PRS techniques for yield estimation, specifically a cost-effective sensor like the Kinect V1.

The nature of our study's datasets resulted in separate data analysis between the bunch- and plant-level datasets. Astounding results were obtained in experiment one ($r^2 = 0.950$; RMSE = 12.458 g). It was evident that the Kinect sensor favours ideal laboratory conditions, allowing accurate yield estimation at bunch-level. The ramification of less-favourable conditions, such as in-situ monitoring, became apparent in experiment two (bunch-level) and three (plant-level).

Although no bunch-level studies have used a Kinect sensor for vineyard yield estimation, a similar approach under laboratory conditions, estimating the volume of cauliflowers, was presented by Andujar et al. [19]. The authors were able to achieve an r^2 of 0.868 when regressing the estimated volume against the known fruit mass. Conversely, our methodology enables the relationship between fruit volume and mass to be modelled, allowing the use of the adjusted mass (calculated from the model) for subsequent yield estimation. A fundamental difference between our methodology and that of Andujar et al. [19] is the method in which the 3-D model was captured. Our methodology captured a 3-D mesh of the fruit, while Andujar et al. [19] captured a 3-D point cloud of the vegetable. This required model reconstruction, constructively producing a 'watertight mesh'. Our improved results at this level ($r^2 = 0.950$) can be accredited to this fundamental difference. Interestingly, both methodologies used the same 3-D reconstruction method—*Screened Poisson Surface Reconstruction* [30] found in MeshLab [29].

The primary limitation of the proposed methodology at bunch-level was a combination of dataset quality and the *Screened Poisson Surface Reconstruction* method [30]. The consequence of a poor-quality mesh became apparent when bunch reconstruction resulted in significant defects, as seen in Figure 14a. Such imperfections directly affect the potential for accurate yield estimation, exemplified by the results of experiment two's FC dataset ($r^2 = 0.020$; in Figure 13b). Future research is necessary to gain a better understanding of this shortfall in the methodology.

This study presented the novel use of a Kinect RGB-D sensor for in-situ vineyard yield estimation at plant-level (experiment three). The presented plant-level methodology produced promising results for the LR treatment ($r^2 = 0.594$; RMSE = 661.739 g), while the effect of the canopy coverage was evident in the FC treatment ($r^2 = 0.487$; RMSE = 673.535 g). Future work should improve on the presented methodology, potentially overcoming several limitations. Additionally, the nature of the Kinect V1 sensor results in the fundamental constraint of being suited to indoor use only, as solar illumination produces excessive interference in the captured imagery [19]. Future work should make use of the Kinect V2 sensor, since several improvements to the sensor have been implemented, effectively allowing improved outdoor imagery to be captured [37]. RGB-D sensors, specifically the Kinect V2, are being incorporated in terrestrial vehicles as cheap sensor alternatives for vineyard modelling and yield estimation, which is demonstrated by [38].

4.3. The Operational Potential of Developed Methodologies

Both the presented 2-D RGB and 3-D RGB-D methodologies achieved acceptable accuracies across the three experiments. Our results (Figures 12 and 13) support the use of PRS technology and techniques for vineyard yield estimation, especially for VSP-trained Shiraz vineyards. The nature of the presented work was conceptualised to assess 2-D and 3-D PRS sensors side-by-side, a novelty in the vineyard yield estimation domain.

Experiment one illustrated the capability of these two techniques for successful yield estimation of individual bunches, where the Kinect RGB-D sensor ($r^2 = 0.950$) outperformed the digital RGB sensor ($r^2 = 0.889$). The suitability of the lighting under laboratory conditions coupled with the Kinect's ability to capture a 3-D model of the bunch both contribute to the success of the Kinect sensor over the RGB sensor. Nonetheless, robust methodologies were established in a controlled environment.

Experiment two tested the established methodologies in situ, under both FC and LR canopy treatments. The produced results of experiment two—and experiment three—confirmed the hypothesis of a superior yield estimation agreement under the LR canopy treatment. If good canopy management practices are established early in the season, better yield estimation results than those obtained under the FC treatment could be achieved. Both sensors produced similar results for FC [approximate $r^2 = 0.61$; using Kinect's modified results (n = 20)] and LR (approximate $r^2 = 0.75$) treatments in experiment two.

The success of the two PRS methodologies can be differentiated at plant-level by the results of experiment three. Unlike experiment one, the RGB methodology outperformed the RGB-D methodology for yield estimation. The RGB results ($r^2 = 0.877$) significantly outclassed the RGB-D results ($r^2 = 0.487$) under the FC treatment, whereas a smaller margin between the RGB ($r^2 = 0.779$) and RGB-D ($r^2 = 0.594$) sensors occurred under the LR treatment. Inference behind the differing results can lead to the following observations. The different lighting conditions influenced the results: the RGB imagery was collected at midday, while the RGB-D imagery was collected immediately before sunset. Additionally, the continuous movement of the Kinect sensor (RGB imagery captured in a stationary position) could have contributed to lowered RGB-D results. Further research is encouraged, using a standardised experimental setup, where feasible, to create more favourable conditions under which both sensors can operate.

For the RGB and RGB-D datasets, the LR treatment yielded higher estimation agreements compared with the FC treatments. The LR treatment was adopted to create an ideal in-situ environment for yield estimation. The commercial feasibility of complete leaf removal in the bunch zone is not practical in some viticulture regions, since there are adverse effects associated with this practice (such as 'sunburn'). However, this practice can be implemented in specific zones of the vineyards, enabling an approximation of the total yield using a monitoring strategy. The FC results provide a better indication of the commercial readiness of the developed methodologies. Future work should determine an optimal canopy coverage that favours both sensor operationality and commercial farming methods.

Overall, the use of the Kinect sensor as a cost-effective RGB-D sensor for vineyard yield estimation, specifically at bunch-level, is supported by the results obtained for experiment one. However, the robustness of the RGB methodology is evident across all three experiments, with substantial plant-level results obtained in situ. A better understanding of the current limitations would allow further improvements to the methodologies. To this end, the results obtained in this study support the potential for operationalisation of both PRS sensors. A refined methodology could see commercially favourable data-acquisition methods implemented on a larger dataset, with fully automated data processing. See for example [15,16]. The commercialisation of such methodologies could become more feasible, due to the simplicity and robustness of the current methodologies, coupled with improved yield estimation accuracy.

5. Conclusions

A novel approach to a side-by-side investigation of 2-D and 3-D PRS techniques for successful vineyard yield estimation has been presented. This study assessed RGB imagery captured by a digital camera and RGB-D imagery captured by a Kinect V1 sensor across three experiments, with in-situ measurements obtained under two canopy treatments. Our results show that the Kinect RGB-D sensor produced the highest yield estimation agreement under laboratory conditions (bunch-level). At bunch-level, RGB and RGB-D PRS techniques performed equally under both canopy treatments for in-situ yield estimation. At plant-level, the best in-situ results were obtained using the RGB imagery, which significantly outperformed the RGB-D results. Both sensors support the use of PRS technology and techniques for vineyard yield estimation, with improved accuracies presented. The results of this study confirm the operational potential of 2-D RGB imagery for accurate yield estimation, with the recommendation that future work should investigate a more automated RGB methodology suitable for operational environments. Regarding the presented RGB-D methodology, the Kinect demonstrates the potential for vineyard yield estimation using 3-D RGB-D imagery. Future work should investigate the use of the Kinect V2 sensor coupled with suitable lighting conditions for in-situ yield estimation.

Author Contributions: C.H.—data analysis, writing and editing of the manuscript; N.P.—content supervision, and editing of the manuscript; N.M.—data collection and analysis; C.P.-E.—conceptualisation, experimental study design, content supervision and editing of the manuscript.

Funding: This research was funded by Winetech, project DVO 07: "Near-real-time characterization of vines for more efficient vineyard management".

Acknowledgments: The authors would like to thank Albert Strever, Berno Greyling, Aloïs Houeto and Alessandro Bellotto for their technical and research support.

Conflicts of Interest: The authors declare no conflicts of interest.

References

1. Arnó, J.; Martínez-Casasnovas, J.A.; Ribes-Dasi, M.; Rosell, J.R. Review. Precision viticulture. Research topics, challenges and opportunities in site-specific vineyard management. *Span. J. Agric. Res.* **2009**, *7*, 779–790. [CrossRef]

2. Blackmore, S. Developing the Principles of Precision farming. In Proceedings of the International Conference on Precision Agriculture, Minneapolis, MN, USA, 14–17 July 2002; p. 5.

3. Matese, A.; Di Gennaro, S.F. Technology in precision viticulture: A state of the art review. *Int. J. Wine Res.* **2015**, *7*, 69–81. [CrossRef]

4. Nuske, S.; Wilshusen, K.; Achar, S.; Yoder, L.; Narasimhan, S.; Singh, S. Automated visual yield estimation in vineyards. *J. Field Robot.* **2014**, *31*, 837–860. [CrossRef]

5. Cunha, M.; Marçal, A.R.S.; Silva, L. Very early prediction of wine yield based on satellite data from vegetation. *Int. J. Remote Sens.* **2010**, *31*, 3125–3142. [CrossRef]

6. Wolpert, J.A.; Vilas, E.P. Estimating Vineyard Yields: Introduction to a Simple, Two-Step Method. *Am. J. Enol. Vitic.* **1992**, *43*, 384–388.

7. De la Fuente, M.; Linares, R.; Baeza, P.; Miranda, C.; Lissarrague, J. Comparison of Different Methods of Grapevine Yield Prediction in the Time Window. *J. Int. Sci. Vigne Vin* **2015**, *49*, 27–35. [CrossRef]

8. Dunn, G.M.; Martin, S.R. Yield prediction from digital image analysis: A technique with potential for vineyard assessments prior to harvest. *Aust. J. Grape Wine Res.* **2004**, *10*, 196–198. [CrossRef]

9. Liu, S.; Marden, S.; Whitty, M. Towards automated yield estimation in viticulture. In Proceedings of the Australasian Conference on Robotics and Automation, Sydney, Australia, 2–4 December 2013.

10. Marinello, F.; Pezzuolo, A.; Cillis, D.; Sartori, L. Kinect 3D reconstruction for quantification of grape bunches volume and mass. In Proceedings of the Engineering for Rural Development, Jelgava, Latvia, 25–27 May 2016; pp. 876–881.

11. Nuske, S.; Achar, S.; Bates, T.; Narasimhan, S.; Singh, S. Yield estimation in vineyards by visual grape detection. In Proceedings of the IEEE International Conference on Intelligent Robots and Systems, San Francisco, CA, USA, 25–30 September 2011; pp. 2352–2358. [CrossRef]

12. Aquino, A.; Millan, B.; Diago, M.P.; Tardaguila, J. Automated early yield prediction in vineyards from on-the-go image acquisition. *Comput. Electron. Agric.* **2018**, *144*, 26–36. [CrossRef]

13. Diago, M.P.; Correa, C.; Millán, B.; Barreiro, P.; Valero, C.; Tardaguila, J. Grapevine yield and leaf area estimation using supervised classification methodology on RGB images taken under field conditions. *Sensors* **2012**, *12*, 16988–17006. [CrossRef]

14. Font, D.; Tresanchez, M.; Martínez, D.; Moreno, J.; Clotet, E.; Palacín, J. Vineyard Yield Estimation Based on the Analysis of High Resolution Images Obtained with Artificial Illumination at Night. *Sensors* **2015**, *15*, 8284–8301. [CrossRef]

15. Liu, S.; Whitty, M. Automatic grape bunch detection in vineyards with an SVM classifier. *J. Appl. Log.* **2015**, *13*, 643–653. [CrossRef]

16. Millan, B.; Velasco-Forero, S.; Aquino, A.; Tardaguila, J. On-the-go grapevine yield estimation using image analysis and boolean model. *J. Sens.* **2018**. [CrossRef]

17. Bengochea-Guevara, J.M.; Andújar, D.; Sanchez-Sardana, F.L.; Cantuña, K.; Ribeiro, A. A low-cost approach to automatically obtain accurate 3D models of woody crops. *Sensors* **2018**, *18*, 30. [CrossRef] [PubMed]

18. Wang, W.; Li, C. Size estimation of sweet onions using consumer-grade RGB-depth sensor. *J. Food Eng.* **2014**, *142*, 153–162. [CrossRef]

19. Andújar, D.; Ribeiro, A.; Fernández-Quintanilla, C.; Dorado, J. Using depth cameras to extract structural parameters to assess the growth state and yield of cauliflower crops. *Comput. Electron. Agric.* **2016**, *122*, 67–73. [CrossRef]

20. Herrero-Huerta, M.; González-Aguilera, D.; Rodriguez-Gonzalvez, P.; Hernández-López, D. Vineyard yield estimation by automatic 3D bunch modelling in field conditions. *Comput. Electron. Agric.* **2015**, *110*, 17–26. [CrossRef]

21. Ivorra, E.; Sánchez, A.J.; Camarasa, J.G.; Diago, M.P.; Tardaguila, J. Assessment of grape cluster yield components based on 3D descriptors using stereo vision. *Food Control* **2015**, *50*, 273–282. [CrossRef]

22. Rose, J.C.; Kicherer, A.; Wieland, M.; Klingbeil, L.; Töpfer, R.; Kuhlmann, H. Towards automated large-scale 3D phenotyping of vineyards under field conditions. *Sensors* **2016**, *16*, 2136. [CrossRef]

23. Conradie, W.; Carey, V.; Bonnardot, V.; Saayman, D.; van Schoor, L. Effect of Different Environmental Factors on the Performance of Sauvignon blanc Grapevines in the Stellenbosch/Durbanville Districts of South Africa. I. Geology, Soil, Climate, Phenology and Grape Composition. *S. Afr. J. Enol. Vitic.* **2002**, *23*, 78–91.

24. Ferreira, J.H.S.; Marais, P.G. Effect of Rootstock Cultivar, Pruning Method and Crop Load on Botrytis cinerea Rot of Vitis vinifera cv. Chenin blanc grapes. *S. Afr. J. Enol. Vitic.* **2017**, *8*, 41–44. [CrossRef]

25. Microsoft. Kinect for Windows SDK 1.8. Available online: https://www.microsoft.com/en-us/download/details.aspx?id=40276 (accessed on 16 May 2019).

26. Labbe, M. RTAB-Map. Available online: http://introlab.github.io/rtabmap/ (accessed on 1 December 2018).

27. The MathWorks Inc. Matlab R2018b v9.5.0.944444. Available online: https://ww2.mathworks.cn/en/ (accessed on 8 January 2019).

28. CloudCompare. CloudCompare Version 2.10.alpha [GPL Software]. Available online: http://www.cloudcompare.org/ (accessed on 29 November 2018).

29. Cignoni, P.; Callieri, M.; Corsini, M.; Dellepiane, M.; Ganovelli, F.; Ranzuglia, G. MeshLab: An Open-Source Mesh Processing Tool. In Proceedings of the Sixth Eurographics Italian Chapter Conference, the Eurographics Association, Pisa, Italy, 2–4 July 2008; pp. 129–136. [CrossRef]

30. Kazhdan, M.; Hoppe, H. Screened poisson surface reconstruction. *ACM Trans. Graph.* **2013**, *32*, 29. [CrossRef]

31. R Core Team. *R: A Language and Environment for Statistical Computing*; R Foundation for Statistical Computing: Vienna, Austria, 2018.

32. Lafarge, T.; Pateiro-Lopez, B. alphashape3d: Implementation of the 3D Alpha-Shape for the Reconstruction of 3D Sets from a Point Cloud. R Packag version 1.3. 2017. Available online: https://rdrr.io/cran/alphashape3d/ (accessed on 20 August 2019).

33. Lafarge, T.; Pateiro-López, B.; Possolo, A.; Dunkers, J.P. R Implementation of a Polyhedral Approximation to a 3D Set of Points Using the α-Shape. *J. Stat. Softw.* **2014**, *56*, 1–19. [CrossRef]

34. Rueda-Ayala, V.; Peña, J.; Höglind, M.; Bengochea-Guevara, J.; Andújar, D. Comparing UAV-Based Technologies and RGB-D Reconstruction Methods for Plant Height and Biomass Monitoring on Grass Ley. *Sensors* **2019**, *19*, 535. [CrossRef] [PubMed]

35. Ribeiro, A.; Bengochea-Guevara, J.M.; Conesa-Muñoz, J.; Nuñez, N.; Cantuña, K.; Andújar, D. 3D monitoring of woody crops using an unmanned ground vehicle. In Proceedings of the 11th European Conference on Precision Agriculture, Advances in Animal Biosciences, Edinburgh, UK, 16–20 July 2017; Volume 8, pp. 210–215. [CrossRef]

36. Kuhn, M. Building predictive models in R using the caret package. *J. Stat. Softw.* **2008**, *28*, 1–26. [CrossRef]

37. Wasenmüller, O.; Stricker, D. Comparison of Kinect V1 and V2 Depth Images in Terms of Accuracy and Precision. In Proceedings of the ACCV Workshops, Taipei, Taiwan, 20–24 November 2016; pp. 34–45. [CrossRef]

38. Lopes, C.; Torres, A.; Guzman, R.; Graca, J.; Reyes, M.; Vitorino, G.; Braga, R.; Monteiro, A.; Barriguinha, A. Using an unmanned ground vehicle to scout vineyards for non-intrusive estimation of canopy features and grape yield. In Proceedings of the 20th GiESCO International Meeting, Mendoza, Argentina, 5–10 November 2017; pp. 16–21.

Article

A Non-Invasive Method Based on Computer Vision for Grapevine Cluster Compactness Assessment Using a Mobile Sensing Platform under Field Conditions

Fernando Palacios [1,2], **Maria P. Diago** [1,2] **and Javier Tardaguila** [1,2,*]

[1]   Televitis Research Group, University of La Rioja, 26006 Logroño (La Rioja), Spain
[2]   Instituto de Ciencias de la Vid y del Vino, University of La Rioja, CSIC, Gobierno de La Rioja, 26007 Logroño, Spain
*   Correspondence: javier.tardaguila@unirioja.es; Tel.: +34-941-299-741; Fax: +34-941-299-721

Received: 8 July 2019; Accepted: 30 August 2019; Published: 2 September 2019

Abstract: Grapevine cluster compactness affects grape composition, fungal disease incidence, and wine quality. Thus far, cluster compactness assessment has been based on visual inspection performed by trained evaluators with very scarce application in the wine industry. The goal of this work was to develop a new, non-invasive method based on the combination of computer vision and machine learning technology for cluster compactness assessment under field conditions from on-the-go red, green, blue (RGB) image acquisition. A mobile sensing platform was used to automatically capture RGB images of grapevine canopies and fruiting zones at night using artificial illumination. Likewise, a set of 195 clusters of four red grapevine varieties of three commercial vineyards were photographed during several years one week prior to harvest. After image acquisition, cluster compactness was evaluated by a group of 15 experts in the laboratory following the International Organization of Vine and Wine (OIV) 204 standard as a reference method. The developed algorithm comprises several steps, including an initial, semi-supervised image segmentation, followed by automated cluster detection and automated compactness estimation using a Gaussian process regression model. Calibration (95 clusters were used as a training set and 100 clusters as the test set) and leave-one-out cross-validation models (LOOCV; performed on the whole 195 clusters set) were elaborated. For these, determination coefficient (R^2) of 0.68 and a root mean squared error (RMSE) of 0.96 were obtained on the test set between the image-based compactness estimated values and the average of the evaluators' ratings (in the range from 1–9). Additionally, the leave-one-out cross-validation yielded a R^2 of 0.70 and an RMSE of 1.11. The results show that the newly developed computer vision based method could be commercially applied by the wine industry for efficient cluster compactness estimation from RGB on-the-go image acquisition platforms in commercial vineyards.

Keywords: image analysis; cluster morphology; RGB; machine learning; non-invasive sensing technologies; proximal sensing; precision viticulture

---

1. Introduction

Grapevine cluster compactness is a key attribute related to grape composition, fruit health status, and wine quality [1,2]. Compactness defines the density of the cluster by the degree of the aggregation of its berries. Highly compacted winegrape clusters can be affected to a greater extent by fungal diseases, such as powdery mildew [3] and botrytis [4], than loose ones [5].

The most prevalent method for assessing cluster compactness was developed by the International Organization of Vine and Wine (OIV) [6] and has been applied in several research studies [7,8]. This OIV method procures cluster compactness assessment by visual inspection in five different classes. This compactness class takes into account several morphological features of the berries and pedicels, which are visually appraised by trained experts. This method and others designed to evaluate compactness on specific varieties [9–11] tend to be inaccurate due to the intrinsic subjectivity of the evaluation linked to the evaluator's opinion. Moreover, these visual inspection methods are laborious and time-consuming, as they may also require the manual measurement of specific cluster morphological parameters. Therefore, alternative methods for objectively and accurately assessing cluster compactness are needed for wine industry applications.

Computer vision and image processing technology enables low-cost, automated information extraction and its analysis from images taken using a digital camera. This technology is being used in viticulture to estimate key parameters such as vine pruning weight [12,13], the number of flowers per inflorescence [14,15], canopy features [16], or yield [17,18], as well as to provide relevant information to grape harvesting robots [19,20].

Automated cluster compactness estimation by computer vision methods was recently attempted by Cubero et al. [21] and Chen et al. [22]. The former involved the automated extraction of image descriptors from red, green, blue (RGB) cluster images taken from different cluster views under laboratory conditions. From these descriptors, a partial least squares (PLS) calibration model was developed to predict their associated OIV compactness rating. In the approach followed by Chen et al. [22], a multi-perspective imaging system was developed, which made use of different mirror reflections that facilitated the simultaneous acquisition of images from multiple views from a single shot. Additionally, the system also included a weighing sensor for cluster mass measurement. Then, a set of image descriptors and features derived from the data provided by the sensing system were automatically extracted and used to calibrate several models. Of these, the PLS model achieved the best results.

Previously developed computer vision methods for cluster compactness assessment provided accurate and objective compactness estimation only working under controlled laboratory conditions, which requires the destructive collection of clusters in the vineyard. This is a laborious and time-consuming practice that precludes the appraisal of cluster compactness as a standard grape quality parameter prior to harvest, thus limiting its industrial applicability. Moreover, to the best of our knowledge, there is no commercial method available to assess grapevine cluster compactness under field conditions in an automated way.

The purpose of this work was to develop a new, non-invasive, and proximal method based on computer vision and machine learning technology for assessing grapevine cluster compactness from on-the-go RGB image acquisition in commercial vineyards.

2. Materials and Methods

2.1. Experimental Layout

The trials were carried out during seasons 2016, 2017, and 2018 in three commercial vineyards planted with four different red grapevine varieties (*Vitis vinifera* L). The vines were trained onto a vertical shoot positioned (VSP) trellis system and were partially defoliated at fruit set.

An overall set of 195 red grape clusters involving five distinct datasets were labeled in the field prior to image acquisition in three commercial vineyards.

- Vineyard site #1: Located in Logroño (lat. 42°27′42.3″N; long. 2°25′40.4″W; La Rioja, Spain) with 2.8 m row spacing and 1.2 m vine spacing, where a set of 95 Tempranillo clusters were imaged and sampled during season 2016, denoted as T16.
- Vineyard site #2: Located in Logroño (lat. 42°28′34.2″N; long. 2°29′10.0″W; La Rioja, Spain) with 2.5 m row spacing and 1 m vine spacing, where a set of 25 Grenache clusters were imaged and sampled during season 2017, denoted as G17.
- Vineyard site #3: Located in Vergalijo (lat. 42°27′46.0″ N; long. 1°48′13.1″ W; Navarra, Spain) with 2 m row spacing and 1 m vine spacing, where three sets of 75 clusters of Syrah, Cabernet

Sauvignon, and Tempranillo (25 per grapevine variety) were imaged and sampled during season 2018, denoted as S18, CS18, and T18, respectively.

The data were divided into a training set, formed by 95 clusters of T16 dataset, and an external validation test set, formed by 100 clusters of G17, S18, CS18, and T18 datasets, in order to test the system performance on new or additional varieties and vineyards.

2.2. Image Acquisition

Vineyard canopy images were taken on-the-go at a speed of 5 Km/h one week prior to harvest using a mobile sensing platform developed at the University of La Rioja. Image acquisition was performed at night using an artificial illumination system mounted onto the mobile platform in order to obtain homogeneity on the illumination of the vines and to separate the vine under evaluation from the vines of the opposite row. An all-terrain vehicle (ATV) (Trail Boss 330, Polaris Industries, Medina, Minnesota, USA) was modified to incorporate all components as described in the work of Diago et al. [23] (Figure 1).

Figure 1. Mobile sensing platform for on-the-go image acquisition: a modified all-terrain vehicle (ATV) incorporating a red, green, blue (RGB) camera, Global Positioning System (GPS), and an artificial illumination system mounted on an adaptable structure.

Additionally, some elements were modified:

- RGB camera: a mirrorless Sony α7II RGB camera (Sony Corp., Tokyo, Japan) mounting a full-frame complementary metal oxide semiconductor (CMOS) sensor (35 mm and 24.3 MP resolution) and equipped with a Zeiss 24/70 mm lens was used for image acquisition in vineyard sites #1 and #2, while a Canon EOS 5D Mark IV RGB camera (Canon Inc. Tokyo, Japan) mounting a full-frame CMOS sensor (35 mm and 30.4 MP) equipped with a Canon EF 35 mm F/2 IS USM lens was used in vineyard site #3.
- Industrial computer: A Nuvo-3100VTC industrial computer was used for image storage and camera parameters setting for the Canon EOS 5D Mark IV using custom software developed, while the parameters of the Sony α7II camera were set in the camera itself and the storage in a Secure Digital (SD)-card.

The camera was positioned at a distance of 1.5 m from the canopy. The camera parameters were manually set at the beginning of the experiment in each vineyard.

2.3. Reference Measurements of Cluster Compactness

After image acquisition, the labeled clusters were manually collected, and their compactness was visually evaluated in a laboratory at the University of La Rioja by a panel composed of 15 experts following the OIV 204 standard [6]. In this reference method, each cluster was classified in one of five discrete classes (Figure 2) ranging from 1, the loosest clusters, to 9, the most compact clusters. In the visual assessment, several aspects related to the morphology of the cluster, such as berries' mobility, pedicels' visibility, and berries' deformation by pressure, were taken into consideration. The average of the evaluators' ratings was used as the reference compactness value for each cluster.

Figure 2. Examples of clusters with different compactness ratings according to the International Organization of Vine and Wine (OIV) 204 standard: class 1 (**a**) very loose clusters; class 3 (**b**) loose clusters; class 5 (**c**) medium compact clusters; class 7 (**d**); compact clusters; class 9 (**e**) very compact clusters.

2.4. Image Processing

Image processing comprised several steps that can be summarized as semi-supervised image segmentation followed by cluster detection and compactness estimation for each detected cluster. While cluster detection and compactness estimation were fully automated, the image segmentation step required the intervention of the user for each dataset. The algorithm was developed and tested using Matlab R2017b (Mathworks, Natick, MA, USA). The flowchart of the algorithm process for a new set of images is described in Figure 3.

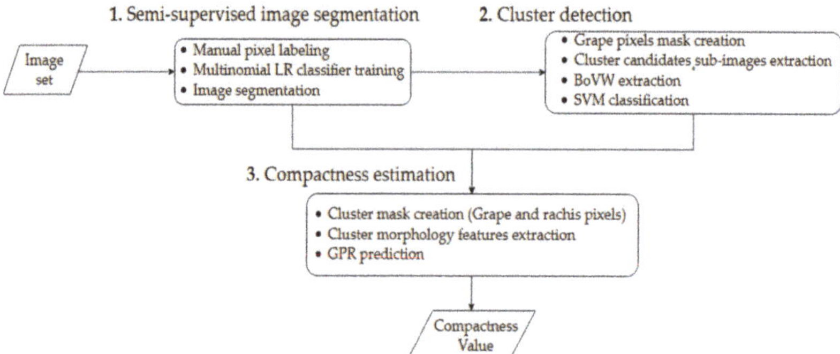

Figure 3. Flow-chart of the full algorithm for a new set of images. First, a set of pixel labeling was required to train a multinomial logistic regression model to segment the whole set. Second, the cluster candidates were extracted and filtered using a bag of visual words model. Finally, compactness features were extracted, and the estimation was performed on each cluster by the Gaussian process regression model.

To ensure consistency in the analysis of the complete algorithm, the classifier used at each step was trained using the training set and evaluated with the output obtained by the classifier of the previous step for the test set, except for the initial image segmentation, where a model was trained on each individual dataset.

2.4.1. Semi-Supervised Image Segmentation

For the proper compactness estimation of every cluster visible in the image, a previous detection of the clusters and their main elements (grape and rachis) was needed.

Most of the red winegrape pixels are easily distinguishable from pixels of other vine elements by their color. Hence, an initial pixelwise color-based segmentation was performed on every image. For this approach, seven classes representing the elements of the grapevines potentially present in their images were defined: "grape", "rachis", "trunk", "shoot", "leaf", "gap", and "trellis". For extracting cluster candidates, only groups of pixels belonging to the first class ("grape") were used, but rachis identification was also relevant for compactness assessment. In summary, an image segmentation considering the seven classes described above as a first step eliminated the necessity of further color segmentation.

A set of 3500 pixels were manually labeled (500 pixels per class), and color features were extracted considering a combination of two color spaces: RGB and CIE L*a*b* (CIELAB) [24]. In this approach, a pixel p was mathematically represented as in Equation (1):

$$p = \left[R_p, G_p, B_p, L_p, a_p, b_p\right] \tag{1}$$

where R_p, G_p, B_p and L_p, a_p, b_p represent the values of p for the three channels of the *RGB* and the *CIELAB* color spaces, respectively. The function rgb2lab from Matlab R2017b was used for the RGB to CIELAB color space conversion.

A multinomial logistic regression was trained with the set described above in order to obtain a pixel wise color based classifier. This classifier, which is a generalization of logistic regression for multiclass problems [25], predicts the probability of each possible outcome for an observation as the relative probability of belonging to each class over belonging to another one chosen arbitrarily as the reference class. Assuming that n is the reference class of a set of $\{1, \ldots, n\}$ classes, the output of the classifier for p is $[\pi_1, \ldots, \pi_n]$, where π_i represents the probability of p belonging to the class i for $i = 1, \ldots, n-1$ [Equation (2)], and π_n represents the probability for the reference class [Equation (3)].

$$\pi_i = \frac{e^{\beta_{i,0} + \sum_{j=1}^{k} \beta_{i,j} x_j}}{1 + \sum_{l=1}^{n-1} e^{\beta_{l,0} + \sum_{j=1}^{k} \beta_{l,j} x_j}} \tag{2}$$

$$\pi_n = \frac{1}{1 + \sum_{l=1}^{n-1} e^{\beta_{l,0} + \sum_{j=1}^{k} \beta_{l,j} x_j}} = 1 - \sum_{l=1}^{n-1} \pi_l \tag{3}$$

where k is the number of predictor variables, x_j the j-th predictor variable, and $\beta_{i,j}$ is the estimated coefficient for the i-th class.

The segmentation was performed by assigning to each pixel the class with the highest probability (Figure 4).

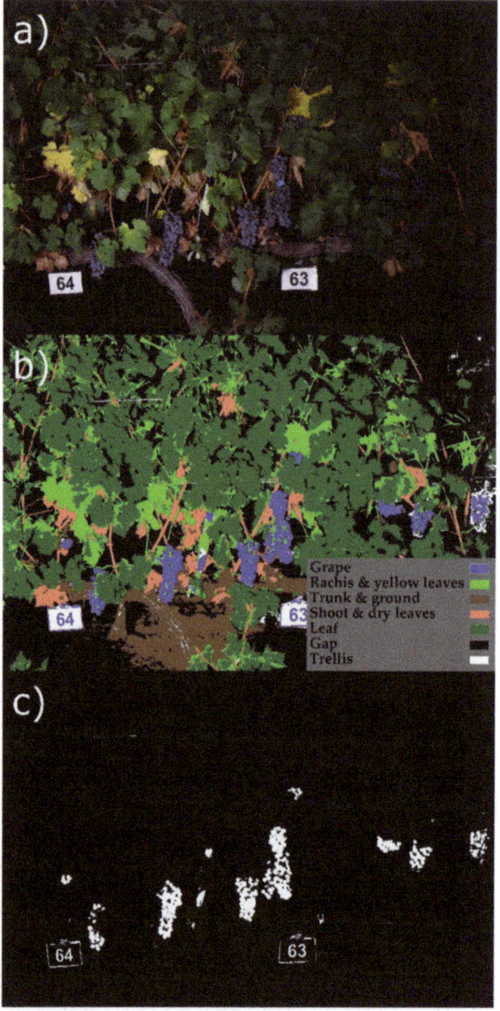

Figure 4. Initial semi supervised segmentation. A set of pixels were manually labeled on the set of the original images (**a**) into seven predefined classes ("grape", "rachis", 'trunk", "shoot", "leaf", "gap", or "trellis"), and the multinomial logistic regression model performed the segmentation on the whole set of images. (**b**) Some pixels of elements without a predefined class were misclassified (e.g., yellow leaves identified as "rachis", dry leaves identified as "shoot", or ground identified as "trunk"). The pixels classified as "grape" and marked in white (**c**) were used for identifying cluster candidates.

2.4.2. Cluster Detection

Using the initial segmentation, a mask of cluster candidates was then generated by selecting those pixels assigned to the "grape" class (Figure 4c).

While the initial color segmentation allowed filtering most of the non-cluster elements presented in the image, a second filtering step was needed to remove those non-cluster groups of pixels with similar color to the "grape" class, which can form objects with different shapes and sizes. This second filtering can be summarized as:

- A morphological opening (morphological erosion operation followed by a dilation) of the clusters' candidates mask using a circular kernel with a radius of three pixels.
- An extraction of a sub-image per minimal bounding box that contains a connected component (groups of connected pixels) in the clusters' candidates mask.
- An extraction of features for each sub-image that represents the information contained on it. For this, the bag-of-visual-words (BoVW) was employed.
- A classification of "cluster" vs. "non-cluster" sub-images.

The bag-of-visual-words (BoVW) model is a concept derived from document classification for image classification and object categorization [26]. In this model, images are treated as documents formed by local features denominated "visual words". These words are grouped to form a "vocabulary" or "codebook". Then, every image is represented by the number of occurrences of every "codeword" in the codebook. In this work, local features were 64-length speeded up robust features (SURF) descriptor vectors [27] clustered by a k-means algorithm [28].

Given a set of n training sub-images, Tr, represented as $Tr = \{tr_1, \ldots, tr_n\}$ and their class $Y = \{y_1, \ldots, y_n\}$ manually labeled into "cluster" (total and partial clusters) and "non-cluster", the process adopted for the training set is the following:

1. To extract SURF points for every sub-image.
2. Cluster SURF points applying k-means. The set of cluster centroids would form the codebook of k codewords.
3. Extraction of the bag-of-words per sub-image:

    a. To assign each SURF point of the image to the nearest centroid of the codebook.
    b. To calculate the histogram by counting the number of SURF points assigned to each centroid.

Then, tr_i had a feature vector $x_i \in \mathbb{R}^k$ that was used to train a support vector machine (SVM) classifier [29]. This is a machine learning algorithm for supervised learning classification or regression tasks that transforms input data into a high-dimensional feature space using a kernel function and finds the hyperplane that maximizes the distance to the nearest training data point of any class. With the classifier trained, only step 3 was performed for new sets of cluster candidates sub-images, and the resulting feature vectors were used to classify each sub-image into "cluster" or "non-cluster" classes, preserving only cluster sub-images for further analysis (Figure 5).

Figure 5. Cluster candidates' extraction and filtering. (**a**) Bounding boxes were extracted from connected components of grape pixels and (**b**) filtered using the bag-of-visual-words (BOVW) model to estimate the cluster compactness of the final non-filtered regions.

2.4.3. Cluster Compactness Estimation

This step involved the extraction of a set of features from the cluster morphology that were related to its compactness. For that purpose, the segmented pixels corresponding to the initial segmentation mask of the sub-images classified as clusters in the previous step had to be extracted. For a given detected cluster sub-image, the next procedure was followed:

- A new mask using only pixels of "grape" and "rachis" classes was created.
- A morphological opening on "grape" pixels using a circular kernel with a radius of two pixels was applied.
- A morphological opening on "rachis" pixels using a circular kernel with a radius of two pixels was also applied.
- A mask containing only the largest connected component formed by "grape" and "rachis" pixels, denoted as mask "A", was created.
- A mask containing the convex hulls of each "grape" pixel's connected component (that can represent several grouped berries on compact clusters or isolated grapes on loose clusters), denoted as mask "B", was created.
- The final mask was created containing "grape" pixels and "rachis" pixels that were in mask "A" and inside the region of the convex hull of mask "B". Those "rachis" pixels in mask "A" that were outside of the convex hull of mask "B" and connected at least two connected components of "grape" pixels were included as well.

The features to estimate the cluster compactness were extracted from the last mask containing only "grape" and "rachis" pixels (Figure 6). These features were the following:

- Ratio of the area of the convex hull body of the cluster corresponding to holes (AH)
- Ratio of the clusters area corresponding to berries (AB)
- Ratio of the area corresponding to "rachis" (AR)
- Average width at $25 \pm 5\%$ of the length of the cluster (W25)
- Average width at $50 \pm 5\%$ of the length of the cluster (W50)
- Average width at $75 \pm 5\%$ of the length of the cluster (W75)
- Ratio between "rachis" and "grape" pixels (RatioRG)
- Roundness of "grape" pixels (RDGrape): $\dfrac{4.0 \times \pi \times A_{Grape}}{P_{Grape}^2}$
- Compactness shape factor of "grape" pixels (CSFGrape): $\dfrac{P_{Grape}^2}{A_{Grape}}$
- Ratio between the maximum width and the length of the cluster (AS)
- Ratio between *W75* and *W25* (RatioW75_W25)
- Proportion of the "rachis" pixels "inside" the cluster (RR_{in})
- Proportion of the "rachis" pixels "outside" the cluster (RR_{out})
- Ratio of the area of the cluster over the mean area of the clusters of its set (R_{AoM})

Where A_{Grape} and P_{Grape} correspond to the area and the perimeter, considering only grape pixels.

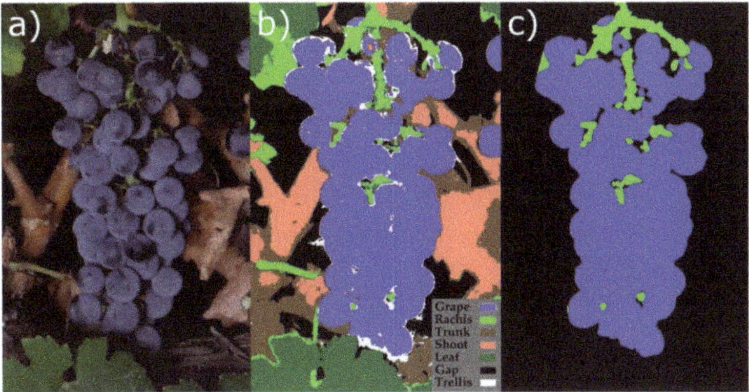

Figure 6. Extraction of the clusters' final masks for compactness estimation. (**a**) Extracted cluster sub-image, (**b**) its corresponding segmentation, and (**c**) a cluster mask obtained using "grape" and "rachis" pixels and morphological operations.

While some of these features were already addressed by Cubero et al. [21] and Chen et al. [22], they had to be adapted to the new environmental situation of field conditions, while others were specifically designed for this study. Features based on clusters' widths and lengths (*W25*, *W50*, *W75*) required a prior rotation of the cluster mask along the longest axis to match the width of the cluster with the horizontal axis, thus the whole set of clusters could have a similar orientation. The features RR_{in} and RR_{out} were calculated as the proportion of "rachis" pixels completely surrounded by "grape" pixels, and the rest of the "rachis" pixels that were not, respectively. The feature R_{AoM} provided a measure about the size of the cluster over the average of the clusters on its set and added robustness by incorporating images of clusters taken at different distances.

The compactness estimation was performed by a Gaussian process regression (GPR) model trained with the data extracted from n clusters $\{\{x_i, y_i\}\}_{i=1}^{n}$ where $x_i \in \mathbb{R}^{14}$ represents the 14-feature vector, and $y_i \in [1,9]$ represents the average of the ratings of the evaluators for the *i*-th cluster.

Gaussian process regression models are probabilistic kernel-based machine learning models that use a Bayesian approach to solve regression problems estimating uncertainty at predictions [30]. A Gaussian process regression model is described in Equation (4):

$$g(x) = f(x) + h(x)^T \beta \tag{4}$$

where $h(x)$ is a vector of basis functions, β is the coefficient of $h(x)$, and $f(x) \sim GP(0, k(x, x'))$ is a zero mean Gaussian process with a $k(x, x')$ covariate function.

2.4.4. Performance Evaluation Metrics

The results obtained for each step were analyzed using a set of metrics corresponding to classification tasks in the case of multinomial logistic regression and support vector machine and regression for the Gaussian process regression. The metrics chosen for classification performance are commonly used for evaluating results of binary classifiers, where a sample can be identified as positive class or negative class. For multinomial logistic regression, the positive and the negative classes were the class under evaluation and the rest of them (e.g., "grape" class vs. "non-grape" classes), while for support vector machine, the "cluster" and the "non-cluster" classes were considered, respectively. The metrics calculated were sensitivity [Equation (5)], specificity [Equation (6)], F1 score [Equation (7)], and intersect over union [IoU; Equation (8)].

$$Sensitivity = \frac{TP}{TP + FN} \tag{5}$$

$$Specificity = \frac{TN}{TN + FP} \tag{6}$$

$$F_1 = 2 \times \frac{Precision \times Sensitivity}{Precision + Sensitivity} \tag{7}$$

$$IoU = \frac{TP}{TP + FP + FN} \tag{8}$$

where TP represents the "true positives" (number of positive samples correctly classified as positive class), FP represents the "false positives" (number of negative samples incorrectly classified as positive class), TN represents the "true negatives" (number of negative samples correctly classified as negative class), and FN represents the "false negatives" (number of positive samples incorrectly classified as negative class). *Precision* was defined as in Equation (9):

$$Precision = \frac{TP}{TP + FP} \tag{9}$$

The area under the receiver operating characteristic (ROC) curve (AUC) [31] was also considered. For regression, the determination coefficient (R^2) and the root mean squared error (RMSE) were selected.

2.4.5. Hyperparameter's Optimization Procedure

Support vector machine and Gaussian process regression are two machine learning algorithms that have a set of hyperparameters that are not learned from the data and need to be set before the training. The most traditional hyperparameter selection method is a brute-force grid search of the best subset of hyperparameters combined with a manually predefined set of values established for each hyperparameter in order to optimize a performance metric. Instead, in this work, a Bayesian optimization algorithm [32] was used for finding the best hyperparameters set, which proved to outperform other optimization algorithms [33]. This algorithm finds the best hyperparameter set that optimizes an objective function (in this context, a performance metric of the machine learning algorithm) using a Gaussian process trained with the objective function evaluations. The Gaussian process is updated with the result of each evaluation of the objective function, and an acquisition function is used to determine the next point to be evaluated in a bounded domain, i.e., the next set of hyperparameters.

The functions fitcsvm and fitrgp of Matlab R2017b were used to train the support vector machine and the Gaussian process regression models, respectively, selecting their hyperparameters with Bayesian optimization. A Gaussian process with automatic relevance determination (ARD) Matérn 5/2 kernel model and the expected-improvement-plus acquisition function were used for this purpose. The ranges considered for each hyperparameter and the final values are shown in Table 1.

Table 1. Hyperparameter range considered for each classifier and the used final values.

	Kernel Function (fixed)	Optimized Hyperparameters Range		Final Values	
SVM	Radial basis function (RBF)	*Box Constraint* $[10^{-3}, 10^3]$	*Kernel scale* $[10^{-3}, 10^3]$	*Box Constraint* 1.4654	*Kernel scale* 24.628
GPR	Exponential	*Sigma* $[10^{-4}, 22.5184]$	*Kernel scale* $[0.1216, 121.6122]$	*Sigma* 0.83194	*Kernel scale* 91.5821

SVM: support vector machine; GPR: Gaussian process regression.

3. Results and Discussion

3.1. Initial Segmentation Performance

The initial segmentation process was a key step towards the accurate assessment of cluster compactness. Therefore, a different segmentation model was applied to each grapevine variety and vineyard to avoid errors associated with slight differences in color and illumination from the images captured from one vineyard to another, which would occur if applying a unique segmentation model. Likewise, five sets of 3500 pixels each were manually labeled (500 pixels per class) for each variety and vineyard, which were used to train five distinct multinomial logistic regression models.

As shown in Table 2, overall, the five models achieved good results in terms of sensitivity, specificity, F1 score, AUC, and IoU when applied to their specific set of images. With regard to the most relevant classes ("rachis" and "grape"), similar and equally good values were obtained for specificity for all sets, while more variable outcomes were obtained for the remaining metrics. In general terms, the T18 model yielded the best results for these two relevant classes, closely followed by the S18 model, in this case only for the "grape" class, and slightly outperformed by the CS18 model for the "rachis" class in terms of sensitivity and AUC. More modest results were obtained for models G17 and T16 (Table 1). Particularly, model T16 yielded values under 0.9 in sensitivity and F1 score metrics for the "rachis" class and in IoU for both "rachis" and "grape" classes.

Table 2. Performance results of each multinomial logistic regression model for segmenting images in their respective dataset using a 10-fold stratified cross validation on the manually labeled pixel sets in terms of sensitivity, specificity, F1 score, area under the receiver operating characteristic (ROC) curve (AUC) and intersect over union (IoU) metrics. Their average was compared with the performance of a single segmentation with all datasets combined using a 5-fold and a 10-fold stratified cross validation.

Vineyard Canopy Class	T16	G17	Dataset S18	CS18	T18	Average	5-Fold CV	10-Fold CV
			Sensitivity					
Trellis	0.9420	0.8760	0.9500	0.9200	0.9760	0.9328	0.5284	0.6700
Gap	0.9740	0.9900	0.9920	0.9960	0.9940	0.9892	0.9868	0.9880
Leaf	0.9760	0.9340	0.9060	0.9680	0.9700	0.9508	0.7476	0.8980
Shoot	0.9440	0.9520	0.9880	1.0000	0.9900	0.9748	0.8992	0.9208
Rachis	0.8640	0.9020	0.9220	0.9660	0.9580	0.9224	0.6912	0.7964
Trunk	0.9020	0.9560	0.9540	0.9660	0.9980	0.9552	0.3620	0.6992
Grape	0.9320	0.9720	0.9700	0.9540	0.9800	0.9616	0.7652	0.8732

Table 2. *Cont.*

Vineyard Canopy Class	T16	G17	Dataset S18	CS18	T18	Average	5-Fold CV	10-Fold CV
			Specificity					
Trellis	0.9897	0.9883	0.9930	0.9883	0.9970	0.9913	0.9554	0.9648
Gap	0.9950	0.9987	0.9967	0.9990	0.9977	0.9974	0.9955	0.9963
Leaf	0.9960	0.9893	0.9897	0.9960	0.9947	0.9931	0.9796	0.9829
Shoot	0.9900	0.9963	0.9983	0.9997	0.9987	0.9966	0.9335	0.9789
Rachis	0.9813	0.9850	0.9810	0.9927	0.9947	0.9869	0.9181	0.9673
Trunk	0.9803	0.9820	0.9923	0.9933	0.9987	0.9893	0.9151	0.9423
Grape	0.9900	0.9907	0.9960	0.9927	0.9963	0.9931	0.9661	0.9751
			F1 Score					
Trellis	0.9401	0.9003	0.9538	0.9246	0.9789	0.9396	0.5884	0.7123
Gap	0.9721	0.9910	0.9861	0.9950	0.9900	0.9868	0.9801	0.9831
Leaf	0.9760	0.9349	0.9207	0.9719	0.9690	0.9545	0.7996	0.8978
Shoot	0.9421	0.9645	0.9890	0.9990	0.9910	0.9771	0.7826	0.8996
Rachis	0.8745	0.9056	0.9057	0.9612	0.9628	0.9220	0.6334	0.7993
Trunk	0.8931	0.9264	0.9540	0.9631	0.9950	0.9463	0.3869	0.6836
Grape	0.9357	0.9586	0.9729	0.9550	0.9790	0.9602	0.7773	0.8634
			AUC					
Trellis	0.9658	0.9322	0.9715	0.9542	0.9865	0.9620	0.7419	0.8174
Gap	0.9845	0.9943	0.9943	0.9975	0.9958	0.9933	0.9912	0.9922
Leaf	0.9860	0.9617	0.9478	0.9820	0.9823	0.9720	0.8636	0.9405
Shoot	0.9670	0.9742	0.9932	0.9998	0.9943	0.9857	0.9164	0.9499
Rachis	0.9227	0.9435	0.9515	0.9793	0.9763	0.9547	0.8047	0.8818
Trunk	0.9412	0.9690	0.9732	0.9797	0.9983	0.9723	0.6386	0.8207
Grape	0.9610	0.9813	0.9830	0.9733	0.9882	0.9774	0.8656	0.9241
			IoU					
Trellis	0.8870	0.8187	0.9117	0.8598	0.9587	0.8872	0.4169	0.5532
Gap	0.9456	0.9821	0.9725	0.9901	0.9803	0.9741	0.9610	0.9667
Leaf	0.9531	0.8778	0.8531	0.9453	0.9399	0.9139	0.6661	0.8146
Shoot	0.8906	0.9315	0.9782	0.9980	0.9821	0.9561	0.6428	0.8175
Rachis	0.7770	0.8275	0.8276	0.9253	0.9283	0.8571	0.4635	0.6657
Trunk	0.8068	0.8628	0.9120	0.9288	0.9901	0.9001	0.2399	0.5193
Grape	0.8792	0.9205	0.9473	0.9138	0.9589	0.9239	0.6358	0.7596

To compare the segmentation performance between individual models on each dataset versus a unique segmentation model, two additional cross-validation methods were applied: a five-fold cross-validation, where at each iteration, the training fold was formed by four datasets and the test fold by the remaining dataset, and a ten-fold stratified cross-validation, where at each iteration, the training and the test folds comprised data at an equal proportion of each dataset and class. The comparison of the results for these two validation methods revealed that better results were obtained for all metrics when the training and the test contained data from the same dataset (ten-fold CV). Comparing both methods with the average of the results obtained by the individual models indicated that applying individual segmentation models produced a substantial improvement in all metrics (except for specificity, for which only a slight increase was recorded) and for all classes (with the exception of the "gap" class, for which similar results were obtained using the three methods). The increase in these metrics for the relevant classes ("grape" and "rachis") and the importance of this step highlights the need for applying individual models for each dataset. Differences between performances could be related to differences in color tonality of the vine elements segmented (e.g., different green tonalities for leaves) between grape varieties and vineyards.

The results show that, given a set of images taken on a vineyard, a multinomial logistic regression model trained with a small subset of pixels manually labeled from the images can be applied to effectively segment vine images in the predefined classes using color information. Also, the pixels needed for compactness estimation ("grape" and "rachis" pixels) can be extracted.

3.2. Cluster Detection Performance

A support vector machine was trained with 600 sub-images manually labeled into 300 "cluster" (total and partial clusters) and 300 "non-cluster" sub-images automatically extracted from the segmentation performed on the T16 set.

The classifier was validated against a set of 800 sub-images automatically extracted from the segmentation performed on sets G17, S18, CS18, and T18 (200 sub-images per set) and manually labeled into 400 "cluster" (total and partial clusters) and 400 "non-cluster" sub-images (100 sub-images of each class per set). A set of k values in a range from 10–200 was chosen for the k-means algorithm, and the performance of the classifier for the "cluster" class was evaluated (Table 3). The model trained with $k = 100$ yielded the best results for all metrics. Similar results were obtained for all k values in terms of sensitivity and F1 score, while for specificity and AUC, the model trained with $k = 10$ performed poorer than the rest of models. The model that yielded the best results ($k = 100$) showed similar values in sensitivity, specificity, and F1 (between 0.76 and 0.8), with specificity being slightly superior, while a higher value was obtained for AUC.

Table 3. Performance results of the support vector machine classifier for cluster detection validated with the external set and performing a 5-fold cross validation on the whole data for several k values of k-means tested in terms of sensitivity, specificity, F1 score, and AUC metrics.

		Sensitivity	Specificity	F1 Score	AUC
	$k = 10$	0.760	0.660	0.724	0.751
	$k = 50$	0.738	0.770	0.750	0.828
Test Set	$k = 100$	0.765	0.795	0.777	0.865
	$k = 150$	0.750	0.770	0.758	0.841
	$k = 200$	0.720	0.765	0.737	0.848
	$k = 10$	0.811	0.678	0.761	0.821
	$k = 50$	0.804	0.788	0.798	0.884
5-Fold CV	$k = 100$	0.821	0.781	0.805	0.903
	$k = 150$	0.790	0.805	0.796	0.888
	$k = 200$	0.799	0.813	0.804	0.902

CV: cross validation.

These results could be improved by incorporating new data, as is visualized in Table 3. A five-fold cross validation (each fold being a different set) was performed to train new support vector machine classifiers. The results show an improvement over all metrics for all k tested values, with the exception of specificity for the $k = 100$ model, whose value was slightly diminished. This model still achieved the best results for all metrics except specificity. The $k = 100$ model was chosen as the final model, as it yielded the best results with the test set in all metrics as well as the best results in almost all metrics at the cross validation.

The performance of the support vector machine proves that this classifier trained with the bag-of-visual-words representation of "cluster" and "non-cluster" sub-images can be applied to classify new sub-images from new datasets previously unknown to the classifier, and therefore, it can be used to filter non-cluster pixel groups before compactness estimation.

3.3. Cluster Compactness Estimation

A Gaussian process regression model was trained on the set of features extracted from the clusters of the T16 set (95 clusters) and validated with the automatically detected clusters of G17, S18, CS18, and T18 sets (100 clusters). A coefficient of determination (R^2) of 0.68 and an RMSE of 0.96 were achieved (Figure 7). This is a remarkable result, considering that the four test sets were totally unknown to the classifier and were taken on different vineyards. Also, S18, CS18, and T18 test sets were photographed with a different RGB camera than the one used to photograph the T16 training set. Even more relevant is that three of the four sets (75 clusters of 100) were formed by varieties different from the one used for training (Tempranillo was used for training, while Grenache, Syrah, Cabernet Sauvignon, and Tempranillo clusters were included in the test set). This outcome paves a way to cluster compactness estimation of winegrape varieties without the requirement of representing variety in the training data—in contrast to the work presented by Cubero et al. [21]—and paves a way to real context application on new varieties and vineyards without the necessity of including specific data from those varieties and vineyards, which would require retrieving new clusters and assessing their compactness by trained experts.

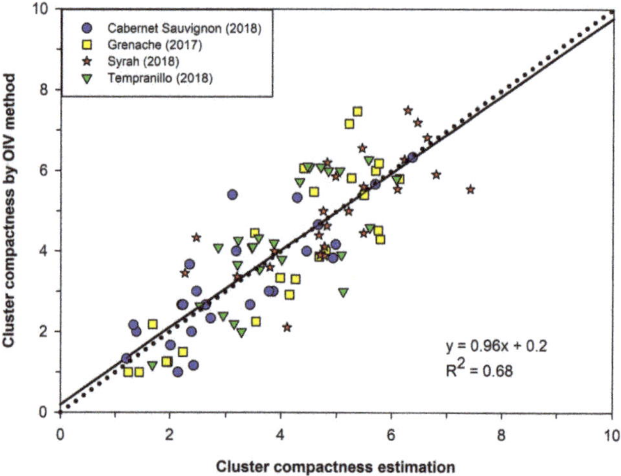

Figure 7. Performance of the GPR model on the test set (100 clusters); correlation between the cluster compactness estimation performed by the model and the OIV ratings (reference method) evaluated visually by the panel of experts.

On the other hand, when leave-one-out cross validation over all datasets (195 clusters) was performed, a coefficient of determination (R^2) of 0.70 and an RMSE of 1.11 were obtained (Figure 8). It can be observed that the algorithm had an accurate performance along most of the compactness range but tended to slightly underestimate highly compacted clusters with an OIV rating close to 9. A feasible reason for this could be that, since this very high compact class was mainly characterized by the deformation of the berries due to the pressure among berries, this feature was difficult to extract by image analysis.

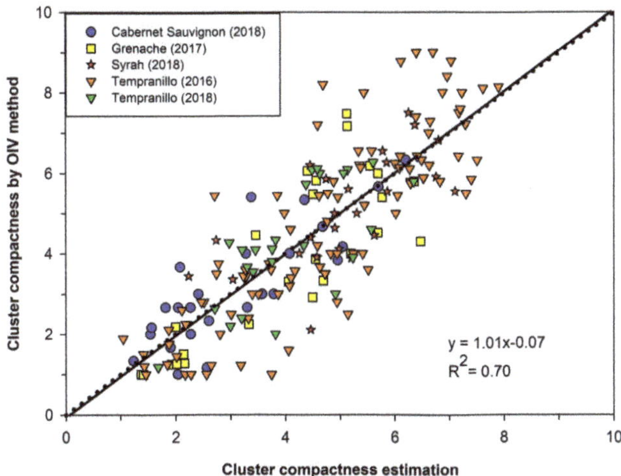

Figure 8. Performance of the GPR model performing leave-one-out cross validation (LOOCV) on the whole data (195 clusters); correlation between the cluster compactness estimation performed by the model and the OIV ratings (reference method) evaluated visually by the panel of experts.

The accuracy of compactness estimation is also meant to be highly influenced by the results obtained in previous steps, i.e., a high misclassification rate in the initial segmentation of cluster and "rachis" pixels would produce wrong shapes in the final cluster mask and a poor feature extraction for the cluster. Also, for estimating the compactness of a given cluster, this has to be previously detected by the BOVW model.

The cluster compactness estimation using the developed methodology in this work could be limited by some experimental in-field conditions, as follows:

- Occlusion of the cluster: the estimation was only performed on the visible region of the cluster. Therefore, a high level of occlusion of the cluster could increase the estimation error. An example of a cluster partially occluded by leaves is shown in Figure 9a and the final mask extracted for compactness estimation in Figure 9b, where the cluster mask presents an anomalous shape that would lead to incorrect compactness estimation.
- Cluster overlapping: highly overlapped clusters would be identified as one, and therefore a unique estimation would be obtained, associated with the set of overlapped clusters. An example is illustrated in Figure 9c, where a set of clusters are overlapped, and in the extracted mask (Figure 9d), the clusters cannot be separated from each other for proper individual compactness estimation.

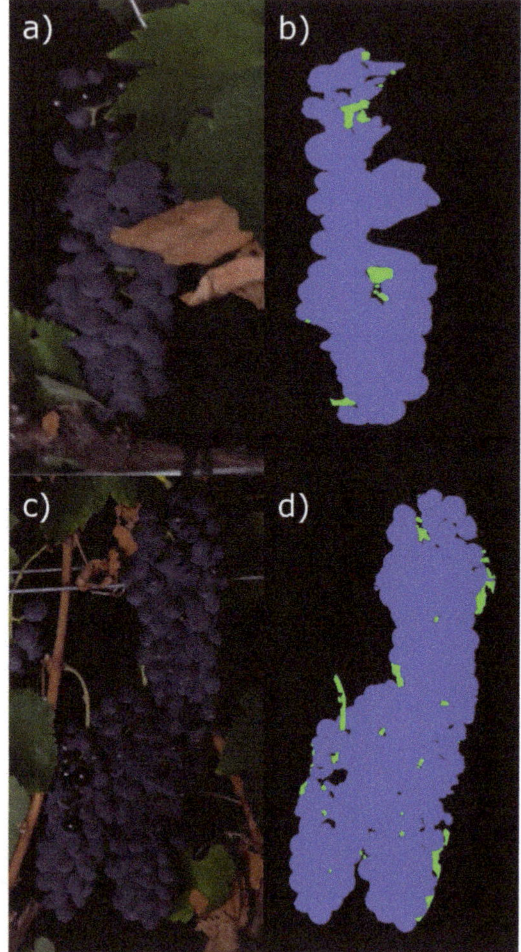

Figure 9. Examples of cluster occlusions and overlapping in commercial vineyards limiting compactness estimation. Cluster partially occluded by leaves (**a**), multiple overlapped clusters (**c**), and final segmented masks used for compactness estimation (**b,d**).

At the current state of the system, the occlusion problem could be overcome by defoliating the side of the vineyard to be photographed. For cluster overlapping, those groups of overlapped clusters could be isolated in the field, or the separation between them could be manually labeled on the images with a clearly different color than "rachis" and "grape" colors (e.g., "trunk" color).

3.4. Commercial Applicability

The developed system can be efficiently used to estimate the cluster compactness in commercial vineyards. The image acquisition carried out by a mobile sensing platform allows the user to take a high number of images in extensive vineyards, which can be automatically geo-referenced. Therefore, the geo-referenced compactness estimations could be used to generate a map that illustrates the spatial variability in cluster compactness, which could be very useful to delineate zones according to cluster compactness and to identify those with similar values. This information could be highly relevant for sorting grapes before harvest, as cluster compactness is often linked to grape quality and health status.

The non-invasive nature of the system could also enable an early identification of very compact clusters before harvest in order to establish strategies against fungal diseases, such as botrytis.

It is also remarkable that the absence of features non-extracted directly from image analysis in the model opens the possibility of a direct application of the algorithm in new vineyards and varieties, in contrast to previous works. Cubero et al [21] introduced, in the PLS model, the cluster winegrape variety as a feature, which requires collecting additional clusters of the variety whose compactness is going to be estimated, evaluating its compactness following the OIV method, and re-training the model. Chen et al. [22] introduced features derived from the cluster mass measured by a weighing sensor, which requires prior harvesting of clusters.

3.5. Future Work

While the current state of the system is capable of estimating compactness in commercial vineyards under uncontrolled field conditions, some improvements still can be made. The algorithm works properly only for red winegrape varieties due to the initial segmentation step, which uses only color information. In this regard, any white grape pixel could be easily misclassified as leaves or "rachis" pixels. A more robust segmentation algorithm could be developed to overcome this problem, combining color and texture information or recurring deep learning techniques. Also, these solutions could help to develop a fully automated system.

The compactness estimation model would also benefit from a more advanced image analysis algorithm capable of extracting features representing the deformation of berries (to increase the accuracy in detecting highly compact clusters), the degree of occlusion of the cluster (to avoid estimations on highly occluded clusters), and to separate overlapped clusters (to enable an individualized estimation on each cluster of the overlapped set).

4. Conclusions

The results of this work show that the developed system was able to estimate winegrape cluster compactness using RGB computer vision on-the-go (at 5 km/h using a mobile sensing platform) and machine learning technology under field conditions. This system enabled a semi-automated, non-invasive method for compactness estimation of a high number of red grapevine clusters under field conditions with low time-consumption. It could be applied to determine the spatial variability of cluster compactness in commercial vineyards, which could be used as new quality input to drive decisions on harvest classification or differential fungicide spraying, for example. The developed methodology constitutes a new tool to improve decision making in precision viticulture, which could be helpful for the wine industry.

Author Contributions: M.P.D. and J.T. conceived and designed the experiments. F.P. developed the algorithm and validated the results. F.P., M.P.D. and J.T. wrote the paper.

Funding: Fernando Palacios would like to acknowledge the research founding FPI grant 286/2017 by Universidad de La Rioja, Gobierno de La Rioja. Dr Maria P. Diago is funded by the Spanish Ministry of Science, Innovation and Universities with a Ramon y Cajal grant RYC-2015-18429.

Acknowledgments: Authors would like to thank Ignacio Barrio, Diego Collado, Eugenio Moreda and Saúl Río for their help collecting field data.

Conflicts of Interest: The authors declare no conflict of interest.

References

1. Hed, B.; Ngugi, H.K.; Travis, J.W. Relationship between cluster compactness and bunch rot in Vignoles grapes. *Plant Dis.* **2009**, *93*, 1195–1201. [CrossRef] [PubMed]
2. Tello, J.; Marcos, J.I. Evaluation of indexes for the quantitative and objective estimation of grapevine bunch compactness. *Vitis* **2014**, *53*, 9–16.
3. Austin, C.N.; Wilcox, W.F. Effects of sunlight exposure on grapevine powdery mildew development. *Phytopathology* **2012**, *102*, 857–866. [CrossRef] [PubMed]

4. Vail, M.; Marois, J. Grape cluster architecture and the susceptibility of berries to Botrytis cinerea. *Phytopathology* **1991**, *81*, 188–191. [CrossRef]

5. Molitor, D.; Behr, M.; Hoffmann, L.; Evers, D. Research note: Benefits and drawbacks of pre-bloom applications of gibberellic acid (GA3) for stem elongation in Sauvignon blanc. *S. Afr. J. Enol. Vitic.* **2012**, *33*, 198–202. [CrossRef]

6. OIV. OIV Descriptor list for grape varieties and Vitis species. *OIV* **2009**, *18*, 178. Available online: http://www.oiv.int/public/medias/2274/code-2e-edition-finale.pdf (accessed on 2 September 2019).

7. Palliotti, A.; Gatti, M.; Poni, S. Early leaf removal to improve vineyard efficiency: gas exchange, source-to-sink balance, and reserve storage responses. *Am. J. Enol. Vitic.* **2011**, *62*, 219–228. [CrossRef]

8. Tardaguila, J.; Blanco, J.; Poni, S.; Diago, M. Mechanical yield regulation in winegrapes: Comparison of early defoliation and crop thinning. *Aust. J. Grape Wine Res.* **2012**, *18*, 344–352. [CrossRef]

9. Zabadal, T.J.; Bukovac, M.J. Effect of CPPU on fruit development of selected seedless and seeded grape cultivars. *HortScience* **2006**, *41*, 154–157. [CrossRef]

10. Evers, D.; Molitor, D.; Rothmeier, M.; Behr, M.; Fischer, S.; Hoffmann, L. Efficiency of different strategies for the control of grey mold on grapes including gibberellic acid (Gibb3), leaf removal and/or botrycide treatments. *OENO One* **2010**, *44*, 151–159. [CrossRef]

11. Tello, J.; Aguirrezábal, R.; Hernáiz, S.; Larreina, B.; Montemayor, M.I.; Vaquero, E.; Ibáñez, J. Multicultivar and multivariate study of the natural variation for grapevine bunch compactness. *Aust. J. Grape Wine Res.* **2015**, *21*, 277–289. [CrossRef]

12. Kicherer, A.; Klodt, M.; Sharifzadeh, S.; Cremers, D.; Töpfer, R.; Herzog, K. Automatic image-based determination of pruning mass as a determinant for yield potential in grapevine management and breeding. *Aust. J. Grape Wine Res.* **2017**, *23*, 120–124. [CrossRef]

13. Millan, B.; Diago, M.P.; Aquino, A.; Palacios, F.; Tardaguila, J. Vineyard pruning weight assessment by machine vision: towards an on-the-go measurement system. *OENO One* **2019**, *53*. [CrossRef]

14. Aquino, A.; Millan, B.; Gutiérrez, S.; Tardáguila, J. Grapevine flower estimation by applying artificial vision techniques on images with uncontrolled scene and multi-model analysis. *Comput. Electron. Agric.* **2015**, *119*, 92–104. [CrossRef]

15. Liu, S.; Li, X.; Wu, H.; Xin, B.; Petrie, P.R.; Whitty, M. A robust automated flower estimation system for grape vines. *Biosystems Eng.* **2018**, *172*, 110–123. [CrossRef]

16. Diago, M.P.; Krasnow, M.; Bubola, M.; Millan, B.; Tardaguila, J. Assessment of vineyard canopy porosity using machine vision. *Am. J. Enol. Vitic.* **2016**, *67*, 229–238. [CrossRef]

17. Nuske, S.; Wilshusen, K.; Achar, S.; Yoder, L.; Narasimhan, S.; Singh, S. Automated visual yield estimation in vineyards. *J. Field Rob.* **2014**, *31*, 837–860. [CrossRef]

18. Millan, B.; Velasco-Forero, S.; Aquino, A.; Tardaguila, J. On-the-Go Grapevine Yield Estimation Using Image Analysis and Boolean Model. *J. Sens.* **2018**, *2018*. [CrossRef]

19. Luo, L.; Tang, Y.; Zou, X.; Ye, M.; Feng, W.; Li, G. Vision-based extraction of spatial information in grape clusters for harvesting robots. *Biosystems Eng.* **2016**, *151*, 90–104. [CrossRef]

20. Luo, L.; Tang, Y.; Lu, Q.; Chen, X.; Zhang, P.; Zou, X. A vision methodology for harvesting robot to detect cutting points on peduncles of double overlapping grape clusters in a vineyard. *Comput. Ind.* **2018**, *99*, 130–139. [CrossRef]

21. Cubero, S.; Diago, M.P.; Blasco, J.; Tardáguila, J.; Prats-Montalbán, J.M.; Ibáñez, J.; Tello, J.; Aleixos, N. A new method for assessment of bunch compactness using automated image analysis. *Aust. J. Grape Wine Res.* **2015**, *21*, 101–109. [CrossRef]

22. Chen, X.; Ding, H.; Yuan, L.-M.; Cai, J.-R.; Chen, X.; Lin, Y. New approach of simultaneous, multi-perspective imaging for quantitative assessment of the compactness of grape bunches. *Aust. J. Grape Wine Res.* **2018**, *24*, 413–420. [CrossRef]

23. Diago, M.P.; Aquino, A.; Millan, B.; Palacios, F.; Tardaguila, J. On-the-go assessment of vineyard canopy porosity, bunch and leaf exposure by image analysis. *Aust. J. Grape Wine Res.* **2019**, *25*, 363–374. [CrossRef]

24. Luo, M.R. CIELAB. In *Encyclopedia of Color Science and Technology*; Springer: Berlin/Heidelberg, Germany, 2014; pp. 1–7.

25. Dobson, A.J.; Barnett, A. *An introduction to generalized linear models*; Chapman and Hall/CRC: New York, NY, USA, 2008.

26. Csurka, G.; Dance, C.; Fan, L.; Willamowski, J.; Bray, C. Visual categorization with bags of keypoints. In Proceedings of the Workshop on statistical learning in computer vision, ECCV, Prague, Czech Republic, 15 May 2004; pp. 1–2.

27. Bay, H.; Tuytelaars, T.; Van Gool, L. Surf: Speeded up robust features. *Springer* **2006**, *3951*, 404–417.

28. Lloyd, S. Least squares quantization in PCM. *IEEE Trans. Inf. Theory* **1982**, *28*, 129–137. [CrossRef]

29. Cortes, C.; Vapnik, V. Support-vector networks. *Mach. Learn.* **1995**, *20*, 273–297. [CrossRef]

30. Williams, C.K.; Rasmussen, C.E. *Gaussian Processes for Machine Learning*; MIT Press: Cambridge, MA, USA, 2006.

31. Bradley, A.P. The use of the area under the ROC curve in the evaluation of machine learning algorithms. *Pattern Recognit.* **1997**, *30*, 1145–1159. [CrossRef]

32. Mockus, J.; Tiesis, V.; Zilinskas, A. The application of Bayesian methods for seeking the extremum. In *Towards Global Optimization*; Elsevier: Amsterdam, The Netherlands, 2014; pp. 117–129.

33. Jones, D.R. A Taxonomy of Global Optimization Methods Based on Response Surfaces. *J. Glob. Optim.* **2001**, *21*, 345–383. [CrossRef]

Article

Spatial Variability of Aroma Profiles of Cocoa Trees Obtained through Computer Vision and Machine Learning Modelling: A Cover Photography and High Spatial Remote Sensing Application

Sigfredo Fuentes [1,*], Gabriela Chacon [1], Damir D. Torrico [1,2], Andrea Zarate [1] and Claudia Gonzalez Viejo [1]

[1]   School of Agriculture and Food, Faculty of Veterinary and Agricultural Sciences, University of Melbourne, Melbourne, VIC 3010, Australia
[2]   Department of Wine, Food and Molecular Biosciences, Faculty of Agriculture and Life Sciences, Lincoln University, Lincoln 7647, New Zealand
*   Correspondence: sfuentes@unimelb.edu.au; Tel.: +61-4245-04434

Received: 5 June 2019; Accepted: 9 July 2019; Published: 11 July 2019

Abstract: Cocoa is an important commodity crop, not only to produce chocolate, one of the most complex products from the sensory perspective, but one that commonly grows in developing countries close to the tropics. This paper presents novel techniques applied using cover photography and a novel computer application (VitiCanopy) to assess the canopy architecture of cocoa trees in a commercial plantation in Queensland, Australia. From the cocoa trees monitored, pod samples were collected, fermented, dried, and ground to obtain the aroma profile per tree using gas chromatography. The canopy architecture data were used as inputs in an artificial neural network (ANN) algorithm, with the aroma profile, considering six main aromas, as targets. The ANN model rendered high accuracy (correlation coefficient (R) = 0.82; mean squared error (MSE) = 0.09) with no overfitting. The model was then applied to an aerial image of the whole cocoa field studied to produce canopy vigor, and aroma profile maps up to the tree-by-tree scale. The tool developed could significantly aid the canopy management practices in cocoa trees, which have a direct effect on cocoa quality.

Keywords: leaf area index; cocoa beans; volatile compounds; artificial neural networks; VitiCanopy app

---

1. Introduction

Cocoa (*Theobroma cacao* L) is considered a major world commodity crop and is the seventh most exported food product [1]. Its primary use is for chocolate manufacture, with an estimate of 3981 thousand tons of cocoa bean production in 2016 globally [2]. However, as with any other crop, the quality of cocoa beans is determinant for consumers' acceptability and end-product cost. The cocoa beans intended for manufacturing must comply with specific quality parameters such as shape, size, color [3], flavor, aroma, cocoa butter, and protein content [4]. These are mainly related to the genotype of the clone, climatic and agricultural conditions, and agricultural practices (e.g., water management, fertilization, canopy management), as well as fermentation, drying, and industrialization processes [5–9]. The cocoa crop and its production are also very important for developing countries. For example, cocoa produce from Ecuador is highly valued internationally due to their recognized aroma intensity development [10].

Cocoa trees are grown in regions with high rainfall (1250–3000 mm/year), average temperatures between 18 and 32 °C, and calm wind conditions to avoid defoliation. Furthermore, shade is an important factor, especially in young trees, as it provides protection from excessive solar radiation and helps enhance aroma profiles that are desirable in the final chocolate product. They are usually planted at a density of 400–2500 trees per hectare and produce an average yield of between 200 and 800 kg of fermented dried cocoa per hectare [4,11,12].

There are three main cocoa tree varieties: (i) Criollo, (ii) Forastero, and (iii) Trinitario, the latter being a hybrid group. Varieties i and iii are considered as fine-quality or flavor-quality cocoa beans, while ii is used to produce bulk-quality cocoa. Specifically, the cocoa quality is mainly influenced by the growing conditions where the tree is grown, such as soil depth, pH (4.5–7), weather changes, and pest and disease control [13]. Some techniques to manage the cocoa trees to assure high-quality beans are: (i) Pest and disease control, (ii) soil nutrients management, (iii) weed control, and (iv) management of shade and canopy pruning [14]. The latter technique, (iv), is important due to the fact that the relationship between light and shade is critical, as the shade provides protection against sunlight and high winds, and lowers the temperature to manage the growth of the cocoa tree [4,11].

In cocoa plantations, some in-field factors that have been reported to affect cocoa quality are the exposure to solar radiation, which has effects on the final fruits' dry weight and the growth temperature, which has, in turn, an influence on the lipid content and melting point [15]. However, no relationship between the aroma quality of the cocoa beans and canopy architecture parameters, such as leaf area index (LAI), has been reported. A complex method proposed by researchers to assess cocoa beans quality from the field is through the genetic identification of the variety by using deoxyribonucleic acid (DNA) fingerprinting and utilizing microsatellite markers containing the information from leaf samples [13,16]. However, this method is time-consuming, non-practical, and cost ineffective. Furthermore, the analysis of the trees at the leaf level has shown to be challenging due to the difficulty of sampling leaves with the same growth conditions (e.g., shade, age) [11] and the complexity of the environment when plants are grown between other tree species (i.e., within a forest).

This paper presents the assessment of canopy architecture parameters of cocoa trees using the VitiCanopy computer application [17] as a ground-truth to validate information extracted from aerial images, to automatically obtain these parameters within a plantation and up to a tree-by-tree resolution. From the same ground-truth cocoa trees, the volatile aroma compounds were assessed from fermented and dried cocoa beans by using gas chromatography–mass spectroscopy (GC–MS) with a solid phase microextraction in the headspace (SPME-HS). The canopy architecture data were used as inputs to develop a machine learning model to predict six of the main volatile compounds in cocoa beans and their associated aromas. Since all the information obtained either through computer vision algorithms (canopy architecture) and machine learning modeling (aroma profiles) was at the tree-by-tree scale, spatial distribution aroma maps were constructed from the cocoa plantation used in this research.

2. Materials and Methods

2.1. Site Description

The cocoa trees used in this research are located in a commercial farm with Papua New Guinea Hybrids (PNH) at Mt Edna Maria Creeks, (17°49′14″ S 146°02′21″ E), North Queensland, Australia. A map showing the location can be found as supplementary material in Figure S1. The soil type is clay loam with a pH of 5, under a tree sprinkler irrigation system with sprinkler discharge of 40 L per hour. Sprinklers were located between tree rows at 4 m intervals, and minimal chemical usage for pests and disease management was applied when required.

Cocoa trees were planted in three blocks using a Tatura trellis pruning system, which requires a continuous shaping of the tree canopy. Measurements were only done in Blocks 1 and 2. Among Blocks 1 and 2, the planting density was 1000 trees ha^{-1} in a double-row layout, with distances of 5 m between rows and 2 m between trees. The blocks and rows were planted in different seasons for pruning test purposes. Block 1 was planted in March 2013, and Block 2 in October 2013. In both blocks, row A was planted first. The pattern area (Figure 1) was excluded from data collection because the trees were established as free-standing trees and were at the flowering stage. On the right side of Block 1 and at the lower lateral of the two blocks, the plantation is surrounded by *Erythrina variegata*, or coral tree, used as a windbreak, while Block 2 is surrounded by common vegetation of the site. The cocoa plantation was grown under full-sunlight conditions without intercropping species or shade species.

Figure 1. Aerial image of a cocoa plantation close to harvest from Mt Edna Maria Creeks, Queensland, Australia (Google Earth Pro™). Yellow dotted lines show the blocks used, numbered from 1 to 3.

The weather data of the site were obtained at an approximate distance of 30 km from the plantation from the Bureau of Meteorology (BOM) weather station located at South Johnstone (Australian BOM Station No. 032037, 17.61° S 146.00° E). Average rainfall of 3289.3 mm per year, and mean values for maximum and minimum temperatures of 28.1 and 19.2 °C, respectively, were recorded in the season studied (2015–2016), which is consistent with the climate classification for the site as a tropical monsoonal region.

2.2. Cocoa Trees and Pods/Beans Sampling

A total of 24 cocoa trees were used for this study. Each tree was measured nine times to use as replicates. The cocoa pods were randomly sampled from the mentioned trees according to the general guidelines on sampling CAC/GL 50-2004, as recommended by the Codex Alimentarius Commission. Two to three cocoa pods were harvested from each tree. Each pod was assigned a label as B1 or B2 according to the blocks from which they were collected (one or two), followed by the number of the tree sampled (e.g., A15) and the letter of the row (A or B). Every tree was geo-located using the VitiCanopy App, which will be described following. The crops for the two blocks from this trial were harvested on the 2 February 2016.

2.3. Image Data Acquisition

For each of the 24 sampled trees, a total of nine upward-looking canopy images were obtained using the VitiCanopy computer application [17] within three consecutive days (three per day). The digital images from the tree canopies were obtained with the front camera of an iPhone 5S (Apple Inc. Corp. Cupertino, CA, USA), which has 1.2 megapixels of resolution. The device was mounted on a Soniq Bluetooth Selfie Stick (SONIQ Australia Pty Ltd., Braeside, Victoria, Australia) facing upwards at 0.5 m height using a tripod. The device was aligned using a bubble level attached to the selfie stick to ensure it was in a horizontal position for each image, obtained at a 0° Zenith angle. It was positioned with its maximum length perpendicular to the direction of the trunk. The images were acquired by locating the device at 0.5 m from the trunk and 0.5 m to the left and right sides (Figure 2). Imagery were obtained four hours before noon to avoid direct sunlight into the lens of the device, as recommended by Poblete-Echeverría et al. [18]. The dates of image acquisition were 29 January to 2 February 2016. All the equipment used for image acquisition were ubiquitous (smartphones), low-cost, and suitable for fieldwork.

Figure 2. (**A**) Distance of image acquisition data in cocoa trees under a Tatura trellis system using the VitiCanopy application (app), and (**B**) image of canopy recorded using the VitiCanopy app.

2.4. Estimation of Canopy Architecture Parameters by the VitiCanopy App

Pictures with direct sunlight, weeds, posts, operators, and canopy from the other trees were considered unsuitable for analysis and discarded. Therefore, a total of 173 images were suitable for the analysis, with six to nine images analyzed from each tree. Canopy architecture parameters were obtained using the following algorithms to obtain the fractions of foliage projective cover (f_f), crown cover (f_c), and crown porosity (Φ), which were calculated from McFarlane et al. [19] as:

$$f_f = 1 - \frac{t_g}{t_p} \tag{1}$$

$$f_c = 1 - \frac{l_g}{t_p} \tag{2}$$

$$\Phi = 1 - \frac{f_f}{f_c} \tag{3}$$

where l_g = large gap pixels; t_g = total pixels in all gaps, and t_p = total gap pixels.

LAI is calculated from Beer's Law:

$$LAI = f_c + \frac{\ln \Phi}{k} \tag{4}$$

where k is the coefficient of light extinction ($k = 0.61$) [20], and the clumping index at the zenith, $\Omega(0)$, was calculated as follows:

$$\Omega(0) = \frac{(1 - \Omega)\ln(1 - f_\text{f})}{\frac{\ln(\Phi)}{f_f}} \tag{5}$$

The clumping index is a correction factor to obtain the effective LAI (LAI_e), which is the product of:

$$LAI_e = LAI_M \times \Omega(0) \tag{6}$$

Equation (5) describes the non-random distribution of canopy elements. If $\Omega(0) = 1$, that means that the canopy displays random dispersion. For $\Omega(0) >$ or < 1, the canopy is defined as clumped.

This innovative app estimates canopy architecture parameters using upward-looking digital images based on a method developed by Fuentes and De Bei et al. [21]. The image can be automatically sub-divided into $m \times n$ (m = number of rows, n = number of columns) sub-images for gap analysis. For this research, a sub-division of 5 × 5 was used for the VitiCanopy App and to analyze each tree sub-image from the aerial image analysis. The algorithms used in the VitiCanopy app were calculated based on the formulas from Macfarlane et al. [19]. In the settings menu of the application, the gap fraction threshold was set as 75%, which means that a big gap was considered when 75% of the pixels from the images corresponded to the sky [21]. A value of the light extinction coefficient (k) of 0.7 was set based on the range of the standard mean values identified from the samples and according to the literature review data of cocoa tree studies made by Yapp [22] and Daymond et al. [23]. Since in these images the sky serves as a background to isolate non-leaf material, the app filters sky and clouds automatically by segmentation of images using the blue channel of the RGB images. In order to record the Global Positioning System (GPS) data of each cocoa tree, the location service of the app was enabled. The images were analyzed with the application individually as they were obtained. The results from the app, which included the canopy architecture parameters, date, GPS location, and settings used were exported via email by tapping the "Export" option, which generates a comma separated values (.csv) file with all the data.

2.5. Aerial Imagery and Processing

The aerial image used for this research was obtained from Google Earth Pro (ver. 7.3.2.5776; Google, Mountain View, CA, USA), which uses a composite between satellite and aerial images. The only available image closer to the harvest was for 27 May 2016. However, it is expected from the biological point of view that the LAI will be similar within this month and that the trends between plants should also be maintained. The image was saved as a maximum high-resolution (4800 × 2679 pixels) Joint Photographic Expert Group (JPEG) file. The aerial image was processed to recognize every single plant and extract information from each one using the methodology proposed and described in previous studies [24,25] through customized codes written in Matlab® (Mathworks Inc., Natick, MA, USA). In general, the automated procedure consists of the following steps:

1. Identification of the area designated by the plant (distance between plants and between rows). This was done by cropping the area for blocks and automatically sub-cropping each tree area for further analysis according to the distance between plants and the distance between rows [24,25] (Figure 3A).
2. Automatic Individual segmentation of tree biomass material using the CIELab (specifically the "a" band) color code to generate a binary image of sub-images analyzed per tree (Figure 3B).
3. Analysis of binary images per area assigned per plant using algorithms described previously (Equations (1)–(6)). Each sub-image is treated as an individual upward-looking image obtained by VitiCanopy, using the sky as a background. A subsequent subdivision of 5 × 5 was used for each tree image for gap analysis, as described by [17,21] (Figure 3C). From the aerial image,

it was assumed to be a downward-looking image at Nadir (0° angle) using the inter-row as a background [17,21].

4. The obtained canopy architecture data were compiled in a matrix form to preserve plant positions and for easier handling for mapping purposes.

5. If canopies were of low cover ($f_c > 0.1$), they were considered as missing trees [24].

6. Tree GPS positions were extracted from the GPS coordinates of the corners of blocks for data mapping purposes.

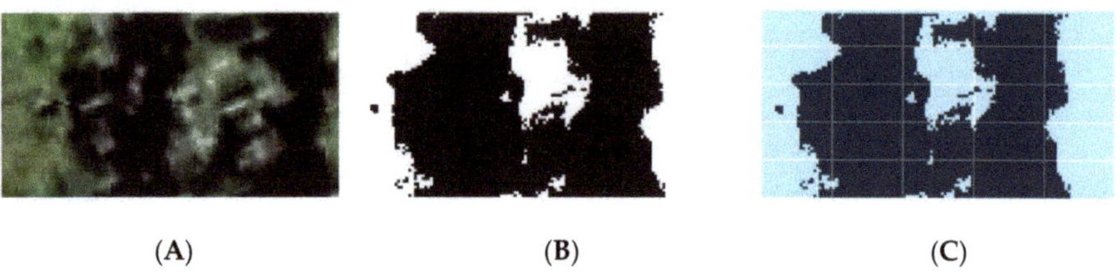

(A)   (B)   (C)

Figure 3. Analysis process from automatically cropped images per tree (**A**) and segmentation using the CieLab color code to obtain a binary image showing main tree biomass (**B**), and sub-division of 5×5 for gap analysis (**C**) and application of image analysis algorithms (Equations (1)–(6)).

2.6. Cocoa Bean Fermentation and Drying Process

The harvested cocoa pods were transported to The University of Melbourne, Australia, and stored for ten days at room temperature between 20 and 25 °C. The storage time allowed a significant increase of the pH related to cocoa flavor potential [26,27]. The seeds were taken from inside each pod sample and mixed with a solution of a commercial yeast *Saccharomyces cerevisiae* (Lowan Whole Foods, Glendenning, NSW, Australia) and water in a plastic container. For the fermentation process, the labeled plastic containers were placed into an incubator I100-G300-D (Thermoline Scientific Equipment Pty Ltd., Wetherill Park NSW, Australia), which was set up at 25 °C (0–12 h), 30 °C (12–24 h), 35 °C (24–36 h), 40 °C (36–48 h), 45 °C (48–72 h), and 48 °C (72–144 h). The temperature was monitored daily, and the seeds were manually flipped every 48 h to ensure aeration and uniform fermentation. After six days of fermentation, a cut test was performed to assess the quality of the process. Subsequently, for the drying process, the beans were set up in a Qualtex Solid State oven, Series 2000 (Qualtex Australasia P/L, Murarrie QLD, Australia) at 45 ± 2 °C, and mixed manually and daily for three days to get a uniform drying.

2.7. Analysis of Volatile Compounds Using Gas Chromatography-Mass Spectroscopy

The cocoa beans from each pod sample were ground in a commercial coffee blender Sunbeam Multigrinder II EM0405 (Sunbeam Products, Boca Raton, FL, USA) for 60 s. One gram of the ground beans was placed into a 15 mL tightly capped vial with 5 mL of saturated salt solution and 100 µL of pyrazine dilution. According to the procedure of De Brito et al. [5], the vials were mixed in a vortex (Vor-Mix VM80) for 1 min and preheated for 45 min at 60 °C, and then placed in a dry block heater and mixed in an orbital mixer OM1 (Ratek Instruments Pty. Ltd., Boronia, Victoria, Australia).

The volatile compounds were extracted by solid phase microextraction in the headspace (SPME-HS) with Supelco fibers 50/30 µm divinylbenzene/carboxene/polydimethylsiloxane (DVB/CAR/PDMS). The extraction of the volatile compounds of each sample took place for 15 min, with 30 min at 60 °C of fiber exposition to the cocoa aromas in the headspace [28]. The fibers were then analyzed by gas chromatography (GC; Agilent Technologies 6850 Series II, Network GC System) coupled to mass spectrometry (MS; Agilent Technologies 5973 Network Mass Selective Detector). The GC–MS was equipped with an Agilent J&W DB-Wax column (30 m × 0.25 mm, 0.25 µm film thickness). The oven temperature was set at 40 °C for 5 min, and then increased to 200 °C at a rate of 10 °C min^{-1}, and finally maintained at 200 °C for 30 min. The carrier gas was helium (Air Liquide or BOC, Ultra-High Purity), with a linear velocity of 36 cm/s and a flow rate of 2.0 mL/min in constant flow mode. The SPME fiber was desorbed in the pulsed splitless injection mode for 5 min at 240 °C and opened after 0.5 min. The temperatures of the selective mass detector quadrupole were MS source 230 °C and MS Quad 180 °C, with an electronic impact ionization system at 70 eV. Compounds identification was performed by comparing the mass spectra of each compound with the National Institute of Standards and Technology MS NIST02.L (National Institute of Standards and Technology, Gaithersburg, MD, USA) library of mass spectra.

2.8. Statistical Analysis and Machine Learning Modeling

An analysis of variance (ANOVA) was conducted to assess significant differences in the LAI values and canopy architectures between the sampled trees, using the SAS® 9.4 software (SAS Institute Inc., Cary, NC, USA). A post-hoc analysis with the Tukey studentized method ($\alpha = 0.05$) was used. Means and standard deviation (SD) were calculated for each parameter.

A machine learning model was developed using the Bayesian Regularization training algorithm after testing 17 different artificial neural network (ANN) algorithms (two backpropagation using Jacobian derivatives, 11 backpropagation using gradient derivatives, and four supervised weight/bias training) using a customized and automatic Matlab® R2018b developed code (unpublished). Six canopy architecture parameters from the canopy measurements ((i) LAI, (ii) LAI_e, (iii) f_f, (iv) f_c, (v) ϕ, and (vi) Ω) were used as inputs to predict six main aroma compounds: (i) Phenethyl acetate, (ii) 2-phenyl-2-butenal, (iii) isoamyl acetate, (iv) tetramethylpyrazine, (v) phenylacetaldehyde, and (vi) 2,3-butanediol. All inputs and target values were normalized from −1 to 1. To develop the model, a random data division was used with 85% (N = 147) of the samples used for training, 15% (N = 26) for testing with a default derivative function, and a performance algorithm based on mean squared error (MSE). The model (Figure 4) was built with a two-layer feedforward network using a tan-sigmoid function in the hidden layer and a linear transfer function in the output layer. A trimming assessment was performed using 3, 5, 7, and 10 neurons to find the model with the best performance and less chance of overfitting. Statistical results are presented for the correlation coefficient (R) and mean squared error values (MSE).

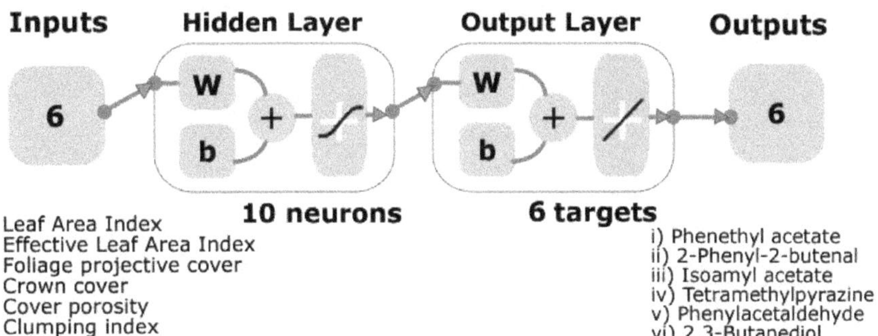

Figure 4. Model diagram showing the two-layer feedforward network with six inputs from the canopy architecture, 10 neurons in the hidden layer with a tan-sigmoid function, and a linear transfer function in the output layer. Six targets/outputs from the volatile compounds were used for the model.

Once developed, the machine learning model was run using the canopy architecture data extracted from the aerial imagery as inputs to obtain the six aromas per tree from the three blocks. Geo-location per tree was obtained from the GPS data from block corners and considering the plantation density through a customized code written in Matlab® R2019a (Fuentes, unpublished). With this information, two maps of LAI_e and the aroma profiles were generated in Matlab, with the first using kriging interpolation algorithms and the second using scattered data representing each tree. A graphical representation of the complete methodology used in this study can be found in the supplementary material (Figure S2).

3. Results

3.1. Canopy Architecture and Volatile Compounds

Table 1 shows the results of the means and SD from the canopy architectures of each of the 24 sampled trees. There were significant differences ($p < 0.05$) between samples for the six parameters. However, for the clumping index, only one sample (B1C4A) was significantly different.

Table 2 shows the means and SD from the results of both blocks, in which significant differences ($p < 0.05$) can be observed for LAI, LAI_e, and crown porosity (Φ). Block 1 had higher mean values of LAI and, although not significantly different, crown cover results compared to Block 2. On the contrary, Block 1 presented lower and significant values of clumping index than Block 2.

Table 3 presents the volatile compounds found in the fermented cocoa beans and the aromas associated with them. It can be observed that most of the associated aromas are related to cocoa, butter, honey, fruity, floral, nutty, and roasted. Figure 5 shows the peak area of each volatile compound found in every sample. It can be seen that most of the samples had phenylethyl alcohol as the most abundant volatile compound, except for the beans of five trees (B1C1A, B1C18A, B1C26B, B2C14A, and B2C16A). Tetramethylpyrazine was present in most of the samples and is higher in those from Block 1 than Block 2. Benzaldehyde and 2,3-butanediol were present in all samples. However, the peak area varied between samples. Pehenthyl acetate and isoamyl acetate were present in all except for two (B1C25A and B2C16A) and one (B1C25A) sample(s), respectively. Other compounds, such as 2-pentanone (fruity) and ethyl acetate (pineapple/fruity/green) were only detected in four (B1C1A, B1C27A, B2C16A, and B2C24A) and six samples (B1C16A, B1C26B, B1C27A, B1C36A, B1C45A, and B2C16A), respectively.

Table 1. Means and standard deviation (SD) of the parameters obtained from the canopy architecture for each sampled tree. Different letters (superscripts) represent statistically significant differences between samples using the Tukey studentized test with $\alpha = 0.05$.

Sample	LAI		LAI_e		Canopy Cover (f_c)		Crown Cover (f_f)		Crown Porosity (Φ)		Clumping Index (Ω)	
	Mean	SD	Mean	SD	Mean	SD	Mean	SD	Mean	SD	Mean	SD
B1C15A	3.08 [b]	0.68	2.79 [a]	0.76	0.836 [a]	0.094	0.962 [a]	0.061	0.134 [a]	0.050	0.951 [a]	0.067
B1C16A	2.78 [b]	0.96	2.55 [a]	1.12	0.755 [a]	0.116	0.879 [a]	0.107	0.142 [a]	0.050	0.851 [a]	0.131
B1C17A	3.60 [a]	0.69	3.19 [a]	0.36	0.868 [a]	0.063	0.957 [a]	0.060	0.094 [b]	0.035	0.904 [a]	0.122
B1C18A	3.47 [a]	0.82	3.18 [a]	1.08	0.843 [a]	0.094	0.932 [a]	0.077	0.097 [b]	0.036	0.887 [a]	0.127
B1C1A	3.62 [a]	0.68	3.32 [a]	0.98	0.863 [a]	0.069	0.945 [a]	0.050	0.088 [b]	0.032	0.888 [a]	0.100
B1C25A	4.56 [a]	0.73	4.08 [a]	0.73	0.939 [a]	0.037	0.994 [a]	0.014	0.056 [b]	0.026	0.986 [a]	0.034
B1C26B	2.55 [b]	0.47	2.33 [a]	0.69	0.730 [a]	0.074	0.863 [a]	0.086	0.153 [a]	0.050	0.814 [a]	0.112
B1C27A	3.31 [a]	0.70	3.04 [a]	0.97	0.834 [a]	0.105	0.928 [a]	0.087	0.104 [b]	0.030	0.894 [a]	0.121
B1C2A	3.50 [a]	1.16	3.20 [a]	1.33	0.827 [a]	0.129	0.915 [a]	0.111	0.099 [b]	0.051	0.875 [a]	0.146
B1C36A	3.68 [a]	0.20	3.32 [a]	0.56	0.867 [a]	0.046	0.946 [a]	0.054	0.080 [b]	0.011	0.872 [a]	0.117
B1C3A	4.29 [a]	1.20	3.92 [a]	1.43	0.881 [a]	0.139	0.935 [a]	0.141	0.059 [b]	0.031	0.905 [a]	0.195
B1C45A	2.65 [b]	0.45	2.41 [a]	0.67	0.772 [a]	0.079	0.913 [a]	0.070	0.156 [a]	0.030	0.886 [a]	0.092
B1C4A	2.40 [b]	0.29	2.18 [b]	0.51	0.668 [b]	0.085	0.775 [b]	0.112	0.135 [a]	0.036	0.723 [b]	0.116
B2C12A	3.06 [b]	0.48	2.84 [a]	0.75	0.847 [a]	0.061	0.980 [a]	0.041	0.136 [a]	0.035	0.972 [a]	0.056
B2C13A	3.33 [a]	0.49	3.03 [a]	0.56	0.875 [a]	0.046	0.991 [a]	0.017	0.117 [b]	0.035	0.986 [a]	0.028
B2C14A	2.61 [b]	0.98	2.46 [b]	1.13	0.727 [b]	0.196	0.853 [a]	0.209	0.152 [a]	0.053	0.863 [a]	0.158
B2C15A	2.30 [b]	0.29	2.12 [b]	0.48	0.720 [b]	0.059	0.888 [a]	0.068	0.188 [a]	0.035	0.864 [a]	0.073
B2C16A	2.81 [b]	0.51	2.59 [a]	0.66	0.791 [a]	0.114	0.919 [a]	0.122	0.140 [a]	0.021	0.902 [a]	0.132
B2C17A	2.88 [b]	0.89	2.63 [a]	0.89	0.770 [a]	0.145	0.885 [a]	0.126	0.134 [a]	0.053	0.871 [a]	0.139
B2C1A	3.27 [a]	0.49	2.99 [a]	0.55	0.867 [a]	0.048	0.984 [a]	0.031	0.119 [b]	0.028	0.974 [a]	0.052
B2C22A	2.91 [b]	0.36	2.69 [a]	0.60	0.816 [a]	0.059	0.947 [a]	0.058	0.138 [a]	0.027	0.913 [a]	0.085
B2C24A	2.56 [b]	0.76	2.36 [b]	0.81	0.719 [b]	0.150	0.849 [a]	0.159	0.153 [a]	0.063	0.828 [a]	0.154
B2C25A	3.02 [b]	0.63	2.80 [a]	0.83	0.796 [a]	0.120	0.901 [a]	0.122	0.118 [b]	0.030	0.865 [a]	0.140
B2C27A	3.13 [a]	0.64	3.13 [a]	0.64	0.849 [a]	0.060	0.976 [a]	0.036	0.131 [a]	0.038	0.962 [a]	0.055

Abbreviations: LAI = Leaf Area Index, LAI_e = effective LAI; sample codes: B1 = Block 1, B2 = Block 2, C# = number of the tree, and A or B = position of the tree (row). Means were taken from six to nine images per tree.

Table 2. Means and standard deviation (SD) of canopy architecture parameters for the two blocks. Different letters represent statistically significant differences between samples using the Tukey studentized test with $\alpha = 0.05$.

Block	LAI		LAI_e		Canopy Cover (f_c)		Crown Cover (f_f)		Crown Porosity (Φ)		Clumping Index (Ω)	
	Mean	SD	Mean	SD	Mean	SD	Mean	SD	Mean	SD	Mean	SD
B1	3.35 [a]	0.93	3.04 [a]	1.01	0.822 [a]	0.11	0.919 [a]	0.095	0.107 [b]	0.048	0.880 [a]	0.126
B2	2.89 [b]	0.67	2.68 [b]	0.76	0.796 [a]	0.12	0.923 [a]	0.116	0.139 [a]	0.043	0.907 [a]	0.116

Abbreviations: LAI = Leaf Area Index, LAI_e = effective LAI, B1 = Block 1, B2 = Block 2.

Table 3. Major volatile compounds found in the fermented cocoa beans, and the associated aromas.

Volatile Compounds	Compound Group	Aroma Associated	References
Phenethyl acetate	*Ester*	*Honey/floral/yeasty/cocoa*	[28,29]
2-Phenyl-2-butenal	*Aldehyde*	*Sweet/cocoa/nutty/beany*	[28,29]
Isoamyl acetate	*Ester*	*Banana/fruity*	[29,30]
Tetramethylpyrazine	*Pyrazine*	*Roasted/cocoa/coffee/nutty/earthy*	[7,29,31]
Phenylacetaldehyde	*Aldehyde*	*Honey/rose/cocoa*	[29,31]
2,3-Butanediol	*Alcohol*	*Cocoa butter/creamy*	[29,30]
Isovaleraldehyde	Aldehyde	Chocolate/nutty/cocoa	[28,29]
2-Pentanone	Ketone	Fruity	[29,32]
Diacetyl/2,3-Butanedione	Ketone	Butter/creamy/caramel	[28,29]
Ethyl acetate	Ester	Pineapple/fruity/green	[28,29]
Acetoin	Ketone	Butter/creamy/dairy/fatty	[28,29]
2-Heptanol	Alcohol	Citrus/fruity	[28,29]
3-Methyl-2-butanol	Alcohol	Cocoa/alcoholic/musty	[29]
Isoamyl alcohol	Alcohol	Malty/banana/pungent	[29,32]
2,3-Dimethylpyrazine	Pyrazine	Caramel/cocoa/coffee/nut skin	[29,33]
2,3-Dimethyl-5-ethylpyrazine	Pyrazine	Burnt/roasted/cocoa	[29]
Benzaldehyde	Aldehyde	Cherry/almond/bitter	[29,33]
Ethyl Phenylacetate	Esters	Honey/rose/dark chocolate	[29]
Benzyl alcohol	Alcohol	Sweet/floral/rose/phenolic	[28]
Phenylethyl alcohol	Alcohol	Honey/bready/floral	[29]

Compounds in italics were used as targets for machine learning modeling.

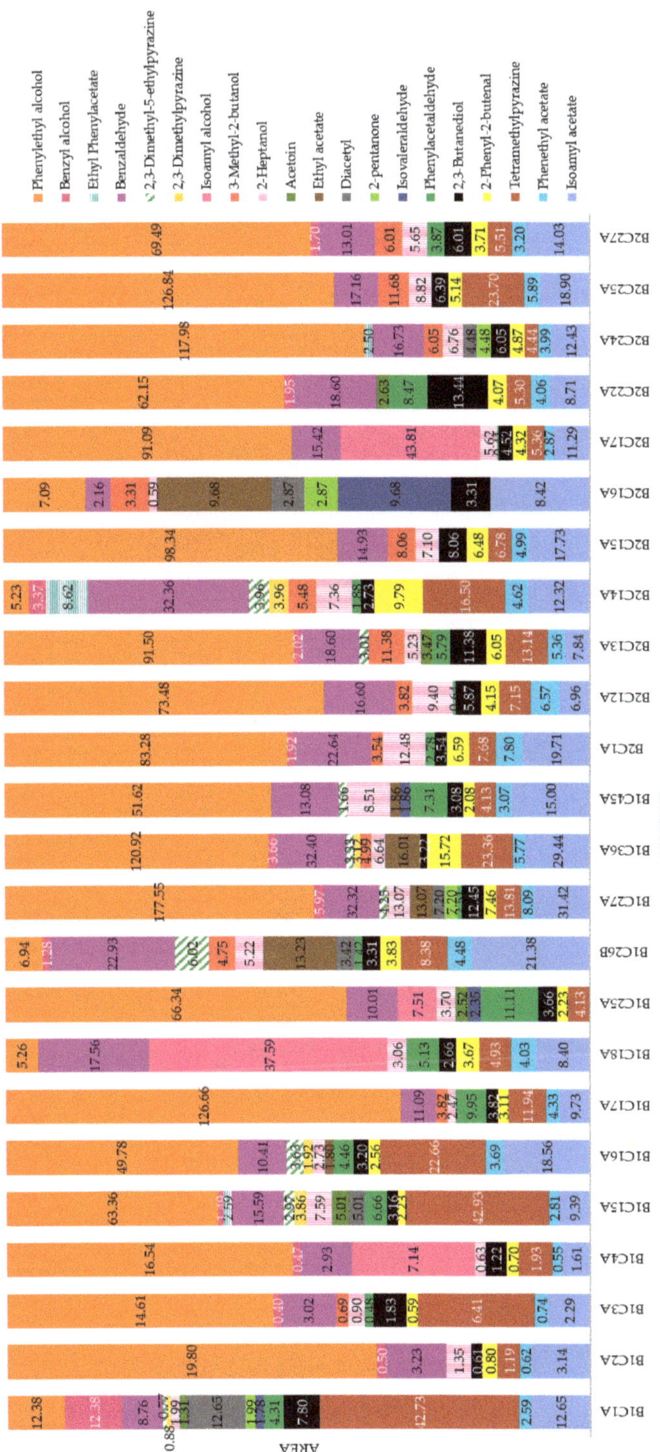

Figure 5. Peak areas of the volatile compounds found in the fermented cocoa bean samples. All values are reported in scientific notation 10^3. Sample codes: B1 = Block 1, B2 = Block 2, C# = number of the tree, and A or B = position of tree (row).

3.2. Machine Learning Modeling

Figure 6 shows the machine learning model to predict six of the main volatile compounds (Figure S3) in cacao beans (phenethyl acetate, 2-phenyl-2-butenal, isoamyl acetate, tetramethylpyrazine, phenylacetaldehyde, and 2,3-butanediol) using the LAI values and canopy architecture data as inputs. The overall model had 4.8% of outliers according to the 95% confidence bounds. In Table 4, it can be observed that the training stage had a correlation coefficient of R = 0.82, while the testing stage and the overall model had correlation coefficients of R = 0.81 and 0.82, respectively. Furthermore, the model had a high performance MSE = 0.09 for the training stage and MSE = 0.11 for the testing stage, which is indicative of an adequate model.

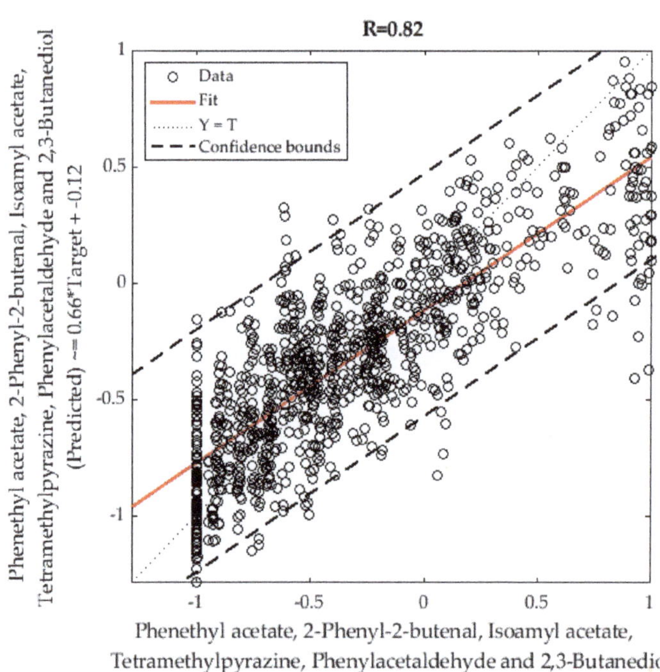

Figure 6. Artificial neural network overall model showing the observed responses (targets) in the x-axis and the predicted values (outputs) in the y-axis of the six volatile compounds. R = correlation coefficient.

Table 4. Results from the artificial neural network model showing the numbers of samples and observations, correlation coefficient (R), and performance based on mean squared error (MSE) for each stage.

Stage	Samples	Observations (Samples × 6 Targets)	R	MSE
Training	147	882	0.82	0.09
Testing	26	156	0.81	0.11
Overall	173	1038	0.82	-

3.3. Implementation of the Machine Learning Model Developed to Map Aroma Compounds from Aerial Imagery

Aerial imagery allowed the extraction of canopy architecture parameters from sub-images per plant that are effectively comparable to upward-looking images from the VitiCanopy App. This was validated using a simple linear regression between the averaged LAI_e parameters from the 24 sampled trees in the field using VitiCanopy (observed) and those obtained from the same plants extracted from the aerial image (correlation coefficient (R) = 0.91; data not shown).

The machine learning model developed and described in Figure 6 and Table 3 was implemented using the canopy architecture parameters obtained from the aerial imagery. Figure 7 shows the LAI_e and aroma profile data from each of the three blocks studied, using a kriging interpolation technique. The denormalized GC area values obtained from the machine learning model implementation were from 0 to 60×10^3, which was used as a scale for all the aroma profiles. Since LAI_e is one of the most important canopy architectural parameters, it was individually mapped with values ranging from 0 (missing plants) to 6. Figure 8 shows the scattered values, which represent the LAI_e and aroma profile for individual plants from the field studied.

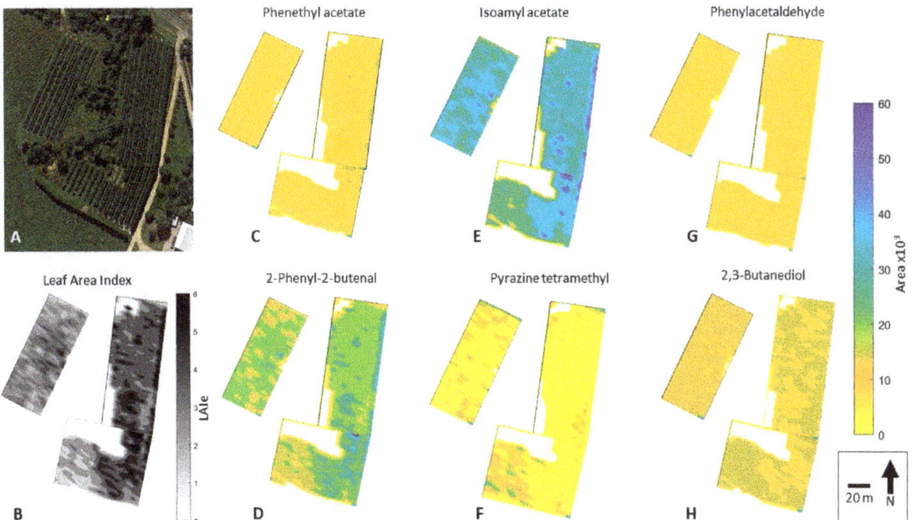

Figure 7. Original aerial image obtained from Google Maps Pro (**A**), and mapping of effective leaf area indexes (LAI_e) (**B**) and aroma profiles for phenethyl acetate (honey/floral/yeasty/cocoa) (**C**); 2-phenyl-2-butenal (sweet/cocoa/nutty/beany) (**D**); isoamyl acetate (banana/fruity) (**E**); pyrazine tetramethyl (roasted/cocoa/coffee/nutty/earthy) (**F**); phenylacetaldehyde (honey/rose/cocoa) (**G**); and 2,3-butanediol (cocoa butter/creamy) (**H**).

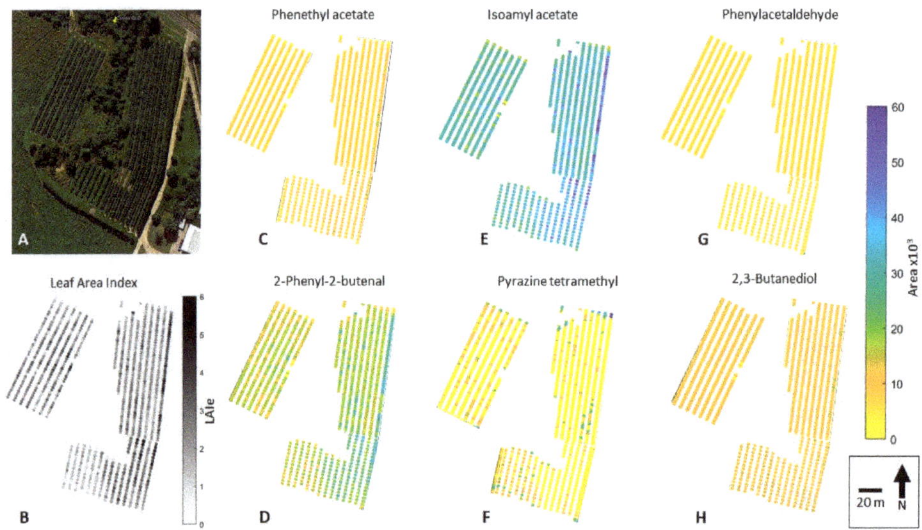

Figure 8. Original aerial image obtained from Google Map Pro (**A**), and plant-by-plant mapping of effective leaf area indexes (LAI$_e$) (**B**) and aroma profiles for phenethyl acetate (honey/floral/yeasty/cocoa) (**C**); 2-phenyl-2-butenal (sweet/cocoa/nutty/beany) (**D**); isoamyl acetate (banana/fruity) (**E**); pyrazine tetramethyl (roasted/cocoa/coffee/nutty/earthy) (**F**); phenylacetaldehyde (honey/rose/cocoa) (**G**); and 2,3-butanediol (cocoa butter/creamy) (**H**).

4. Discussion

4.1. Canopy Architecture and Volatile Compounds

The mean values of LAI and LAI$_e$ for all trees fall within those reported by Daymond et al. [23] (2.2–4.5) for different clones of cacao in Brazil. The range of the mean values of canopy cover found in the present study was between 67% and 94%, which is similar to the range (72%–91%) obtained by Bos et al. [34] in Criollo and Forastero hybrids of cacao trees from Indonesia. The significant differences found between the two blocks for LAI, LAI$_e$, and crown porosity may be due to the fact that Block 1 was planted in March 2013, while Block 2 was planted in October 2013, and therefore, their maturity stages were different at the time of measurements, giving an average LAI of 2.89 for Block 2, which is lower than that of Block 1 (3.35). High-resolution aerial images could be obtained using Unmanned Aerial Vehicle (UAV) platforms and high-resolution visible cameras. It is expected that by using the latter type of high-resolution images, the correlation between VitiCanopy and the aerial extraction of canopy architecture features will improve due to a more accurate gap analysis.

All volatile compounds found in the cocoa beans have been previously reported in other studies. Ten out of 20 of these compounds provide aromas related to nuts, roasted, cocoa, and chocolate, while six compounds provide fruity aromas including banana, cherry, and pineapple. Acetoin and 2,3-butanediol, which elicit creamy and butter aromas, were also found. According to Aprotosoaie et al. [35], the alcohols found in cocoa are due to fermentation and decrease with drying and roasting. However, 2-heptanol and phenylethyl alcohol, which give fruity and floral aromas, are commonly found in cacao. In the present study, phenylethyl alcohol was the most abundant compound in most of the samples. Furthermore, aldehydes and ketones, followed by esters and pyrazines, are the most important to determine cocoa quality [28,35], and in this study, seven compounds from aldehydes and ketones, four esters, and three pyrazines were found (Table 2). According to Rodriguez-Campos et al. [28], phenethyl acetate is an important ester in fermented cocoa beans, and higher concentrations of this compound are an indicator of high cocoa quality. In the present study, this volatile compound was identified in 22 out of 24 samples. On the other hand, similar to the present study, tetramethylpyrazine has been reported in other studies as the most abundant pyrazine and is one of the most important volatiles for cocoa flavor [36]. This was found in all except one sample from this study (Figure 5).

4.2. Machine Learning Modeling to Predict the Aroma Profile

Besides fermentation and drying, other factors that contribute to differences in aromas and flavors in cacao are the genotype, location and growth conditions, ripening, and amount of shade [37]. Therefore, the machine learning model developed in this study based on canopy architecture parameters has a strong basis on the physiological factors affecting the aroma profile. Hence, the high accuracy and performance obtained (R = 0.82; MSE = 0.09) using the canopy architecture information, which is related to the amount of shade, accurately represents the six volatile compounds commonly found in cocoa and that are determinant for its quality (Table 2). Up to date, there are no known studies that have managed to predict the volatile compounds in cacao using the canopy architecture. However, there was a study in grapevines in which flavonol contents were predicted using crown porosity and LAI using partial least squares (PLS) regression [38].

The model presented in this paper may be useful by producers to modify the canopy architecture of the cocoa trees according to the desired aroma characteristics on the final product. This would allow them to avoid financial loses or predict the end use of the final product by anticipating the quality based on the aroma profile. This model may be fed with more data from samples of other regions and genotypes to create a general predictive model to be used by any grower. On the other hand, similar models may be developed for other crops to predict and improve the quality at early stages.

4.3. Implementation of the Machine Learning Model Developed to Map Aroma Compounds from Aerial Imagery

Spatial distribution of LAI_e and aroma profiles were possible due to high-resolution aerial images, which can be effectively used to implement the canopy cover photography algorithms described (Equations (1)–(6)). A similar approach was implemented for grapevines using f_c, with similar results from imagery obtained using unmanned aerial vehicles [24]. From the kriging interpolated maps (Figure 7) it can be seen that some compounds are highly dependent on vigor characteristics of plants, such as isoamyl acetate, with clear differences of higher areas found for the far-right row in Block 1 (Figure 7E). These trees corresponded to almost double LAI_e values (~5) compared to the average of trees in Blocks 1 and 2 (Figure 7A,B). Other compounds, such as phenethyl acetate, phenylacetaldehyde, and pyrazine tetramethyl, do not have many spatial differences among the three blocks (Figure 7C,G,F, respectively). The latter results confirm that the pattern of the machine learning models was effectively identified, and they are independent for each aroma related to canopy architecture parameters.

The differences in the spatial distribution of growth and aroma profiles can be more clearly seen using a scattering plot, which represents the tree-by-tree scale (Figure 8). This spatial distribution map could offer growers a powerful tool to perform tailored canopy management strategies to manage aroma profile outputs that could be considered desirable or to obtain a uniform quality of pods and, finally, the chocolate produced.

5. Conclusions

The data analysis tools presented in this paper could contribute significantly to adapting management strategies from growers, specifically in relation to canopy management, to obtain desired aroma profiles at the tree-by-tree scale. Specific practical applications could be related to homogenize the aroma quality profile within plants and the entire plantation or to increase desired and specific aromas to change the style of the cocoa volatiles. These management strategies can be implemented within the critical stages of the growing period of cocoa trees, or close to harvest. By implementing VitiCanopy and the machine learning modeling strategies presented, these potential changes in the volatile composition of beans related to canopy architecture changes due to canopy management can be obtained in real time at the fingertips of growers through their portable electronic devices (smartphones and tablet PCs). The latter will enhance crop management capabilities from major industries, but more importantly, by growers in developing countries.

Supplementary Materials: The following are available online at http://www.mdpi.com/1424-8220/19/14/3054/s1, Figure S1. Map showing the site location where the study was conducted. The image was obtained from Google Earth (Google Technology company, Menlo Park, CA, USA); Figure S2. Diagram representing the methods used in the study, where (A) represents the canopy architecture assessment at ground level using the VitiCanopy app, (B) depicts the cacao pods processing and assessment of volatile compounds using gas chromatography, mass spectroscopy, (C) the development of the machine learning modelling using (A) as inputs and (B) as targets/outputs, (D) represents the canopy architecture assessment from aerial images using a customized Matlab® code, and (E) the development of spatial distribution maps using the outputs from (C) using the data from (D) as inputs; Figure S3. Peak areas of the six volatile compounds found in the fermented cocoa bean samples used to develop the machine learning model. All values are reported in scientific notation 10^3. Sample codes: B1 = Block 1, B2 = Block 2, C# = number of the tree, and A or B = position of the tree (row).

Author Contributions: Conceptualization, S.F.; data curation, S.F. and C.G.V.; formal analysis, S.F. and C.G.V.; investigation, S.F., G.C., D.D.T., and A.Z.; methodology, S.F., G.C., D.D.T., A.Z., and C.G.V.; project administration, S.F.; software, S.F.; supervision, S.F.; validation, C.G.V.; visualization, S.F. and C.G.V.; writing—original draft, S.F., G.C., D.D.T., A.Z., and C.G.V.; writing—review & editing, S.F. and C.G.V.

Funding: This research was partially funded by the Ecuadorian government through the academic-award scholarships program granted to GC. This research was supported by the Digital Viticulture program funded by the University of Melbourne's Networked Society Institute, Australia.

Acknowledgments: The authors would like to acknowledge the support from Yan Diczbalis, Principal Horticulturist from the Department of Agriculture and Fisheries, Queensland, Australia. Furthermore, the authors acknowledge Pangzhen Zhang for his support in the use of GC-MS in the chemistry laboratory of The University of Melbourne, Australia.

Conflicts of Interest: The authors declare no conflict of interest.

References

1. Dand, R. Chapter 9: Cocoa Bean Processing and the Manufacture of Chocolate. In *The International Cocoa Trade*; MacFadyen, H., Ed.; Woodhead Publishing: Cambridge, MA, USA, 2011.

2. ICCO. *Production of Cocoa Beans*; International Cocoa Organization: London, UK, 2017.

3. Astika, I.; Solahudin, M.; Kurniawan, A.; Wulandari, Y. Determination of cocoa bean quality with image processing and artificial neural network. In *The Quality Information for Competitive Agricultural Based Production System and Commerce, Proceedings of the AFITA 2010 International Conference, Bogor, Indonesia, 4–7 October 2010*; Asian Federation for Information Technology in Agriculture: Bogor, Indonesia, 2010.

4. Beckett, S.T.; Fowler, M.S.; Ziegler, G.R. *Beckett's Industrial Chocolate Manufacture and Use*; John Wiley & Sons: Hoboken, NJ, USA, 2017.

5. De Brito, E.S.; García, N.H.P.; Amancio, A.C.; Valente, A.L.; Pini, G.F.; Augusto, F. Effect of autoclaving cocoa nibs before roasting on the precursors of the Maillard reaction and pyrazines. *Int. J. Food Sci. Technol.* **2001**, *36*, 625–630. [CrossRef]

6. Ilangantileke, S.; Wahyudi, T.; Bailon, M.G. Assessment methodology to predict quality of cocoa beans for export. *J. Food Qual.* **1991**, *14*, 481–496. [CrossRef]

7. Afoakwa, E.O.; Paterson, A.; Fowler, M.; Ryan, A. Flavor formation and character in cocoa and chocolate: A critical review. *Crit. Rev. Food Sci. Nutr.* **2008**, *48*, 840–857. [CrossRef] [PubMed]

8. Elwers, S.; Zambrano, A.; Rohsius, C.; Lieberei, R. Differences between the content of phenolic compounds in Criollo, Forastero and Trinitario cocoa seed (*Theobroma cacao* L.). *Eur. Food Res. Technol.* **2009**, *229*, 937–948. [CrossRef]

9. Hansen, C.E.; Mañez, A.; Burri, C.; Bousbaine, A. Comparison of enzyme activities involved in flavour precursor formation in unfermented beans of different cocoa genotypes. *J. Sci. Food Agric.* **2000**, *80*, 1193–1198. [CrossRef]

10. Saravia-Matus, S.L.; Rodríguez, A.G.; Saravia, J.A. Determinants of certified organic cocoa production: Evidence from the province of Guayas, Ecuador. *Org. Agric.* **2019**, 1–12. [CrossRef]

11. Department of Agriculture and Fisheries. *Growing Cocoa*; Government of Queensland: Brisbane, Australia, 2015.

12. Trognitz, B.; Cros, E.; Assemat, S.; Davrieux, F.; Forestier-Chiron, N.; Ayestas, E.; Kuant, A.; Scheldeman, X.; Hermann, M. Diversity of cacao trees in Waslala, Nicaragua: Associations between genotype spectra, product quality and yield potential. *PLoS ONE* **2013**, *8*, e54079. [CrossRef]

13. Smulders, M.; Esselink, D.; Amores, F.; Ramos, G.; Sukha, D.; Butler, D.; Vosman, B.; Van Loo, E. Identification of cocoa (*Theobroma cacao* L.) varieties with different quality attributes and parentage analysis of their beans. *IGENIC Newsl.* **2008**, *12*, 1–13.

14. Olujide, M.; Adeogun, S. Assessment of cocoa growers' farm management practices in Ondo State, Nigeria. *Span. J. Agric. Res.* **2006**, *4*, 173–179. [CrossRef]

15. Daymond, A.; Hadley, P. Differential effects of temperature on fruit development and bean quality of contrasting genotypes of cacao (*Theobroma cacao*). *Ann. Appl. Biol.* **2008**, *153*, 175–185. [CrossRef]

16. Saunders, J.A.; Mischke, S.; Leamy, E.A.; Hemeida, A.A. Selection of international molecular standards for DNA fingerprinting of *Theobroma cacao*. *Theor. Appl. Genet.* **2004**, *110*, 41–47. [CrossRef] [PubMed]

17. De Bei, R.; Fuentes, S.; Gilliham, M.; Tyerman, S.; Edwards, E.; Bianchini, N.; Smith, J.; Collins, C. VitiCanopy: A free computer app to estimate canopy vigor and porosity for grapevine. *Sensors* **2016**, *16*, 585. [CrossRef]

18. Poblete-Echeverría, C.; Fuentes, S.; Ortega-Farias, S.; Gonzalez-Talice, J.; Yuri, J.A. Digital cover photography for estimating leaf area index (LAI) in apple trees using a variable light extinction coefficient. *Sensors* **2015**, *15*, 2860–2872. [CrossRef] [PubMed]

19. Macfarlane, C.; Hoffman, M.; Eamus, D.; Kerp, N.; Higginson, S.; McMurtrie, R.; Adams, M. Estimation of leaf area index in eucalypt forest using digital photography. *Agric. For. Meteorol.* **2007**, *143*, 176–188. [CrossRef]

20. Miyaji, K.-I.; Da Silva, W.S.; Alvim, P.D.T. Longevity of leaves of a tropical tree, *Theobroma cacao*, grown under shading, in relation to position within the canopy and time of emergence. *New Phytol.* **1997**, *135*, 445–454. [CrossRef]

21. Fuentes, S.; Palmer, A.R.; Taylor, D.; Zeppel, M.; Whitley, R.; Eamus, D. An automated procedure for estimating the leaf area index (LAI) of woodland ecosystems using digital imagery, MATLAB programming and its application to an examination of the relationship between remotely sensed and field measurements of LAI. *Funct. Plant Biol.* **2008**, *35*, 1070–1079. [CrossRef]

22. Yapp, J.H.; Hong, H. *A Study into the Potential for Enhancing Productivity in Cocoa (Theobroma cacao L.) through Exploitation of Physiological and Genetic Variation*; University of Reading: Reading, UK, 1992.

23. Daymond, A.; Hadley, P.; Machado, R.; Ng, E. Canopy characteristics of contrasting clones of cacao (*Theobroma cacao*). *Exp. Agric.* **2002**, *38*, 359–367. [CrossRef]

24. Baofeng, S.; Jinru, X.; Chunyu, X.; Yuyang, S.; Fuentes, S. Digital surface model applied to unmanned aerial vehicle based photogrammetry to assess potential biotic or abiotic effects on grapevine canopies. *Int. J. Agric. Biol. Eng.* **2016**, *9*, 119–130.

25. Xue, J.; Fan, Y.; Su, B.; Fuentes, S. Assessment of canopy vigor information from kiwifruit plants based on a digital surface model from unmanned aerial vehicle imagery. *Int. J. Agric. Biol. Eng.* **2019**, *12*, 165–171. [CrossRef]

26. Biehl, B.; Meyer, B.; Crone, G.; Pollmann, L.; Said, M.B. Chemical and physical changes in the pulp during ripening and post-harvest storage of cocoa pods. *J. Sci. Food Agric.* **1989**, *48*, 189–208. [CrossRef]

27. Selamat, J.; Mordingah Harun, S. Formation of methyl pyrazine during cocoa bean fermentation. *Pertanika* **1994**, *17*, 27.

28. Rodriguez-Campos, J.; Escalona-Buendía, H.; Contreras-Ramos, S.; Orozco-Avila, I.; Jaramillo-Flores, E.; Lugo-Cervantes, E. Effect of fermentation time and drying temperature on volatile compounds in cocoa. *Food Chem.* **2012**, *132*, 277–288. [CrossRef] [PubMed]

29. The Good Scents Company. The Good Scents Company Information System. Available online: http://www.thegoodscentscompany.com (accessed on 3 March 2019).

30. Ramos, C.L.; Dias, D.R.; PedrozoMiguel, M.G.D.; FreitasSchwan, R. Impact of different cocoa hybrids (*Theobroma cacao* L.) and S. cerevisiae UFLA CA11 inoculation on microbial communities and volatile compounds of cocoa fermentation. *Food Res. Int.* **2014**, *64*, 908–918. [CrossRef] [PubMed]

31. Owusu, M.; Petersen, M.A.; Heimdal, H. Assessment of aroma of chocolate produced from two Ghanaian cocoa fermentation types. In Proceedings of the 12th International Weurman Flavour Research Symposium, Interlaken, Switzerland, 1–4 July 2008.

32. Rodriguez-Campos, J.; Escalona-Buendía, H.; Orozco-Avila, I.; Lugo-Cervantes, E.; Jaramillo-Flores, M.E. Dynamics of volatile and non-volatile compounds in cocoa (*Theobroma cacao* L.) during fermentation and drying processes using principal components analysis. *Food Res. Int.* **2011**, *44*, 250–258. [CrossRef]

33. Bonvehí, J.S. Technology. Investigation of aromatic compounds in roasted cocoa powder. *Eur. Food Res. Technol.* **2005**, *221*, 19–29. [CrossRef]

34. Bos, M.M.; Steffan-Dewenter, I.; Tscharntke, T. Shade tree management affects fruit abortion, insect pests and pathogens of cacao. *Agric. Ecosyst. Environ.* **2007**, *120*, 201–205. [CrossRef]

35. Aprotosoaie, A.C.; Luca, S.V.; Miron, A. Flavor chemistry of cocoa and cocoa products—An overview. *Compr. Rev. Food Sci. Food Saf.* **2016**, *15*, 73–91. [CrossRef]

36. Ramli, N.; Hassan, O.; Said, M.; Samsudin, W.; Idris, N.A. Influence of roasting conditions on volatile flavor of roasted Malaysian cocoa beans. *J. Food Process. Preserv.* **2006**, *30*, 280–298. [CrossRef]

37. Kongor, J.E.; Hinneh, M.; Van de Walle, D.; Afoakwa, E.O.; Boeckx, P.; Dewettinck, K. Factors influencing quality variation in cocoa (*Theobroma cacao*) bean flavour profile—A review. *Food Res. Int.* **2016**, *82*, 44–52. [CrossRef]

38. Martínez-Lüscher, J.; Brillante, L.; Kurtural, S.K. Flavonol profile is a reliable indicator to assess canopy architecture and the exposure of red wine grapes to solar radiation. *Front. Plant Sci.* **2019**, *10*, 10. [CrossRef]

 *sensors*

Article

Non-Invasive Tools to Detect Smoke Contamination in Grapevine Canopies, Berries and Wine: A Remote Sensing and Machine Learning Modeling Approach

Sigfredo Fuentes [1,*], Eden Jane Tongson [1], Roberta De Bei [2], Claudia Gonzalez Viejo [1], Renata Ristic [2], Stephen Tyerman [2] and Kerry Wilkinson [2]

[1] School of Agriculture and Food, Faculty of Veterinary and Agricultural Sciences, The University of Melbourne, Parkville, VIC 3010, Australia
[2] School of Agriculture, Food and Wine, The University of Adelaide, PMB 1, Glen Osmond, SA 5064, Australia
* Correspondence: sfuentes@unimelb.edu.au; Tel.: +61-3-9035-9670

Received: 16 July 2019; Accepted: 28 July 2019; Published: 30 July 2019

Abstract: Bushfires are becoming more frequent and intensive due to changing climate. Those that occur close to vineyards can cause smoke contamination of grapevines and grapes, which can affect wines, producing smoke-taint. At present, there are no available practical in-field tools available for detection of smoke contamination or taint in berries. This research proposes a non-invasive/in-field detection system for smoke contamination in grapevine canopies based on predictable changes in stomatal conductance patterns based on infrared thermal image analysis and machine learning modeling based on pattern recognition. A second model was also proposed to quantify levels of smoke-taint related compounds as targets in berries and wines using near-infrared spectroscopy (NIR) as inputs for machine learning fitting modeling. Results showed that the pattern recognition model to detect smoke contamination from canopies had 96% accuracy. The second model to predict smoke taint compounds in berries and wine fit the NIR data with a correlation coefficient (R) of 0.97 and with no indication of overfitting. These methods can offer grape growers quick, affordable, accurate, non-destructive in-field screening tools to assist in vineyard management practices to minimize smoke taint in wines with in-field applications using smartphones and unmanned aerial systems (UAS).

Keywords: bushfires; infrared thermography; near-infrared spectroscopy; smoke taint; artificial intelligence

---

1. Introduction

A recent report from the Victorian government of Australia concluded that bushfires have increased in number and severity since the 1970s across the east and south of the country [1]. The main contributing factor to this environmental disaster is climate change, specifically the increased frequency of recurrent heat waves (i.e., prolonged periods of hotter weather) and drought conditions, which have increased the window of risk for bushfires, as well as their number, and severity. Recently, Chile (central region), USA (California), Greece, South Africa (Stellenbosch) and Australia (various states) have suffered some of the worst bushfires experienced in each country's history. These countries are major producers of wines, and their grape growers and winemakers are similarly affected by global warming with detrimental effects in drought, vine phenological changes, shifting of suitable grapevine growing regions towards the north and south, and increased bush fire events near wine growing regions [2–4].

When bushfires occur in close proximity to vineyards, smoke can contaminate leaves and fruit. One of the main physiological effects of bush fire smoke in grapevines is the reduction of stomatal conductance (g_s) [5]. Decreased g_s may be explained by the combination of the main smoke components carbon dioxide (CO_2) and carbon monoxide (CO), with water vapor (100% RH in the substomatal cavity) producing carbonic acid (H_2CO_3) that reduces pH in the stomata, thereby causing partial or complete stomatal closure [5]. In berries, smoke contamination results in adsorption of smoke-derived volatile phenols (which accumulate in glycoconjugate forms), that are extracted into the final wine during the winemaking process [6]. Several mitigating measures have been evaluated to minimize smoke taint in berries or to remove volatile phenols (and their glycoconjugates) from wine, including defoliation [7] or foliar application of kaolin [8] in the vineyard, reverse osmosis treatment [9] and the addition of fining agents in wines [10]. However, implementation of defoliation or kaolin applications are often indiscriminate and broadly applied irrespective of the degree of grapevine exposure to smoke. Furthermore, the removal of smoke taint from wine is not selective and may inadvertently remove important wine compounds, thereby affecting the desirable organoleptic characteristics of wine.

Physiological assessment of control (non-smoked) and smoke affected grapevine cultivars have shown that some cultivars are more susceptible than others in terms of photosynthesis and stomatal conductance, in particular, Merlot and Cabernet Sauvignon. In contrast, Sauvignon Blanc was not significantly affected by smoke contamination from a physiological perspective. However, berries exposed to smoke resulted in wines with significantly higher concentrations of volatile phenols and guaiacol glycoconjugates compared to wines made from uncontaminated fruit [6].

The need to assess smoke contaminated fruit and wine has led to the implementation of new laboratory-based analytical methods [11], including liquid chromatography-tandem mass spectrometry for quantification of volatile phenol glycoconjugates [12,13]. However, these techniques require expensive laboratory instrumentation, and specialized technical skills to prepare samples (i.e., to extract the analytes of interest), operate the instruments and data analysis. Spectral methods, both mid-infrared (MIR) reflectance spectroscopy and chemometric techniques have been evaluated for rapid detection of smoke-taint in grapes [8] and bottled wines [14], but also with limitations. Spectral reflectance measurements of berries were affected by fruit maturity, while MRI-based classification of wines was influenced by cultivar, oak maturation and the level of smoke taint. Thus, reliable and rapid in-field techniques available to determine whether vines and fruits have been contaminated with smoke from bushfires are not yet available. This paper presents pattern recognition and regression models based on machine learning algorithms developed for the identification of smoke contaminated grapevine canopies and fruit. The first machine learning model generated used infrared thermography data from grapevine canopies as inputs to predict smoke contamination, as a target for four grapevine cultivars. Near infrared (NIR) spectroscopy readings from berries were used as inputs for regression machine learning algorithms to assess specific smoke-related compounds in berries and final wines from seven cultivars. These models combined with affordable geo-referenced NIR spectroscopy measurements of berries could allow growers to map contaminated areas of a vineyard to facilitate decision making at harvest. Finally, potential applications of these models using proximal and mid-range remote sensing using unmanned aerial systems (UAS) are discussed.

2. Materials and Methods

2.1. Experimental Site and Application of Smoke to Grapevines

Grapevine smoke exposure experiments were conducted in the 2009/10 season using seven different cultivars grown at two locations: (i) Sauvignon Blanc, Pinot Gris, Chardonnay and Pinot Noir grown in a commercial vineyard in Adelaide Hills region, South Australia, Australia (35°00′ S, 138°49′ E) and (ii) Shiraz, Cabernet Sauvignon and Merlot vines grown in a vineyard located at the University of Adelaide's Waite campus in Adelaide, South Australia (34°58′ S, 138°38′ E). Grapevines (three per replicate) were exposed to smoke (for 1 h) using a purpose-built smoke tent and experimental conditions described previously (Figure 1) [15]. Smoke was applied to vines at a phenological stage corresponding to approximately seven days post-veraison; when total soluble solids (TSS) concentrations were approximately 15 Brix, determined using a digital handheld refractometer (PAL-1, Atago, Tokyo, Japan).

Experiment 1 consisted of the physiological assessment of smoke contamination at the canopy level using porometry, infrared thermography, and pattern recognition machine learning using four cultivars two hours after smoke exposure: Chardonnay, Merlot, Sauvignon Blanc, and Shiraz. Experiment 2 assessed smoke taint in berries and wine at harvest for all seven cultivars. In this experiment, control (unsmoked) and smoke-affected berry samples (two berries taken from the mid-section of two bunches from two replicates per treatment, per cultivar; $n = 112$ berries) were collected at harvest. Morphometry of berries was measured using a caliper to obtain diameter (equatorial length in cm); length (cm) and calculated radius (cm), area (cm^2), and perimeter (cm).

Figure 1. Grapevines were enclosed in a tent, and smoke from combustion of straw was blown into the tent using a fan. Pre-installation of the tent (**A**), and installed and operational tent (**B**). Photos obtained from the 2009/10 trial in Adelaide, Australia.

2.2. Experiment 1

2.2.1. Physiological Measurements Using Leaf Porometry

Leaf conductance to water vapor was measured as stomatal conductance (g_s) using a porometer (AP4, Delta-T Devices, Cambridge, UK). Porometer readings used were obtained from the cultivars Shiraz, Sauvignon Blanc, Chardonnay, and Merlot. Measurements were performed two hours after smoke treatments using nine mature, fully expanded sunlit leaves, for each of the two middle vines of two replicates per treatment per cultivar ($n = 72$) under natural leaf orientation with natural light intensity. Leaves were chosen to ensure measurements were performed on three leaves from the top, middle, and bottom parts of the canopies from each vine in a 3 × 3 matrix arrangement.

2.2.2. Infrared Thermal Imagery of Canopies

Thermal images were acquired from grapevine canopies using an infrared thermal camera FLIR® T-series (Model B360) (FLIR Systems, Portland, OR, USA), with a resolution of 320 × 240 pixels. The camera measures temperature in the range of −20 to +1200 °C. The thermal sensitivity of the camera is <0.08 °C @ +30 °C/80 mK with a spatial resolution of 1.36 milliradians. Each pixel is considered an effective temperature reading in degrees Celsius (°C). Infrared thermal images were acquired from the same side and in parallel to porometer measurements (shaded side of the canopy to reduce variability) in the estimation of the infrared index (I_g), which is proportional to g_s [16]. One thermal image from the canopy of each of the middle vines of two replicates per treatment per cultivar was obtained from a constant distance of 2.5 m perpendicular to the row direction (distance between rows being 3 m; Figure 2A). The calculated infrared thermal index (I_g) was compared with porometry measurements acquired immediately after obtaining each thermal image from corresponding vines. All thermal images were acquired on a clear day. The smoke treatments were applied with minimal wind; a requirement for undertaking the field trials implemented to avoid the risk of fire spreading from accidental burning of interrow dry plant material and to secure representativeness of thermal images [16,17].

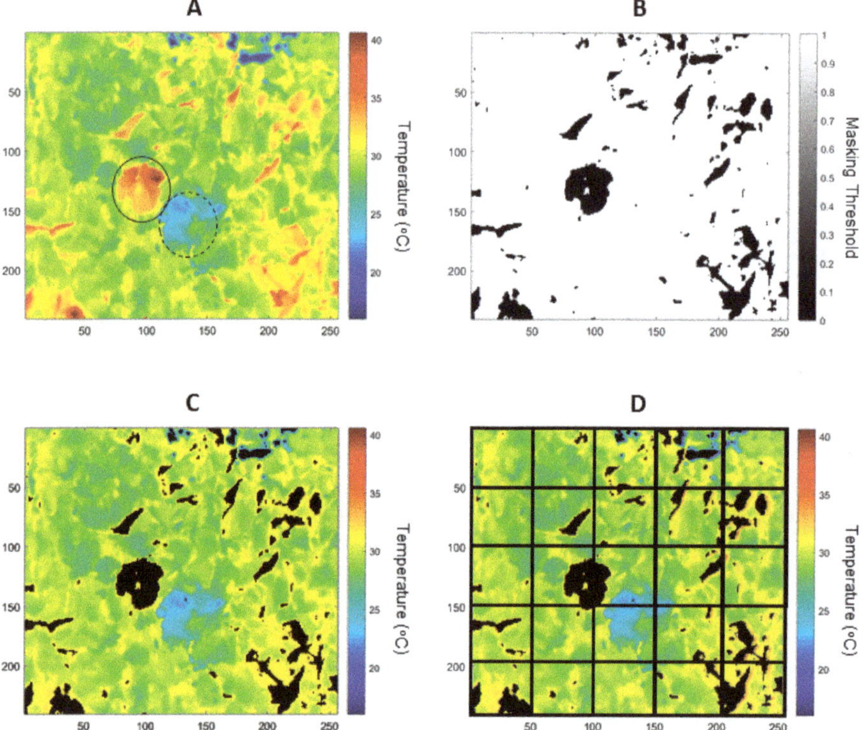

Figure 2. Examples of a radiometric thermal image (**A**) processing for data extraction of T_{dry} (A, solid circle) and T_{wet} (A, dotted circle) by painting leaves with petroleum jelly and water, respectively. Binary image obtained by thresholding T_{dry} and T_{wet} (**B**); masked radiometric image extracting non-leaf material, such as overheated elements and sky (**C**); and subdivision of thermal image to extract information from sections of the canopy in a 5 × 5 sub-division (**D**).

2.2.3. Algorithms Used to Calculate Crop Water Stress Indices (CWSI) and Infrared Index (I_g)

Crop water stress index (CWSI) was calculated using the following equation, after determining T_{dry} and T_{wet} [18]:

$$\text{CWSI} = \frac{T_{canopy} - T_{wet}}{T_{dry} - T_{wet}} \tag{1}$$

where T_{canopy} is the actual canopy temperature extracted from the thermal image at determined positions, and T_{dry} and T_{wet} are the reference temperatures (in °C), obtained using the method of painting both sides of reference leaves with petroleum jelly and water, respectively [16].

An infrared index (I_g), proportional to leaf conductance to water vapor transfer (g_s), can be obtained using the relationship as follows [19]:

$$I_g = \frac{T_{canopy} - T_{wet}}{T_{dry} - T_{wet}} = g_s \left(r_{aw} + \left(\frac{s}{\gamma} \right) r_{HR} \right) \tag{2}$$

where r_{aw} = boundary layer resistance to water vapor, r_{RH} = the parallel resistance to heat and radiative transfer, Υ = psychrometric constant and s = slope of the curve relating saturation vapor pressure to temperature [17,19].

2.2.4. Infrared Thermography Data Extraction

The T_{dry} and T_{wet} values were obtained on a per image basis using a customized code written in Matlab® R2019a (Mathworks Inc. Natick, MA, USA) to crop the radiometric data from the areas within the respective painted leaves with water (T_{wet}) and petroleum jelly (T_{dry}) (Figure 2A). To filter non-leaf material from the radiometric image using the determined threshold, a second customized code was written in Matlab®to binarize a masked image (Figure 2B) and to extract these values from the original image (Figure 2C). For automatic extraction of data within a canopy, a pre-defined subdivision of $3 \times 3 = 9; 5 \times 5 = 25; 7 \times 7 = 49$ and $10 \times 10 = 100$ was automatically implemented (Figure 2D; for the case of 5×5). From these subdivisions, data were extracted for T_{canopy} per image (Figure 2D), I_g Equation (2) and CWSI Equation (1).

The image sub-divisions (Figure 2D) represent the matrix (A) with $m \times n$ (m = rows and n = columns) extraction points represented as per the following matrix:

$$A(m,n) = \begin{pmatrix} T_{1,1} & \cdots & T_{1,n} \\ \vdots & \ddots & \vdots \\ T_{m,1} & \cdots & T_{m,n} \end{pmatrix} \tag{3}$$

Since m, n represent the pre-determined subdivision for automatic cropping sections from the infrared thermal image (A), every sub-image is processed for automatic canopy extraction by filtering non-leaf temperatures using the T_{dry} and T_{wet} values extracted (Figure 2B) as minimum and maximum possible temperatures for the canopy. The calculated T value then corresponds to the averaged T_{canopy} for each sub-division.

2.2.5. Pattern Recognition of Infrared Thermal Imagery using Machine Learning for Smoke Contamination Prediction

Pattern recognition models were developed using a customized Matlab®code, which is able to test 17 different training algorithms, two from Backpropagation with Jacobian derivatives, 11 from Backpropagation with gradient derivatives and four from Supervised weight and bias training functions, in loop to select the best model. This model was constructed using the infrared thermal image output values as inputs to classify the samples into smoked and non-smoked (control). The infrared thermal images were analyzed with the methodology described in Figure 2 to obtain T_{canopy}, I_g, and CWSI data obtained using Equations (1) and (2) with sub-divisions of 3×3 ($n = 27$ per image); 5×5 ($n = 75$, per image; 7×7 ($n = 147$, per image) and 10×10 ($n = 300$ per image). All algorithms tested used a random data division. However, for the algorithms such as scaled conjugate gradient, which consist of three stages—training, validation and testing, the data was divided as 60% ($n = 28$ images) for training, 20% ($n = 10$ images) for validation with a cross-entropy performance algorithm, and 20% ($n = 10$ images) for testing with a default derivative function. For the algorithms such as sequential order weights and bias, which only consist of training and testing stages, the data was divided as 70% ($n = 34$) for training and 30% ($n = 14$) for testing with a cross-entropy performance algorithm. A trimming exercise was conducted using 3, 7 and 10 neurons to select the best model with no signs of overfitting (Figure 3).

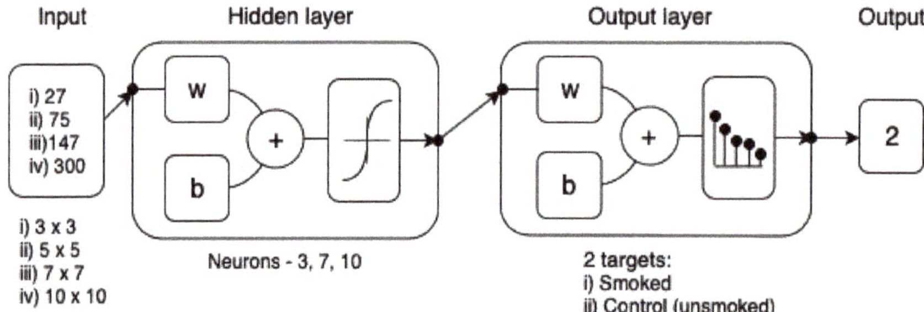

Figure 3. Diagram of the two-layer feedforward network with a tan-sigmoid function in the hidden layer and a Softmax transfer function in the output layer. For hidden and output layers, w = weights and b = biases. Input volume and neuron trimming exercises are included in the diagram.

2.3. Experiment 2

2.3.1. Berry Near Infrared (NIR) Spectroscopy Measurements

Full berries were scanned using a spectrophotometer (ASD FieldSpec®3, Analytical Spectral Devices, Boulder, CO, USA) equipped with the ASD contact probe, built for contact measurements, attached by fiber optic cable to the instrument. A total of 112 berries collected at harvest from seven cultivars (16 berries per cultivar) were scanned by putting the probe's lens in contact with the berries and a total of 401 spectra were recorded for each berry. The instrument records spectra with a resolution of 1.4 nm for the region 350–1000 nm and 2 nm for the region 1000–1850 nm. The instrument was used in reflectance mode and data was then transformed into absorbance values (absorbance = log (1/reflectance)). A reference tile (Spectralon®, Analytical Spectral Devices, Boulder, CO, USA) was used as a white reference, for scatter correction. A new reference was taken every ten spectra acquisitions.

2.3.2. Winemaking and Chemical Analysis of Berries and Wine

Small scale winemaking of control and smoke-affected fruit from this trial has been described previously in detail by Ristic et al. [6]. Guaiacol glycoconjugates were measured in fruit and wine by HPLC–MS/MS using a stable isotope dilution analysis (SIDA) method developed by Dungey et al. (2011) [12]. Volatile phenols, including guaiacol, were determined in berries and wine by the Australian Wine Research Institute's (AWRI) Commercial Services Laboratory (Adelaide, Australia). Volatile phenols were measured by GC–MS according to SIDA methods reported previously [13].

2.3.3. Fitting Modeling of Near-Infrared (NIR) Spectroscopy of Berries Using Machine Learning Modeling to Predict Smoke Taint in Berries and Wine

A regression model was developed using a customized Matlab® code, which is able to test 17 different training algorithms, two from Backpropagation with Jacobian derivatives, 11 from Backpropagation with gradient derivatives and four from Supervised weight and bias training functions, in loop to select the best model. NIR absorbance values corresponding to the range of wavelengths within 700 and 1100 nm with a second derivative transformation, which were used as inputs in the machine learning algorithms, since that range corresponds to alcohol and alcohol-based compounds to predict (i) guaiacol glycoconjugates in berries (μg Kg^{-1}), (ii) guaiacol glycoconjugates in wines (μg L^{-1}) and iii) guaiacol in wine (μg L^{-1}). All algorithms tested used a random data division. However, for the algorithms, which consist of three stages—training, validation and testing, the data was divided as 60% (n = 28) for training, 20% (n = 10) for validation with a means squared error (MSE) performance algorithm and 20% (n = 10) for testing with a default derivative function (data not shown). For the algorithms such as sequential order weights and bias, which only consist of training and testing stages, data was divided as 70% (n = 34) for training and 30% (n = 14) for testing with a means squared error performance algorithm. A trimming exercise was conducted using 3, 7, and 10 neurons to select the best model with no signs of overfitting (Figure 4).

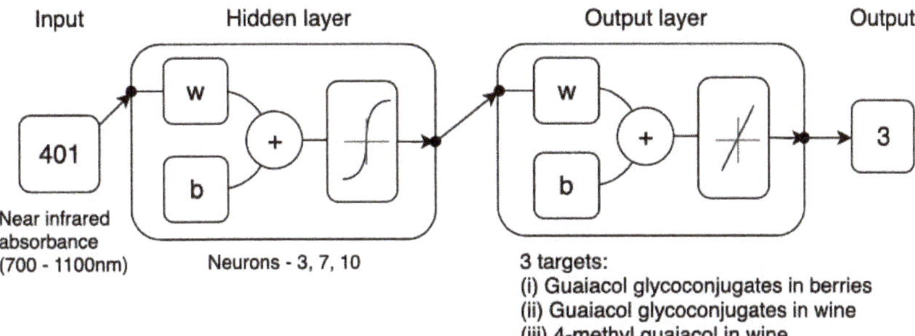

Figure 4. Diagram of the two-layer feedforward network with a tan-sigmoid function in the hidden layer and a linear transfer function in the output layer. For hidden and output layers, w = weights and b = biases. Neuron trimming exercise is included in the diagram.

2.4. Statistical Analysis

Data from chemometry and morphometry of berries, wine compounds, and I_g and g_s were analyzed through ANOVA using SAS® 9.4 software (SAS Institute Inc., Cary, NC, USA) with Tukey's studentized range test (HSD; $p < 0.05$) as post-hoc analysis for multiple comparisons to assess significant differences. Statistical data such as means and standard deviation (SD) were obtained from the replicates of each cultivar and treatment.

3. Results

3.1. Experiment 1

3.1.1. Grapevine Physiological Data Relationships between Porometry and Infrared Thermal Imagery

Table 1 shows the mean values for g_s and I_g with respective standard deviations (SD) for the four cultivars from Experiment 1. The general trend for the mean values of the control treatments follows a positive linear relationship ($R^2 = 0.99$; $I_g = 0.0027\ g_s$). On the contrary, the trend for the mean values of the smoke treatments have lower linearity and relationship, but still showed a positive linear pattern ($R^2 = 0.23$; $I_g = 0.0023\ g_s$; data not shown). In the control samples, the mean Ig values per cultivar did not show significant change, as reflected by the SD values, but Merlot showed a significantly higher I_g ($p < 0.05$). This trend was similar for the mean g_s values showing Merlot with the highest mean ($p < 0.05$). The mean I_g values for smoked samples were more variable, while g_s showed higher mean values, except Sauvignon Blanc, and more variability as reflected by the higher SD values compared to control. The I_g mean values for both treatments were not very sensitive, as seen in Table 1.

Figure 5 shows the relationships between g_s and I_g for different sections of canopies (top, middle, and bottom) of the grapevines monitored for both non-smoked (control) and smoked treatments. The graph (Figure 5A) shows a strong and significant linear relationship between g_s and I_g ($R^2 = 0.85$; $I_g = 0.0026\ g_s$). However, there was no relationship observed for smoke treatments, with the data presenting high variability, which is consistent with results shown in Table 1. Figure 5A shows that regardless of the measurement position within the canopy for I_g, there is a broader distribution of values between top, middle, and bottom of the canopy along the linear relationship found. On the contrary, Figure 5B shows that the bottom readings for g_s are more concentrated towards the lower values (<200 mmol m^2 s^{-1}). Furthermore, the Ig values become less sensitive (spread between 0 and 1). The same pattern can be seen for most of the top readings with the middle readings having a wider spread distribution.

Table 1. Means and standard deviation (SD) per variety and treatment for the infrared index (Ig, unitless) and stomatal conductance (g_s in mmol m^2 s^{-1}) calculated for all the images without sub-divisions.

Variety	I_g		g_s (mmol m^2 s^{-1})		I_g		g_s (mmol m^2 s^{-1})	
	Control				Smoked			
	Mean	SD	Mean	SD	Mean	SD	Mean	SD
Chardonnay	0.32 [b]	0.22	112.66 [c]	60.55	0.60 [a]	0.34	203.02 [ab]	145.72
Merlot	1.06 [a]	0.29	384.93 [a]	102.68	0.43 [ab]	0.28	251.00 [a]	131.44
Sauvignon Blanc	0.34 [b]	0.18	130.60 [c]	60.12	0.32 [b]	0.23	135.89 [b]	46.49
Shiraz	0.52 [b]	0.26	211.40 [b]	88.97	0.59 [a]	0.10	235.35 [ab]	148.71

Means followed by different superscript letters are statistically significant between treatments based on Tukey's studentized range test (HSD, $p < 0.05$).

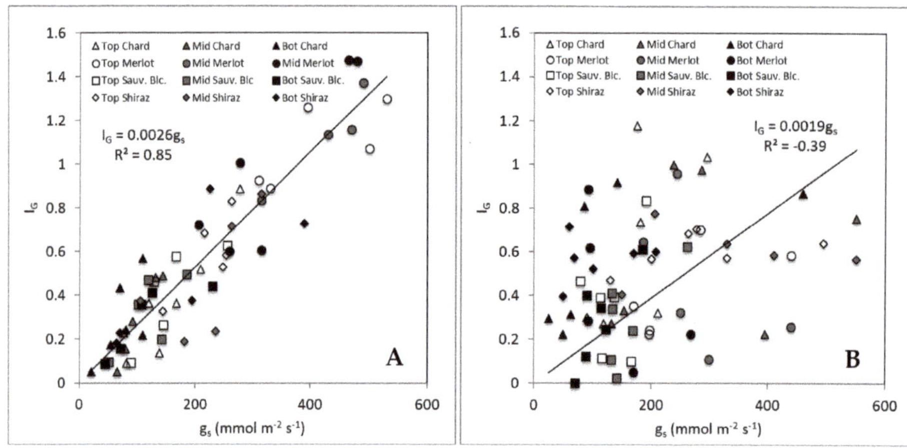

Figure 5. Relationship between g_s and I_g for Experiment 1 in the four cultivars with data separated between canopy sections: top (Top), middle (Mid) and bottom (Bot) measurements for control treatments (**A**) and smoked treatments (**B**).

3.1.2. Pattern Recognition Using Machine Learning Modeling of Physiological and Infrared Thermal Data

Table 2 shows the results of the pattern recognition modeling for the data extracted from infrared thermal images from the canopies of four different cultivars combined for Experiment 1. The best performing algorithm for the 3 × 3 sub-division and extraction of T_{canopy}, I_g, and CWSI used as inputs and classification of smoked and non-smoked as target was the scaled conjugate gradient algorithm. The training, validation, and testing procedures (using 10 neurons) resulted in an overall model with 94% accuracy. In the case of the data extracted using a 5 × 5 sub-division, the overall best model (sequential order weight and bias) resulted in an accuracy of 88% (using 10 neurons) in the classification of smoked and non-smoked canopies. For the 7 × 7 sub-division, the best algorithm (also the sequential order weight and bias) resulted in an accuracy of 94% (using 7 neurons) in the classification. Finally, the 10 × 10 was the best performing algorithm overall (sequential order weight and bias) resulted in an accuracy of 96% (using 3 neurons). Furthermore, the performance of training was lower than the one for testing, and testing accuracy was close to that from the training stage, which are evidence of no overfitting [20,21].

Table 2. Best pattern recognition model developed for each set of inputs showing the best training algorithm and number of neurons to predict whether canopies are smoked or non-smoked (control). Inputs corresponds to data extracted from infrared thermal images for T_{canopy}, I_g and crop water stress index (CWSI) in matrix arrangement of 3×3 ($n = 27$), 5×5 ($n = 75$), 7×7 ($n = 147$) and 10×10 ($n = 300$) data points per thermography. Performance reported is based on cross-entropy.

Inputs	Algorithm	Neurons	Stage	Samples	Accuracy	Performance
3×3	Scaled conjugate gradient	10	Training	28	100%	0.03
			Validation	10	90%	0.16
			Test	10	80%	0.44
			Overall	48	94%	-
5×5	Sequential order weight and bias	10	Training	34	85%	0.37
			Test	14	93%	0.43
			Overall	48	88%	-
7×7	Sequential order weight and bias	7	Training	34	94%	0.72
			Test	14	93%	0.71
			Overall	48	94%	-
10 × 10	**Sequential order weight and bias**	**3**	**Training**	**34**	**97%**	**0.45**
			Test	**14**	**93%**	**0.47**
			Overall	**48**	**96%**	**-**

Figure 6 shows the Receiver Operating Characteristic (ROC) for the best performing model found to predict smoke contamination in grapevine canopies (10×10 sub-division; Table 2). The figure shows that results for both smoke and control pattern recognition using infrared thermography data as inputs are projected in a similar trend to the True Positive Rate prediction axis of the graph.

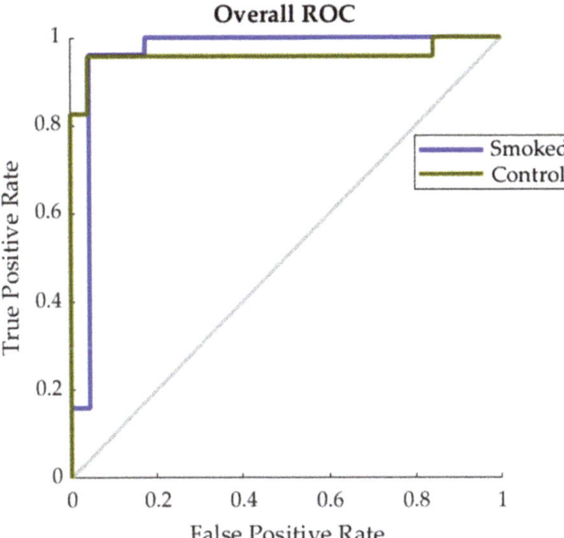

Figure 6. Receiver Operating Characteristic (ROC) showing the false positive rate (*x*-axis) and true positive (*y*-axis) for control and smoked treatments for the best performing classification model found in Table 3.

Table 3. Morphometric data obtain from berries for all seven cultivars consisting in Perimeter (P in cm), Equatorial Diameter (D in cm), calculated Area (A in cm2) and Radius (D/2 in cm). For chemometry, Total Soluble Solids (TSS) represented by Brix and Near Infrared (NIR) absorbance at 982 nm corresponding to H–O–H and O–H chemical bonds.

	P (cm)		D (cm)		Area (cm^2)		R (cm)		Brix (°)		NIR982 (nm)	
	C	S	C	S	C	S	C	S	C	S	C	S
Merlot	4.7 [a]	4.6 [a]	1.4 [a]	1.3 [a]	1.5 [a]	1.5 [a]	0.3 [c]	0.3 [bc]	23.9 [a]	24.2 [ab]	0.4 [abc]	0.3 [ab]
Shiraz	4.1 [b]	3.9 [bc]	1.3 [ab]	1.2 [b]	1.5 [a]	1.2 [c]	0.4 [ab]	0.3 [cd]	24.9 [a]	25.4 [a]	0.3 [cd]	0.4 [a]
PinGr	4.0 [bc]	4.2 [b]	1.3 [bc]	1.3 [ab]	1.4 [ab]	1.5 [a]	0.3 [bc]	0.4 [a]	18.7 [c]	19.8 [d]	0.3 [abc]	0.4 [a]
Char	4.0 [bcd]	3.8 [cd]	1.4 [a]	1.3 [ab]	1.5 [a]	1.3 [bc]	0.4 [a]	0.3 [abc]	19.7 [cd]	18.6 [de]	0.5 [a]	0.4 [a]
PinNoir	3.8 [cd]	3.8 [cd]	1.2 [c]	1.3 [ab]	1.2 [bc]	1.3 [abc]	0.3 [c]	0.3 [ab]	17.2 [d]	18.2 [e]	0.3 [bcd]	0.4 [a]
CabSauv	3.7 [cd]	3.6 [d]	1.1 [d]	1.1 [c]	1.1 [c]	1.0 [d]	0.3 [d]	0.3 [d]	24.1 [a]	23.1 [b]	0.4 [ab]	0.4 [a]
SauvBl	3.7 [d]	3.8 [c]	1.3 [bc]	1.3 [ab]	1.3 [b]	1.4 [ab]	0.4 [ab]	0.4 [a]	20.8 [b]	21.5 [c]	0.2 [d]	0.2 [b]

Abbreviations: C = Control, S = Smoke, PinGr = Pinot Gris, PinNoir = Pinot Noir, Char = Chardonnay, CabSauv = Cabernet Sauvignon, SauvBl = Sauvignon Blanc. Different superscript letters are statistically significant between treatments based on Tukey's studentized range test (HSD, $p < 0.05$).

3.2. Experiment 2

3.2.1. Berry Morphology and NIR Peak within the 700–1100 nm

Table 3 shows the average data of morphometric and chemometric measurements obtained from berry samples for all the seven cultivars for Experiment 2. Even though there are some significant differences between morphometric measurements of berries for the different cultivars comparing smoke and non-smoked (Control) treatments, they do not affect results and models developed.

3.2.2. Smoke-Related Compounds Found in Berries and Wines

Data for smoke-related compounds have been previously reported by Ristic et al. (2016) [6], and comprised of volatiles with statistical differences between control (non-smoked) and smoked treatments. Specifically, for purposes of modeling, guaiacol glycoconjugates found in berries (μg Kg^{-1}), guaiacol glycoconjugates found in wines (μg L^{-1}) and guaiacol found in wines (μg L^{-1}) were used since these are the primary compounds identified by the industry to contribute to smoke taint. In berries, the guaiacol glycoconjugates average concentration ranged for control between 37 and 602 μg kg^{-1} and from 253 to 2452 μg kg^{-1} for smoke-affected treatments. The guaiacol glycoconjugates concentrations in wines ranged from 8 to 334 μg L^{-1} for control and from 111 to 1480 μg L^{-1} for smoke-affected treatments. In the case of guaiacol concentration in wines, values ranged from 0 (not detected) to 9 μg L^{-1} for control and from 0 (not detected) to 26 μg L^{-1} [6].

3.2.3. Near-Infrared (NIR) Spectroscopy from Berries and Smoke Taint Compounds Found

Figure 7 shows the main average spectra for berries from smoke and non-smoked (control) treatments for red (Figure 7A) and white cultivars (Figure 7B). There were no significant differences in the averaged spectra between smoked and non-smoked berries for red cultivars. On the contrary, there appears to be a consistent difference for white cultivars of around 0.05 in absorbance, especially from 820 to 1100 nm for the range considered for this study. Smoke-related compounds for this trial and used for the machine learning model reported here have been previously reported by Ristic et al. [6]. In this study, statistically significant differences in the main smoke taint compounds were reported for all the seven cultivars included in Experiment 2.

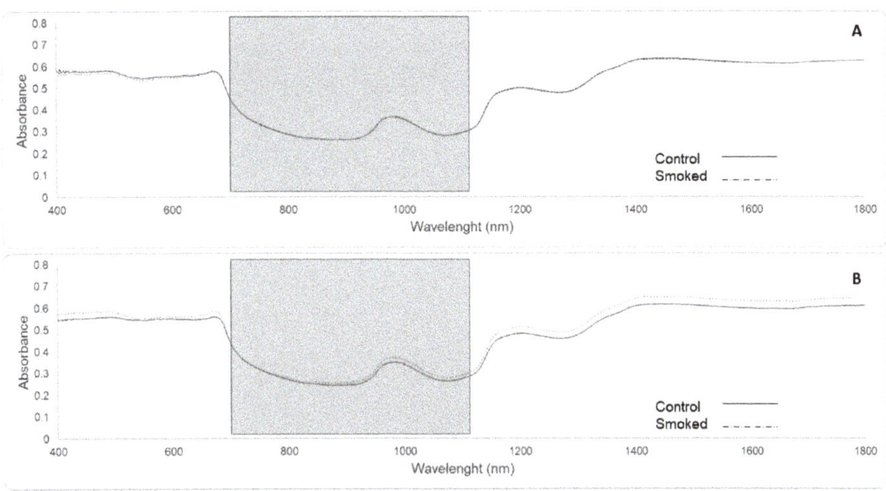

Figure 7. Average spectra for control (solid line) and smoke-affected berries (dashed line) from red cultivars (**A**): Merlot, Shiraz, Pinot Noir, Cabernet Sauvignon and white cultivars (**B**): Pinot Gris, Chardonnay and Sauvignon Blanc. The grey rectangles represent the wavelengths used for machine learning fitting modeling (700–1100 nm), with the main peak at 982 nm.

3.2.4. Machine Learning Modeling Based on NIR Spectra to Estimate Smoke Taint Compounds in Berries and Wine

Table 4 shows the best machine learning regression model obtained for the NIR data from berries (700–1100 nm using the second derivative transformation; Sequential Order Weights and Bias) as inputs and smoke taint compounds measured in berries and wine. The correlation between the estimated and observed values was R = 0.97 and slope b = 0.93 (close to unity). The same correlations and similar slopes were found for the training and the test stages. The overall model can also be seen in Figure 8, in which most of the point cloud data fits in the 1:1 line representing the accuracy of predicted versus observed data. Based on the 95% confidence bounds, the overall model had 3.6% of outliers. The performance of training was lower than the one for testing, and testing accuracy was the same as that from the training stage, which are evidence of no overfitting [20,21].

Table 4. Regression model using machine learning (Sequential Order Weights and Bias) for NIR data from berries of seven grapevine cultivars showing the correlation coefficient (R) and performance based on mean squared error (MSE) for each stage.

Stage	Samples	Observations	R	Slope	Performance (MSE)
Training	78	234	0.97	0.91	0.86
Test	33	99	0.97	0.96	0.91
Overall	111	333	0.97	0.93	-

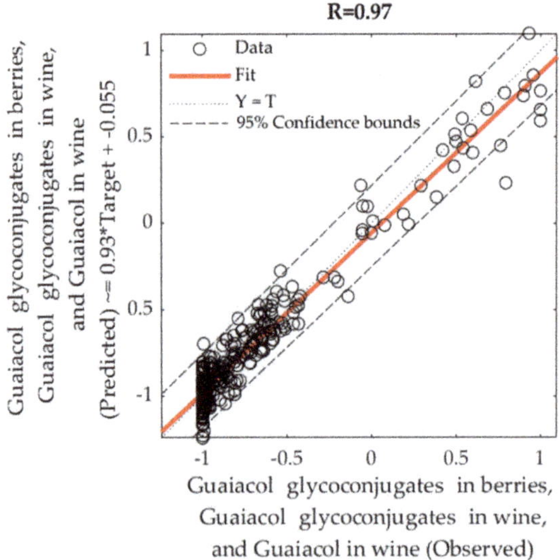

Figure 8. Overall fitting model using machine learning (Sequential Order Weights and Bias) using NIR spectra (700–1100 nm; second derivative transformation) of berries from seven grapevine cultivars as inputs and main smoke taint compounds found in berries and wine as targets.

4. Discussion

4.1. Physiological Changes within Grapevine Canopies Due to Smoke Contamination

The relationship between the I_g thermal index and g_s is linear, as shown in Table 1 and Figure 5A for non-smoked vines. These results are consistent with other studies showing the same relationships for grapevines [16,17], coffee plants [22] and olive trees [23], which are tree-like or bushy canopies. However, this relationship was not observed for smoked canopies of the four cultivars from Experiment 1 (Figure 5B). Smoke contamination is an external signal to the plant which is composed mainly of CO, CO_2 and other gases, which cause acidification of the sub-stomatal cavity due to the production of carbonic acid (H_2CO_3) when combined with water, with the resulting pH reduction causing partial or complete stomata closure [5]. This effect could explain the increased variability within g_s data amongst individual leaves that was detected in porometry data (Table 1 and Figure 5B). The reported I_g data from the whole infrared thermal images (Table 1) did not have significant differences in the variability of the data, which can be explained by the unrepresentativeness of means when using this type of high-resolution information.

It is important to note that the comparison between g_s and I_g for Figure 5 was made in this case using the methodology proposed in Figure 2 and with a sub-division of 3 × 3 for comparison purposes. Since every image was taken from 2.5 m distance, the field of view from infrared thermal images was around 140 × 110 cm of the canopy, which divided by nine gives a sub-area of 47 × 37 cm (area = 1739 cm^2). Considering that the area of an average leaf (data not shown) is of around 50–80 cm^2 [24], the I_g values represent the average of an area of approximately 25-fold of single leaves, in which porometry was conducted. This difference may explain the lower sensitivity of I_g to g_s, especially for smoked canopies with higher g_s variability expected even at the leaf level (patchy stomata behavior).

The extraction of I_g values from infrared thermal images require a T_{dry} and T_{wet} reference temperatures. In this study, the painted leaves method was implemented for more accuracy in the determination of reference temperature thresholds to separate leaf from non-leaf material in the analysis. However, this method is manual and hinders the possibility of automation. Alternatively, the leaf energy balance method could be implemented using micrometeorological weather data such as temperature, relative humidity, and solar radiation to calculate T_{dry} and T_{wet} on-the-go, while obtaining the infrared thermal images. It is common nowadays to access cheap sensor technology to measure these micrometeorological variables and dataloggers or access to the Internet of Things (IoT) for data transmission and processing. Previous research has shown that these reference temperatures can be calculated with high accuracy ($R^2 = 0.95$; RMSE = 0.85; $p < 0.001$) [16]. Furthermore, there is the requirement for infrared thermal images to be explored and assessed more in-depth at higher subdivisions and using machine learning modeling to assess the pattern variability and use it as a predictor of smoke contamination levels.

4.2. Pattern Recognition of Smoke Contamination Using Machine Learning Modeling

Considering the sub-division of infrared thermography data, the field of view of canopies and size of single leaves for this study, it is not surprising that the best pattern recognition model (96% accuracy) using machine learning (Sequential order weight and bias) was obtained with the 10×10 subdivision. This sub-division will render comparison areas within the canopy of 154 cm^2, which is only 2.2-fold compared to a single leaf area (70 cm^2). Furthermore, from the neuron trimming analysis, a highly accurate model was obtained for the classification of smoked and non-smoked canopies with three neurons, which makes the model more efficient and less susceptible to overfitting. The latter is also supported by the performance value obtained by this model. Results shown in this paper from pattern recognition modeling using machine learning to asses smoke contamination of canopies have excellent potential for the use in short and mid-range remote sensing based on Unmanned Aerial Vehicles (UAVs) platforms. From Figure 5B, it can be seen that the main variability within g_s values is in the bottom and top parts of the canopies, which validates obtaining infrared thermal imagery using UAVs at 0° Nadir angle. Furthermore, models developed in this study should be tested using UAV with infrared cameras that could render a 15×15-pixel resolution, which corresponds to an area of 225 cm^2, which is close to the 154 cm^2 area used for machine learning modeling here.

This kind of remote sensing tool can render spatial distribution maps of contaminated areas within vineyards that could aid growers to apply differential management strategies discussed before to mitigate smoke contamination of the fruit. Spatial maps of smoke contamination can also help to achieve differential harvests to avoid mixing fruit with smoke-tainted fruit. Hence, a system is proposed using these methods, which is depicted in Figure 9 for proximal and mid-distance remote sensing using infrared cameras and UAV platforms. For proximal remote sensing, the algorithms developed in this study can be implemented in smartphone devices as computer applications (Apps) connected to portable and affordable infrared thermal cameras (i.e., FLIR One®, FLIR Systems, Portland, OR, USA) and near-infrared spectroscopy devices (i.e., Lighting Passport®, AsenseTek, Taipei, Taiwan).

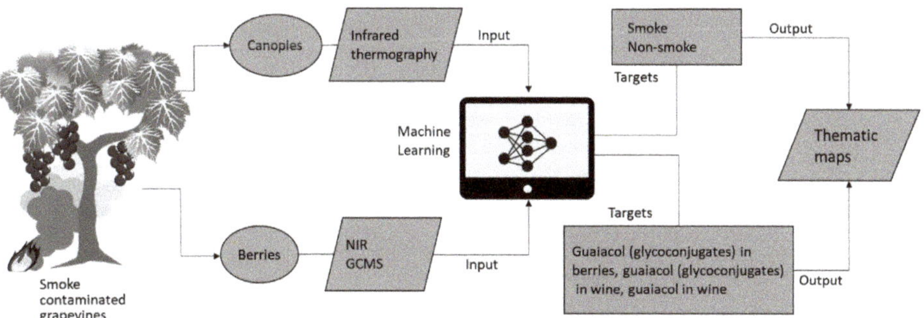

Figure 9. Diagram showing the implementation of machine learning modeling strategies proposed in this paper for proximal (using smartphones and portable infrared thermal cameras and NIR devices) and mid-distance remote sensing using unmanned aerial system (UAS) platforms.

4.3. Near-Infrared (NIR) Spectroscopy of Berries

Since NIR spectroscopy was obtained from full berries, the tool proposed in this paper is non-destructive. Furthermore, it has been shown that a higher concentration of smoke-related compounds after contamination can be found in the skin of berries, which is higher than in the pulp and higher than the seeds [12]. Furthermore, the range of 700–1100 nm was chosen since most of the available NIR instrumentation in this range can be affordable for growers compared to the instrument used in this study which can cost around 45 times more. The 982 nm overtone is associated with the OH overtone band and 1100 for the CH bands, which corresponds to alcohol and phenolic compounds [25].

The model reported using machine learning fitting algorithms can be of great assistance to growers and winemakers to obtain chemometry data in real time using the proposed methodology shown in Figure 9. Currently, growers do not have sophisticated tools to assess potential smoke contamination of berries bunches and wines. The only option available is collection of samples within a vineyard for compositional analysis by an accredited laboratory using GC-MS or HPLC-MS/MS. This process is destructive, expensive, and takes a long time, which makes it less ideal for the implementation of mitigation strategies and/or decision making before harvest. Furthermore, it may minimize smoke taint by the information provided through a spatial assessment of the contamination either through canopies or berries for informed decision making regarding palliative measures (as presented in this paper) or differential harvest.

The models developed in this study were able to predict smoke contamination in canopies, berries and wines, regardless of the cultivar. Hence, the models could be applied as a universal methodology. Further studies and data acquired could be added to models to include more cultivars. However, the seven cultivars included in this study were some of the most commercially important in Australia. Finally, it is important to note that the levels of smoke-taint compounds present in wine are in part related to the winemaking process (i.e., duration of skin contact time during fermentation), hence this model will need to be adjusted for different winemaking techniques, which can influence the extraction of smoke-related compounds from the berry.

5. Conclusions

This paper showed two main advancements for tools to detect smoke contamination in grapevine canopies and smoke-related compounds in berries and wine using remote sensing techniques. This study is the first to apply machine learning modeling techniques to assist growers confronted with vineyard exposure to smoke from bushfires, an issue which has been exacerbated in prominent wine regions around the world due to climate change. Furthermore, this paper has proposed an affordable method to implement these novel techniques using smartphones, portable thermal imagery and NIR spectroscopy devices. More research is required to assess the usage of these affordable devices in the future using the models proposed.

Author Contributions: S.F. conceived the machine learning modeling idea and practical applications; S.F., E.J.T. and C.G.V. analyzed the data and created the machine learning models; K.W., S.T. and S.F. were awarded funding for the study; K.W. and R.R. performed field trials, laboratory analysis and winemaking; S.F. and R.D.B. acquired the physiological and NIR data. All authors contributed to the writing of the paper.

Funding: This research received no external funding.

Acknowledgments: This research was supported under the Australian Research Council's Linkage Projects funding scheme (LP0989138); the financial contributions of industry partners are also gratefully acknowledged. The machine learning modeling research was supported by the Digital Viticulture program funded by the University of Melbourne's Networked Society Institute, Australia.

Conflicts of Interest: The authors declare no conflict of interest.

References

1. Hughes, L.; Alexander, D. Climate Change and the Victoria Bushfire Threat: Update 2017. Climate Council Report. 2017. Available online: http://www.climatecouncil.org.au/uploads/98c26db6af45080a32377f3ef4800102.pdf (accessed on 10 July 2019).
2. Webb, L.; Whetton, P.; Bhend, J.; Darbyshire, R.; Briggs, P.; Barlow, E. Earlier wine-grape ripening driven by climatic warming and drying and management practices. *Nat. Clim. Chang.* **2012**, *2*, 259–264. [CrossRef]
3. Webb, L.B. Climate change and winegrape quality in Australia. *Clim. Res.* **2008**, *36*, 99–111. [CrossRef]
4. Webb, L.B.; Whetton, P.H.; Barlow, E.W.R. Modelled impact of future climate change on the phenology of winegrapes in Australia. *Aust. J. Grape Wine Res.* **2007**, *13*, 165–175. [CrossRef]
5. Su, B.; Xue, J.; Xie, C.; Fang, Y.; Song, Y.; Fuentes, S. Digital surface model applied to unmanned aerial vehicle based photogrammetry to assess potential biotic or abiotic effects on grapevine canopies. *Int. J. Agric. Biol. Eng.* **2016**, *9*, 119–130.
6. Ristic, R.; Fudge, A.L.; Pinchbeck, K.A.; De Bei, R.; Fuentes, S.; Hayasaka, Y.; Tyerman, S.D.; Wilkinson, K.L. Impact of grapevine exposure to smoke on vine physiology and the composition and sensory properties of wine. *Theor. Exp. Plant Physiol.* **2016**, *28*, 67–83. [CrossRef]
7. Ristic, R.; Pinchbeck, K.; Fudge, A.; Hayasaka, Y.; Wilkinson, K. Effect of leaf removal and grapevine smoke exposure on colour, chemical composition and sensory properties of Chardonnay wines. *Aust. J. Grape Wine Res.* **2013**, *19*, 230–237. [CrossRef]
8. van der Hulst, L.; Munguia, P.; Culbert, J.A.; Ford, C.M.; Burton, R.A.; Wilkinson, K.L. Accumulation of volatile phenol glycoconjugates in grapes following grapevine exposure to smoke and potential mitigation of smoke taint by foliar application of kaolin. *Planta* **2019**, *249*, 941–952. [CrossRef]
9. Fudge, A.; Ristic, R.; Wollan, D.; Wilkinson, K.L. Amelioration of smoke taint in wine by reverse osmosis and solid phase adsorption. *Aust. J. Grape Wine Res.* **2011**, *17*, S41–S48. [CrossRef]
10. Fudge, A.; Schiettecatte, M.; Ristic, R.; Hayasaka, Y.; Wilkinson, K.L. Amelioration of smoke taint in wine by treatment with commercial fining agents. *Aust. J. Grape Wine Res.* **2012**, *18*, 302–307. [CrossRef]
11. Wilkinson, K.; Ristic, R.; Pinchbeck, K.; Fudge, A.; Singh, D.; Pitt, K.; Downey, M.; Baldock, G.; Hayasaka, Y.; Parker, M. Comparison of methods for the analysis of smoke related phenols and their conjugates in grapes and wine. *Aust. J. Grape Wine Res.* **2011**, *17*, S22–S28. [CrossRef]
12. Dungey, K.A.; Hayasaka, Y.; Wilkinson, K.L. Quantitative analysis of glycoconjugate precursors of guaiacol in smoke-affected grapes using liquid chromatography–tandem mass spectrometry based stable isotope dilution analysis. *Food Chem.* **2011**, *126*, 801–806. [CrossRef]
13. Hayasaka, Y.; Parker, M.; Baldock, G.A.; Pardon, K.H.; Black, C.A.; Jeffery, D.W.; Herderich, M.J. Assessing the impact of smoke exposure in grapes: Development and validation of a HPLC-MS/MS method for the quantitative analysis of smoke-derived phenolic glycosides in grapes and wine. *J. Agric. Food Chem.* **2012**, *61*, 25–33. [CrossRef] [PubMed]
14. Fudge, A.L.; Wilkinson, K.L.; Ristic, R.; Cozzolino, D. Classification of smoke tainted wines using mid-infrared spectroscopy and chemometrics. *J. Agric. Food Chem.* **2011**, *60*, 52–59. [CrossRef] [PubMed]
15. Kennison, K.R.; Gibberd, M.R.; Pollnitz, A.P.; Wilkinson, K.L. Smoke-derived taint in wine: The release of smoke-derived volatile phenols during fermentation of Merlot juice following grapevine exposure to smoke. *J. Agric. Food Chem.* **2008**, *56*, 7379–7383. [CrossRef] [PubMed]
16. Fuentes, S.; De Bei, R.; Pech, J.; Tyerman, S. Computational water stress indices obtained from thermal image analysis of grapevine canopies. *Irrig. Sci.* **2012**, *30*, 523–536. [CrossRef]
17. Jones, H.G.; Stoll, M.; Santos, T.; Sousa, C.D.; Chaves, M.M.; Grant, O.M. Use of infrared thermography for monitoring stomatal closure in the field: Application to grapevine. *J. Exp. Bot.* **2002**, *53*, 2249–2260. [CrossRef] [PubMed]
18. Moran, M.S.; Inoue, Y.; Barnes, E. Opportunities and limitations for image-based remote sensing in precision crop management. *Remote Sens. Environ.* **1997**, *61*, 319–346. [CrossRef]
19. Jones, H.G. Use of infrared thermometry for estimation of stomatal conductance as a possible aid to irrigation scheduling. *Agric. For. Meteorol.* **1999**, *95*, 139–149. [CrossRef]
20. Beale, M.H.; Hagan, M.T.; Demuth, H.B. *Deep Learning Toolbox User's Guide*; The Mathworks Inc.: Herborn, MA, USA, 2018.

21. Gonzalez Viejo, C.; Torrico, D.D.; Dunshea, F.R.; Fuentes, S. Development of Artificial Neural Network Models to Assess Beer Acceptability Based on Sensory Properties Using a Robotic Pourer: A Comparative Model Approach to Achieve an Artificial Intelligence System. *Beverages* **2019**, *5*, 33. [CrossRef]

22. Craparo, A.; Steppe, K.; Van Asten, P.J.; Läderach, P.; Jassogne, L.T.; Grab, S. Application of thermography for monitoring stomatal conductance of Coffea arabica under different shading systems. *Sci. Total Environ.* **2017**, *609*, 755–763. [CrossRef]

23. Egea, G.; Padilla-Díaz, C.M.; Martinez-Guanter, J.; Fernández, J.E.; Pérez-Ruiz, M. Assessing a crop water stress index derived from aerial thermal imaging and infrared thermometry in super-high density olive orchards. *Agric. Water Manag.* **2017**, *187*, 210–221. [CrossRef]

24. Fuentes, S.; Hernández-Montes, E.; Escalona, J.; Bota, J.; Viejo, C.G.; Poblete-Echeverría, C.; Tongson, E.; Medrano, H. Automated grapevine cultivar classification based on machine learning using leaf morpho-colorimetry, fractal dimension and near-infrared spectroscopy parameters. *Comput. Electron. Agric.* **2018**, *151*, 311–318. [CrossRef]

25. Wang, X.; Bao, Y.; Liu, G.; Li, G.; Lin, L. Study on the best analysis spectral section of NIR to detect alcohol concentration based on SiPLS. *Procedia Eng.* **2012**, *29*, 2285–2290. [CrossRef]

Article

Vibration Monitoring of the Mechanical Harvesting of Citrus to Improve Fruit Detachment Efficiency

Sergio Castro-Garcia [1,*], Fernando Aragon-Rodriguez [1], Rafael R. Sola-Guirado [1], Antonio J. Serrano [2], Emilio Soria-Olivas [2] and Jesús A. Gil-Ribes [1]

[1] Department of Rural Engineering, Universidad de Cordoba, 14004 Cordoba, Spain; g92arrof@uco.es (F.A.-R.); ir2sogur@uco.es (R.R.S.-G.); gilribes@uco.es (J.A.G.-R.)

[2] IDAL, Intelligent Data Analysis Laboratory, Universidad de Valencia, 46100 Valencia, Spain; ajserran@uv.es (A.J.S.); emilio.soria@uv.es (E.S.-O.)

* Correspondence: scastro@uco.es; Tel.: +34-957-218-548

Received: 13 March 2019; Accepted: 9 April 2019; Published: 12 April 2019

Abstract: The introduction of a mechanical harvesting process for oranges can contribute to enhancing farm profitability and reducing labour dependency. The objective of this work is to determine the spread of the vibration in citrus tree canopies to establish recommendations to reach high values of fruit detachment efficiency and eliminate the need for subsequent hand-harvesting processes. Field tests were carried out with a lateral tractor-drawn canopy shaker on four commercial plots of sweet oranges. Canopy vibration during the harvesting process was measured with a set of triaxial accelerometer sensors with a datalogger placed on 90 bearing branches. Monitoring of the vibration process, fruit production, and branch properties were analysed. The improvement of fruit detachment efficiency was possible if both the hedge tree and the machinery were mutually adjusted. The hedge should be trained to facilitate access of the rods and to encourage external fructification since the internal canopy branches showed 43% of the acceleration vibration level of the external branches. The machine should be adjusted to vibrate the branches at a vibration time of at least 5.8 s, after the interaction of the rod with the branch, together with a root mean square acceleration value of 23.9 m/s^2 to a complete process of fruit detachment.

Keywords: *Citrus sinensis* L. Osbeck; mechanical harvesting; acceleration sensor; vibration time; logistic regression

---

1. Introduction

Citrus fruits, whether for fresh consumption or industrial processing, are mainly harvested by hand. Worldwide, 147 million tonnes of citrus were produced in 2017, including orange, grapefruit, lemon, mandarin, and other citrus fruits [1]. Spain is the sixth largest producer of citrus fruit in the world, with an approximate production in 2017–2018 of more than seven million tonnes. In Spain, the predominant citrus orchards are trained for manual harvesting, with an orientation towards the fresh market. Manually harvested orchards experience problems due to the availability of labour and the high cost of operation.

Within the citrus production process, harvesting is a phase of enormous economic importance due to its high impact on the final cost of production. The manual harvesting in Southern Spain requires an average of 95 days' work per hectare and represents between 25% and 35% of the final cost of production [2]. Roka and Hyman [3] stated that, under Florida conditions, the application of mechanical harvesting for industrial processing could provide a 50% cost reduction, while increasing labour productivity by ten per cent. With the current approach to citrus production, the high costs of manual harvesting could compromise the profitability of the activity and the future of plantations in the long term [4].

Since the 1970s, the development of mechanical citrus harvesting systems for the juice industry has mainly taken place in Florida. However, none of the mechanised systems have been able to match the flexibility and fruit selection capabilities of manual harvesting [5]. The foremost mechanical harvesting systems are the trunk and canopy shaker systems that were applied to and developed for citrus fruits, and which reach high values of harvesting efficiency ranging between 84–95% and 55–95%, respectively [6]. However, the detachment of immature fruitlets has been identified in both systems; this occurs particularly in the late varieties that are of special interest to the juice industry and, in addition to a reduction in the working capacity (ha/h), it could represent an obstacle to the adoption of these mechanical harvesting systems by farmers. Roka et al. [7] showed yield reduction values for the use of these machines compared to manual harvesting of 20–50% according to their use and regulation. In parallel, the development and testing of abscission agents have permitted an increase in the working capacity and an improvement in the detachment of mature fruit with these harvesting technologies [8]. It was shown that a moderate reduction in fruit detachment force, through the application of an abscission agent, was enough to significantly increase the harvesting efficiency [9]. Subsequently, it was demonstrated that the use of an abscission agent together with an adjustment of the vibration parameters, fundamentally time and frequency of vibration, allowed high percentages of mature fruit detachment to be achieved without significantly affecting the following season [10]. This result was confirmed based on the different frequency responses of mature fruit and immature fruitlets to mechanical harvesting [11].

Canopy shaker systems allow a continuous vibration of the tree row; the rods penetrate the canopy and achieve a high value of fruit detached in areas where there is direct contact of rods with branches. The use of a canopy shaker can generate a greater fall of leaves, shoots, and branches than manual harvesting. The fall of these organs is considered as tree damage, which could have negative implications in the yield, the productive life, and the cost of transport from the orchard to the industry [12]. For this reason, the improvement of these machines has been based on a dual objective; to increase the efficiency of mature fruit removal and to reduce the damage caused to trees. In order to improve the mechanical harvesting process, Savary et al. [13] developed a canopy shaker simulation based on finite element methods to predict and evaluate the interaction between the tree and the machine. Then, Savary et al. [14] evaluated the effect of vibration on the tree canopy according to the distribution of forces and accelerations in branches and fruits. Subsequently, machine improvement proposals were based on mathematical models and prototype tests with a combination of machine operating parameters such as frequency and amplitude of vibration [15,16], and the configuration and material of the rods [17]. The rods have been shown to play an essential role in the process of tree shaking, both in terms of fruit detachment and the possible generation of tree and fruit damage. Liu et al. [17] indicated that rod material affected the vibratory response of trees with respect to the acceleration peaks in branches. They recommended that the rods have high stiffness values, but that their surface should be smooth to reduce damage. The shape of the rods revealed that arc-shaped flexible rods had better performance with respect to fruit detachment efficiency and a lower rate of damage to trees than the free end rods [18]. In a further attempt to adapt the vibration process to tree requirements, Pu et al. [18] designed and tested a two-section independent shaker system in order to minimise tree damage and maximise harvesting efficiency. The adaptation of the machine to the variability of the tree canopy made it possible to reduce tree damage compared to other canopy shaker systems.

During the harvesting process, the machine rods penetrate the tree but do not usually reach all parts of the canopy, so vibration must be transmitted via the branches in order to detach the inner fruit. Whitney et al. [19] indicated that a greater fruit detachment efficiency value was obtained by harvesting smaller rather than larger canopy trees. Canopy shaker systems generally have a limited capacity to detach internal fruits from the tree and yet can remove almost all external fruits. The harvesting process of fruit inside the canopy may cause damage to the outermost parts due to a reduction in machine ground speed or an increment in vibration frequency value and may also require additional manual harvesting.

This study works on the hypothesis that mechanised harvesting processes can be carried out with a high value of harvesting efficiency, where subsequent hand-harvesting is not necessary, and that the possible, but moderate tree damage does not affect the productive life of the orchard. The objective of this work is to analyse the interaction of the canopy shaker rods with the tree branches from the data collected continuously during the mechanical harvesting process of citrus trees. First, the fruit-bearing branches with and without direct contact with the machine rods were characterised in the position and production. Then, the canopy vibration process during mechanical harvesting was monitored. Finally, the vibration parameters necessary to detach the fruit were determined, and recommendations were given on the harvest parameters necessary to increase the fruit removal efficiency.

2. Materials and Methods

Mechanical harvest tests were carried out in Cordoba (Spain) in 2017 during the sweet orange (*Citrus sinensis* L. Osbeck cv. Valencia) harvest season for juice production, during four weeks from the end of flowering to before the natural fall of immature fruitlets in June. Table 1 shows the main characteristics of the four, mechanically harvested citrus orchards. Trees had been planted in wide hedges over 0.4 m ridges and had wide row distances to allow machine manoeuvrability and the use of canopy shaker harvesters (Figure 1).

Table 1. Characteristics of citrus orchards mechanically harvested with the canopy shaker system.

	Plot 1	Plot 2	Plot 3	Plot 4
Date planted	2006	2005	2007	2005
Plot area (ha)	54.7	38.0	33.1	57.3
Trees per ha	440	330	440	330
Tree distance (m)	7×3	7×4	7×3	7×4
Hedge height (m)	4.0	4.0	4.3	4.4
Hedge width (m)	3.9	4.1	4.5	4.6

Figure 1. Example of a tree row trained in width hedge for mechanised harvesting.

Mechanical harvesting was carried out with a lateral tractor-drawn continuous canopy shaker system (Oxbo 3210, Byron, New York, NY, USA), working under regular conditions, with a ground speed range between 1 and 1.5 km/h (0.28–0.42 m/s), and a vibration frequency close to 4.5 Hz, which caused the fruit to fall to the ground. Harvesting tests were carried out to ensure close contact of the shaker system with the tree canopy (Figure 2). The machine harvested both sides of the hedge in independent passes, with an approximate working capacity of 0.4–0.5 ha/h. Subsequently, the fruit remaining in the canopy was hand-picked and collected together with the fruit from the ground and loaded into a container.

Figure 2. Lateral tractor-drawn continuous canopy shaker system (Oxbo, 3210) used in citrus harvesting tests. (**a**) front view before the harvesting process; (**b**) rear view after the harvesting process.

Interaction of the harvesting system with the tree canopy was studied for each side of the tree hedges (Figure 3). Because the tree hedges differed between orchards, a representative cross-section of the tree canopy was selected. The cross-section ranged from 4–5 m², and was composed of a distance from the line of the trunks to the outside of the canopy of up to 2.0–2.5 m, and of a height ranging from 0.5–2.5 m. This cross-section was selected because it differentiated an external area of the canopy with direct contact of the machine rods with the branches, for a rod length of 1.4 m, and an internal area of the canopy without direct contact with the rods. Furthermore, the cross-section was representative of the tree canopy, with high yield, avoiding the effect of the lower pendulous branches and allowing analysis of the vibration process in a homogeneous canopy area between the tested plots. The cross-section was divided into 16–20 sectors according to the width of the hedge, at intervals of 0.5 m both horizontally and vertically. In each sector, the values of vibration, the properties of the branches, and the number of fruits were recorded. The canopy area that had direct contact with the rods was sampled with 10 sectors, while the canopy area without direct contact was sampled with 6 or 10 sectors.

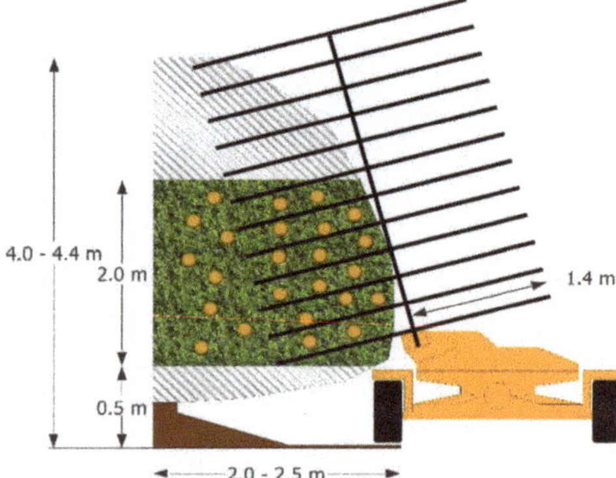

Figure 3. Cross-section of the tree hedge and canopy areas with and without direct contact with the canopy shaker system rods.

Before the mechanical harvesting process, a total of 90 fruit-bearing branches were selected that had mature fruit and were distributed in different sectors. The statistical design established a stratified random sampling, each cluster was a tested plot, and in each plot, 18–24 fruit-bearing branches were randomly selected. The sample guaranteed at least three measurements in each sector. Each branch was assigned a position value in the cross-section at a point close to the fruit that was able to support an acceleration sensor, but which had a diameter less than 10 mm. The fruit detachment ratio was determined by the number of fruits removed from each branch before and after the harvesting process.

Branch vibration measurements were recorded with a triaxial MEMS accelerometer sensor (Gulf Coast Data Concepts LLC X200-4, Waveland, MS, USA) with a measurement range of ±2000 m/s^2, a 16-bit resolution, a sensitivity of 0.06 m/s^2, and a sampling frequency of 400 Hz. Figure 4 shows the placement of the sensor on the fruit-bearing branch.

Figure 4. Location of the acceleration sensor on a fruit-bearing branch on the outermost part of the canopy.

Analysis of the acceleration signals in the time domain and statistical analysis was performed using the R open software (R Core Team, 2016) and in the frequency domain using the NV Gate v8.0 software, with a fast Fourier transformation with 401 lines in a frequency range of 0–156.2 Hz. In the time domain, the resultant acceleration value (A_r) was determined as the module of the vector sum of the three measurement axes in each sensor. Figure 5 shows a sample of A_r in a fruit-bearing branch.

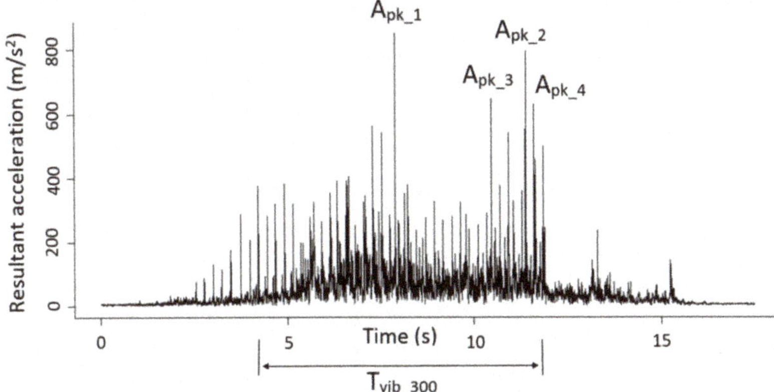

Figure 5. Example of the resultant acceleration (A_r) in the time domain measured in a fruit-bearing branch. T_{vib_300}—time elapsing between the first and the last event with an A_r value of 300 m/s^2; A_{pk_n}— n maximum peak value of A_r.

In the time domain, the vibration variables studied were:

- Vibration time (T_{vib}): time (s) elapsing between the first and last value of A_r measured on the branch, ranging from A_r values of 20 m/s^2 up to 600 m/s^2.
- Mean peak acceleration (A_{pk}): the mean value of the 10 maximum peak values of A_r (m/s^2) for T_{vib_20}.

In the frequency domain, the vibration variables studied were:

- Frequency: number of cycles per second (Hz) of rod movement in the canopy.
- RMS acceleration (A_{RMS}): vector sum of the Root Mean Square values (RMS) of each accelerometer axis at the vibration frequency.

The statistical analysis focused on predicting the fruit detachment ratio according to the vibration variables and the branch measurements as the predicted parameters. Logistic regression was used with a K = 2-fold cross-validation method.

3. Results and Discussion

Most of the fruit (72.7%) was located in the canopy area that had direct contact with the rods, in the height range of 1–2 m from the ground, and in a range of 0.5–2 m from the trunk. Gupta et al. (2015) stated that the area with the highest fructification is in the primary branches of the intermediate zone of the canopy at a height of 1.14–2.29 m, and at a distance of 0.78–0.83 m from the canopy exterior. However, the distribution of fruit in the canopy also depends on planting distances and tree height. A reduced distance between trees can generate a greater percentage of fruit in the upper canopy parts but can reduce the number of fruits inside [19].

The diameter of the branch at the vibration measurement point was 7.9 ± 2.4 mm (mean ± sd), with a variation that ranged from 10.2 mm for the branches closest to the trunk and the ground to 5.53 mm for the outermost and tallest branches of the canopy. Each tested branch carried a mean value of 3.7 ± 1.8 fruits. The results showed a high variability in the distribution of fruit in the canopy and in the morphology of the branches. This variability is important for the outcome of mechanical harvesting systems and was considered by Gupta et al. [15,16] for modelling the tree and simulating the harvesting process in order to improve the canopy shaker system. However, the current harvesting systems based on canopy shakers do not contemplate the variability of branches and fruit within the tree canopy. In an attempt to improve the machine adaptation to the tree, Pu et al. [18] designed and tested a canopy shaker system capable of applying different vibration parameters to the upper and lower parts of the tree. These authors showed the need to use different harvesting parameters and were able to achieve a high fruit detachment ratio (82.6%) with low tree damage.

Canopy shaker systems continuously harvest fruit as they move along the row of trees. Before the machine comes into contact with a branch, the branch may vibrate due to contact with other branches or due to the transmission of vibration from the trunk. Then, the branch vibrates due to contact with the machine rods, and finally, the branch vibrates freely when the machine has passed. Table 2 shows the results of this vibration process measured in branches, both in the canopy area with and without direct contact with the machine rods. In order to define the beginning and end of the vibration process, the acceleration values produced only by natural sources, mainly by wind and gravity, and without machine interaction were recorded. The vibration time where the branch was excited by the machine was defined as the time elapsing between resultant acceleration values greater than 18 m/s^2 (T_{vib_18}). The average vibration time (T_{vib_18}) of the branches was 14.3 ± 2.8 s. No significant differences were found (Student's t, $p > 0.05$) between the vibration time of the branches located in the canopy area with or without direct contact with rods. This indicated that all branches vibrated at the same time, but not all at the same level of acceleration. The length of the machine's shaking system, i.e., the length of the rods and drums, together with the machine ground speed, determined the vibration time during which the rods had direct contact with the branches, whereas tree-training and canopy density could define the vibration time during which the branch vibrated without direct contact with the rods. For the field test conditions, i.e., a rod length of 1.4 m and ground speed ranging from 1–1.5 km/h (0.28–0.42 m/s), the canopy shaker systems were able to maintain direct contact with the branches within the range of 7.7–11.5 s. This indicated that between 20% and 46% of the vibration time may correspond to the transmission of vibration before and after the passage of the machine.

Table 2. Vibration parameters measured on branches with and without direct contact with the rods during mechanical harvesting with the canopy shaker systems.

	Branches with Direct Contact with Rods	Branches without Direct Contact with Rods	Mean Value
Vibration time (s)	14.8 ± 2.8 [a]	13.8 ± 2.9 [a]	14.3 ± 2.8
Frequency (Hz)	4.1 ± 0.2 [a]	4.0 ± 0.3 [a]	4.1 ± 0.5
A_{RMS} (m/s^2)	29.6 ± 10.2 [a]	12.8 ± 6.4 [b]	26.5 ± 13.6
Acceleration peak (m/s^2)	616.7 ± 283.3 [a]	268.1 ± 164.6 [b]	495.1 ± 270.9
Fruit detachment ratio (%)	84.7 ± 30.5 [a]	25.1 ± 22.2 [b]	69.1 ± 40.7

Values shown are mean ± standard deviation, n = 90. The same superscript letters in the same row are not significantly different (Student's *t*, $p < 0.05$; Wilcoxon–Mann–Whitney test, $p < 0.05$).

During the harvesting process, the branches in contact with the rods experienced a process of forced vibration. The rods, powered with an alternating and rotating movement, penetrated the canopy, producing an impulse excitation in the branches of the fruit of the outermost part of the canopy [20]. The branches showed a mean vibration frequency value of 4.1 ± 0.5 Hz. The value of the vibration frequency did not correlate with the position of the branch in the canopy (Pearson = 0.135, $p > 0.05$). The frequency value used was within the range recommended for citrus detachment with canopy shakers. Liu et al. [17] established that the value of 5 Hz was appropriate for the detachment of fruit without increasing the damage caused to the tree. Similarly, it was demonstrated that a frequency value of 4.8 Hz was adequate to produce a high acceleration in the branches when using rigid rods [18]. The frequency of the vibration in combination with the design of the rods used and the state of the immature fruits [10] play an important role in the damage caused to the trees. Although tree damage is an important element in the process of improving mechanised harvesting, it has not been considered in this study because the damage that occurred was minor and similar to that of previous years, mechanically harvested plantations had not shown any problems in tree development or yield compared to previous seasons.

The values of A_{RMS} measured in the branches was positively related to the fruit detachment ratio. The branch vibration was characterised by an A_{RMS} value of 26.5 ± 13.6 m/s^2 for the machine vibration frequency. However, there was a significant variation in the mean A_{RMS} values in branches depending on their position in the canopy (Figure 6). Branches with direct contact with rods showed a significantly higher mean A_{RMS} value (29.6 ± 10.2 m/s^2) (Student *t*, $p < 0.05$; Wilcoxon–Mann–Whitney, $p < 0.05$) than branches without direct contact (12.8 ± 6.4 m/s^2). Pu et al. (2018) showed that the highest values of acceleration in the branches (31.4 m/s^2) were provided by contact with the machine rods and these branches reached the highest values of fruit removal efficiency.

During the harvesting process, the acceleration produced in the branch was transmitted to the fruits for their detachment. The internal branches, without direct contact with the rods, showed 43% of the A_{RMS} vibration level of the external branches. A similar result was achieved by Liu et al. [17], whose results showed a reduction of the acceleration in the inner branches of the canopy of 42% with respect to the outer branches. In field trials with measurements inside the fruit before its detachment with canopy shaker systems, Castro–Garcia et al. [20] recorded mean A_{RMS} values between 38.8 and 51.3 m/s^2. These values indicated that there was an amplification of the acceleration values from the branch to the fruit. Savary et al. [13] evaluated the acceleration produced in the branches and indicated that the resultant acceleration values were higher at the ends of the branches, especially on the thinnest and outermost branches. These same authors pointed out that the trunk region of unbranched trees had very low acceleration values, whereas the acceleration values began to be noticeably higher from the first branch of the trunk region.

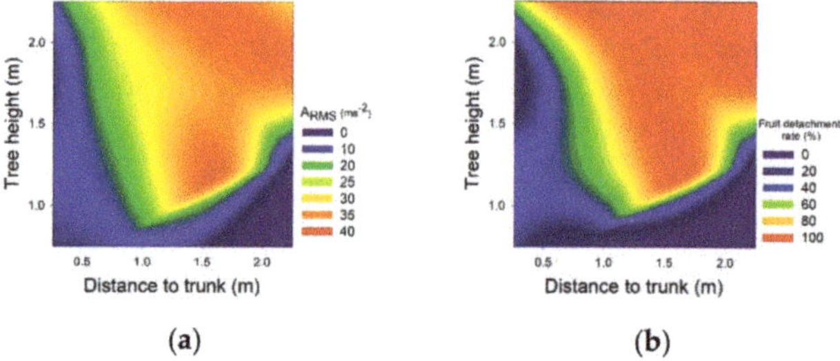

Figure 6. (**a**) Distribution of the A_{RMS} values (m/s^2); and (**b**) fruit detachment rate (%) produced with a canopy shaker system and measured in fruit-bearing branches in the cross-section of the tree canopy.

The interaction of the rods with the branches was characterizsed by a succession of impacts with a high acceleration value in accordance with the vibration frequency of the machine. These impacts presented a mean A_{pk} value of 495.1 ± 270.9 m/s^2. Similarly to Figure 6, the A_{pk} values in branches with direct contact with rods (616.7 ± 283.3 m/s^2) were higher (Student's t, $p < 0.05$) than branches without direct contact (268.1 ± 164.6 m/s^2). The A_{pk} and A_{RMS} values showed a positive linear correlation (Pearson = 0.70, $p < 0.05$) in the tree canopy. In both cases, the direct contact of the rod represented an increment of 2.3 times the acceleration values reached in the branch.

The canopy shaking system achieved a mean fruit detachment value of 69.1 ± 40.7%. However, this variable presented high variability within the cross-section of the tree hedge (Figure 6). As expected, this variable reached its highest values in the branches with direct contact with rods. The fruit detachment ratio was reduced from an average value of 84.7 ± 30.5% in branches with direct contact with rods to 25.1 ± 22.2% for branches without direct contact. The fruit detachment value of 100% was reached in all branches located between 2 and 2.5 m in height. A similar result was reported by Whitney et al. [6] who found that working with small canopies that were accessible to rods achieved a fruit detachment of 96%, while wider canopies achieved reduced values of 55%. Savary et al. [14] reached a fruit detachment value of 88% on the outside of the canopy, while on the inside this figure was reduced to 57%. However, in order to improve the harvesting efficiency, it is not only necessary for the rod to penetrate the canopy, but also for it to interact with the branch. Liu et al. [17], analysing fruit detachment according to the point of contact of the rod with the branch, determined that the operation was more effective when the rod impacted at 30% of the distance to the free end of the branch.

The results obtained from the cross-section of the tree canopy have shown a high variability, both in vibration and in fruit values. Reducing this variability and improving the mechanised harvesting process requires knowledge of the requirements to detach fruit from the tree. The measurements in the canopy showed a high linear correlation between variables, which indicated that fruit detachment prediction could not have a single solution. Analysis of the data focused on the discretisation of quantitative variables that could discriminate whether there was a fruit detachment with a value of 100% and reasonable success. Due to its simplicity and efficiency, logistic regression was used. A_{RMS} and time elapsed between an acceleration greater than 300 m/s^2 (T_{vib_300}) were significant variables to discriminate the events of 100% fruit detachment. The result was defined as a straight line that separates the conditions at which a fruit detachment of 100% was obtained with a precision measured as the area under the ROC curve of 0.95 in the validation set. Equation (1) shows the values obtained and Figure 7 is the graphical representation.

$$ln\left(\frac{\text{Prob of complete fruit detachment}}{1 - \text{Prob of complete fruit detachment}}\right) = 7.13417 - 0.52754\,T_{vib_{300}} - 0.17206\,A_{RMS} \qquad (1)$$

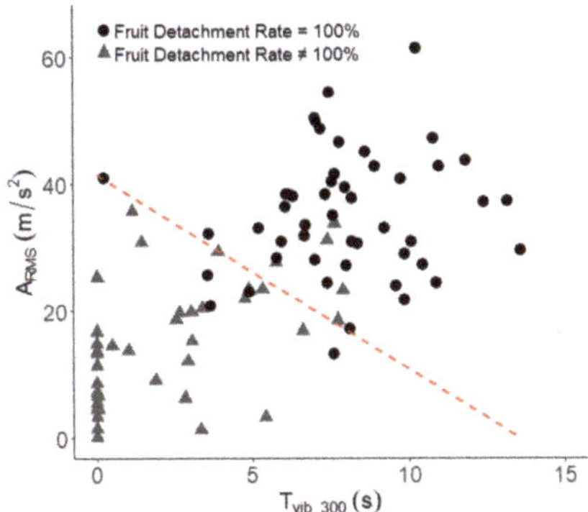

Figure 7. Distribution of the fruit detachment rate values according to A_{RMS} (m/s^2) and T_{vib_300} (s).

Currently, canopy shaker systems for citrus harvesting can employ various types of vibration systems, with variations in the frequency or amplitude of movement, different machine ground speeds or rods with different designs, or mechanical properties. In all cases, the machine produces a forced vibration of the branches with the aim of detaching fruit and avoiding major damage to the tree or fruit. Monitoring of the forced vibration process of the tree canopy showed that it was possible to achieve a 100% fruit detachment ratio based on a combination of acceleration levels and vibration times in branches (Figure 7). Under the conditions of the field tests performed, we propose a combination of a vibration time of at least 5.8 s, after the interaction of the rod with the branch (T_{vib_300}), together with an A_{RMS} value of 23.9 m/s^2. With these harvest parameters, a complete process of fruit detachment was achieved in 88.9% of the branches tested. Although these values could be modified if another type of tree formation, citrus variety, or harvesting machine was considered, the fruit detachment process can be estimated by vibration time and branch acceleration. Both parameters are of great importance, not only for training trees to facilitate the rod penetration in the canopy but also for the design of new canopy shaker systems.

4. Conclusions

The monitoring of the vibration process in the tree during mechanised harvesting with the canopy shaker showed a great variability in results depending on different parts of the canopy. Branches that had direct contact with the machine rods showed a higher mean value of fruit detachment ratio (84.7%) than non-contact branches (25.1%). Vibration transmission from the external branches to the internal branches in the canopy was not effective to remove internal canopy fruit. During the harvesting process, values of 100% fruit detachment ratio could be achieved with a combination of harvest parameters in the branch. Achieving a complete process of fruit detachment is possible if both the tree canopy and machinery are mutually adjusted to facilitate the contact of the shaking system and the necessary vibration time.

Author Contributions: All the authors made significant contributions to the manuscript. S.C.G., F.A.-R., and R.R.S.-G. designed the experiment and wrote the manuscript; S.C.G. and F.A.-R. performed the data acquisition; F.A.-R. and R.R.S.-G. performed the field component; E.S.-O. and A.J.S. performed the statistical analysis; J.A.G.-R. and A.J.S. scientifically supported and reviewed the paper; and S.C.G and E.S.-O. managed the project.

Funding: This research was funded by National Institute for Agricultural and Food Research and Technology (INIA, Spain) financed by FEDER funds, from research projects RTA2014-00025-C05-03 and RTA2014-00025-C05-05.

Conflicts of Interest: The authors declare no conflict of interest. The funders had no role in the design of the study, in the collection, analyses, or interpretation of data, in the writing of the manuscript, or in the decision to publish the results.

References

1. FAOSTAT. Food and Agriculture Data. Available online: http://www.fao.org/faostat/en/#data/QC (accessed on 25 March 2019).
2. Junta de Andalucía. Average Production Costs. 2016–17 Season. Available online: http://www.juntadeandalucia.es/agriculturaypesca/observatorio (accessed on 25 March 2019).
3. Roka, F.M.; Hyman, B.R. Mechanical harvesting of sweet oranges for juice processing. *Acta Hortic.* **2012**, *965*, 241–244. [CrossRef]
4. Brotons-Martínez, J.M.; Martin-Gorriz, B.; Torregrosa, A.; Porras, I. Economic evaluation of mechanical harvesting of lemons. *Outlook Agric.* **2018**, *47*, 44–50. [CrossRef]
5. Sanders, K.F. Orange harvesting systems review. *Biosyst. Eng.* **2005**, *90*, 115–125. [CrossRef]
6. Whitney, J.D. Field test results with mechanical harvesting equipment in Florida oranges. *Appl. Eng. Agric.* **1999**, *15*, 205–210. [CrossRef]
7. Roka, F.M.; Burns, J.K.; Buker, R.S. Mechanical harvesting without abscission agents-Yield impacts on late season 'Valencia' oranges. *Proc. Fla. State Hortic. Soc.* **2005**, *118*, 25–27.
8. Burns, J.K.; Buker, R.S.; Roka, F.M. Mechanical harvesting capacity in sweet orange is increased with an abscission agent. *HortTechnology* **2005**, *15*, 758–765. [CrossRef]
9. Hartmond, U.; Whitney, J.D.; Burns, J.K.; Kender, W.J. Seasonal variation in the response of 'Valencia' orange to two abscission compounds. *HortScience* **2000**, *35*, 226–229. [CrossRef]
10. Burns, J.K.; Roka, F.M.; Li, K.T.; Pozo, L.; Buker, R.S. Late-season 'Valencia' orange mechanical harvesting with an abscission agent and low-frequency harvesting. *HortScience* **2006**, *41*, 660–663. [CrossRef]
11. Castro-Garcia, S.; Blanco-Roldán, G.L.; Ferguson, L.; González-Sánchez, E.J.; Gil-Ribes, J.A. Frequency response of late-season 'Valencia' orange to selective harvesting by vibration for juice industry. *Biosyst. Eng.* **2017**, *155*, 77–83. [CrossRef]
12. Spann, T.M.; Danyluk, M.D. Mechanical harvesting increases leaf and stem debris in loads of mechanically harvested citrus fruit. *HortScience* **2010**, *45*, 1297–1300. [CrossRef]
13. Savary, S.K.J.U.; Ehsani, R.; Schueller, J.K.; Rajaraman, B.P. Simulation study of citrus tree canopy motion during harvesting using a canopy shaker. *Trans. ASABE* **2010**, *53*, 1373–1381. [CrossRef]
14. Savary, S.K.J.U.; Ehsani, R.; Salyani, M.; Hebel, M.A.; Bora, G.C. Study of force distribution in the citrus tree canopy during harvest using a continuous canopy shaker. *Comput. Electron. Agric.* **2011**, *76*, 51–58. [CrossRef]
15. Gupta, S.K.; Ehsani, R.; Kim, N.H. Optimization of a citrus canopy shaker harvesting system: Properties and modeling of tree limbs. *Trans. ASABE* **2015**, *58*, 971–985. [CrossRef]
16. Gupta, S.K.; Ehsani, R.; Kim, N.H. Optimization of a citrus canopy shaker harvesting system: Mechanistic tree damage and fruit detachment models. *Trans. ASABE* **2016**, *59*, 761–776. [CrossRef]
17. Liu, T.H.; Luo, G.; Ehsani, R.; Toudeshki, A.; Zou, X.J.; Wang, H.J. Simulation study on the effects of tine-shaking frequency and penetrating depth on fruit detachment for citrus canopy-shaker harvesting. *Comput. Electron. Agric.* **2018**, *148*, 54–62. [CrossRef]
18. Pu, Y.; Toudeshki, A.; Ehsani, R.; Yang, F.; Abdulridha, J. Selection and experimental evaluation of shaking rods of canopy shaker to reduce tree damage for citrus mechanical harvesting. *Int. J. Agric. Biol. Eng.* **2018**, *11*, 48–54. [CrossRef]
19. Whitney, J.D.; Wheaton, T.A. Tree spacing affects citrus fruit distribution and yield. *Proc. Fla. State Hortic. Soc.* **1984**, *97*, 44–47.
20. Castro-Garcia, S.; Sola-Guirado, R.R.; Gil-Ribes, J.A. Vibration analysis of the fruit detachment process in late-season 'Valencia' orange with canopy shaker technology. *Biosyst. Eng.* **2018**, *170*, 130–137. [CrossRef]

 *sensors*

Article

Predicting Forage Quality of Warm-Season Legumes by Near Infrared Spectroscopy Coupled with Machine Learning Techniques

Gurjinder S. Baath [1,*], Harpinder K. Baath [2], Prasanna H. Gowda [3], Johnson P. Thomas [2], Brian K. Northup [4], Srinivas C. Rao [4] and Hardeep Singh [1]

[1] Department of Plant and Soil Sciences, Oklahoma State University, 371 Agricultural Hall, Stillwater, OK 74078, USA; hardeep.singh@okstate.edu

[2] Department of Computer Science, Oklahoma State University, 219 MSCS, Stillwater, OK 74078, USA; hbaath@okstate.edu (H.K.B.); jpthomas@okstate.edu (J.P.T.)

[3] USDA-ARS, Southeast Area Branch, 114 Experiment Station Road, Stoneville, MS 38776, USA; prasanna.gowda@usda.gov

[4] USDA-ARS, Grazinglands Research Laboratory, 7207 W. Cheyenne St., El Reno, OK 73036, USA; brian.northup@usda.gov (B.K.N.); srinivas.rao@usda.gov (S.C.R.)

* Correspondence: gbaath@okstate.edu; Tel.: +15-756428027

Received: 20 December 2019; Accepted: 4 February 2020; Published: 6 February 2020

Abstract: Warm-season legumes have been receiving increased attention as forage resources in the southern United States and other countries. However, the near infrared spectroscopy (NIRS) technique has not been widely explored for predicting the forage quality of many of these legumes. The objective of this research was to assess the performance of NIRS in predicting the forage quality parameters of five warm-season legumes—guar (*Cyamopsis tetragonoloba*), tepary bean (*Phaseolus acutifolius*), pigeon pea (*Cajanus cajan*), soybean (*Glycine max*), and mothbean (*Vigna aconitifolia*)—using three machine learning techniques: partial least square (PLS), support vector machine (SVM), and Gaussian processes (GP). Additionally, the efficacy of global models in predicting forage quality was investigated. A set of 70 forage samples was used to develop species-based models for concentrations of crude protein (CP), acid detergent fiber (ADF), neutral detergent fiber (NDF), and in vitro true digestibility (IVTD) of guar and tepary bean forages, and CP and IVTD in pigeon pea and soybean. All species-based models were tested through 10-fold cross-validations, followed by external validations using 20 samples of each species. The global models for CP and IVTD of warm-season legumes were developed using a set of 150 random samples, including 30 samples for each of the five species. The global models were tested through 10-fold cross-validation, and external validation using five individual sets of 20 samples each for different legume species. Among techniques, PLS consistently performed best at calibrating (R^2_c = 0.94–0.98) all forage quality parameters in both species-based and global models. The SVM provided the most accurate predictions for guar and soybean crops, and global models, and both SVM and PLS performed better for tepary bean and pigeon pea forages. The global modeling approach that developed a single model for all five crops yielded sufficient accuracy (R^2_{cv}/R^2_v = 0.92–0.99) in predicting CP of the different legumes. However, the accuracy of predictions of in vitro true digestibility (IVTD) for the different legumes was variable (R^2_{cv}/R^2_v = 0.42–0.98). Machine learning algorithms like SVM could help develop robust NIRS-based models for predicting forage quality with a relatively small number of samples, and thus needs further attention in different NIRS based applications.

Keywords: partial least square; support vector machine; Gaussian processes; soybean; pigeon pea; guar; tepary bean

---

1. Introduction

Perennial warm-season grasses, such as bermudagrass (*Cynodon dactylon*), old world bluestems (*Bothriochloa* spp.), and bahiagrass (*Paspalum notatum*), serve as major summer forage resources for grazing stocker cattle in the southern United States (US). While capable of producing large amounts of biomass, these perennial grasses often show a decline in forage quality with their maturation towards the mid-late summer growing season and do not meet the nutritional needs of grazing stocker cattle for the entire season [1,2]. Legumes, being high-quality forages, can be adopted to offset the summer slump in forage quality, and enhance the efficiency of forage-based beef production systems. Further, the continued increase in the cost of nitrogen fertilizers has added to the interest of producers in utilizing legume crops as forage in many regions across the US. In response, extensive research in the southern US over the last decade has focused on evaluating warm-season annual legumes as summer forage resources that can be grown in rotation with winter-wheat (*Triticum aestivum* L.) [3–6]. In more recent years, several legumes have received increased attention due to their capabilities of generating high amounts of biomass under the limited moisture conditions that prevail in the southern US [7].

Quantifying the quality of forage in pastures is crucial for both agriculture research and forage management, including cattle grazing and harvesting. However, the determination of the different parameters of forage quality, such as crude protein (CP), neutral detergent fiber (NDF), acid detergent fiber (ADF), and in vitro true digestibility (IVTD), by classical analytical techniques is time-consuming and expensive, especially when numerous samples are required. The vast evolution of computers and multivariate statistical techniques has enabled the use of near infrared spectroscopy (NIRS) in assessing the quality parameters of many forages. The NIRS method is quick, inexpensive, and facilitates timely decision-making related to grazing periods. The technique is based on interactions between light reflectance in the wavelength ranging between 750–2500 nm and organic compounds in the plant biomass [8]. The method of applying NIRS to predict forage quality involves analyzing a particular forage with both traditional lab analysis and NIRS, and then developing a predictive equation by pairing the information in a calibration dataset (Figure 1). The NIRS has been widely used in forage quality predictions of crops including alfalfa (*Medicago sativa*) [9], maize (*Zea mays*) [10], ryegrass (*Lolium multiflorum*) [11], tall fescue (*Festuca arundinacea*) [12], and other species. However, the technique has been underutilized to provide predictions of forage quality for many warm-season legumes.

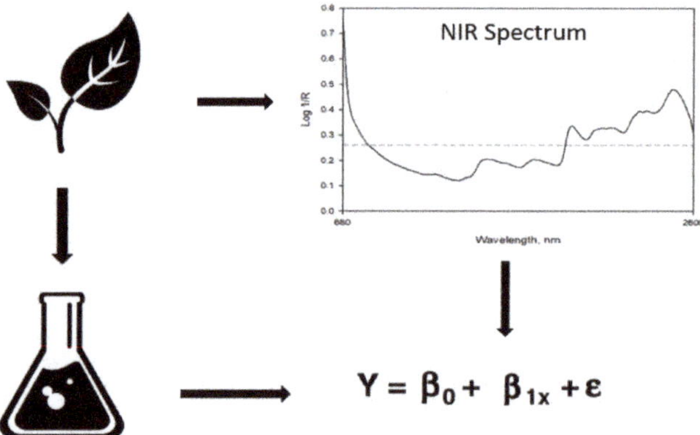

Figure 1. Illustration of the procedure used for applying near infrared spectroscopy (NIRS) technique in forage quality predictions.

As developed for other forage crops, well-calibrated NIRS species-based models for warm-season legumes could be useful tools to quickly asses the forage quality of different legume species grown under a range of environmental or management settings, or harvested at different stages of growth, or cutting or grazing height. Therefore, it is necessary to examine the effectiveness of NIRS in predicting forage characteristics of some important warm-season legumes. This work includes species such as guar (*Cyamopsis tetragonoloba*), tepary bean (*Phaseolus acutifolius*), soybean (*Glycine max*), and pigeon pea (*Cajanus cajan*), given past research, and the potential for expansion of use of these species across the southern US and other similar environments. There are also other warm-season legumes that may be capable of providing high-quality forage for summer grazing [3,13,14]. However, developing NIR calibration equations for every species can become challenging for public or private laboratories that test forage quality. Generally, accurate chemical analyses of a large number of samples is not readily available or feasible to develop calibrations, especially when novel legume species are involved. In response to challenges related to developing species-based calibrations, global models developed from samples of ranges of different warm-season legumes can prove useful, if such calibrations provide sufficiently accurate predictions.

Several calibration techniques are known to perform well in the application of NIRS in estimating forage quality and are generally available in most chemometric packages [15]. Partial least squares (PLS) is among the most commonly used methods, where least square algorithms are used to compute regressions [16]. In contrast, a comparatively novel and robust machine learning algorithm, support vector machine (SVM), has been gaining attention for NIRS calibrations [15]. Further, the Gaussian processes (GP) have provided better calibration results than PLS and SVM, in some cases [17,18]. However, tests of these calibration techniques on wide ranges of common and more novel legumes are required to define their function.

The combination of NIRS and machine learning calibration techniques could serve as an effective tool to streamline the monitoring efforts in warm-season legumes by eliminating the need for classical forage analytical methods. Therefore, the objective of this research was to evaluate the performance of NIRS in predicting the forage quality of four warm-season legumes (guar, tepary bean, pigeon pea, and soybean), using three different calibration techniques—PLS, SVM, and GP—on individual species bases. Additionally, the efficacy of global calibrations of these techniques, developed by combining datasets of all four species and mothbean (*Vigna aconitifolia*), was tested using different independent datasets of five species.

2. Materials and Methods

2.1. Materials

Forage samples used in the study ($n = 410$) were collected as parts of two different field experiments conducted at the USDA-ARS Grazinglands Research Laboratory near El Reno, Oklahoma, US (35.57° N, 98.03° W, elevation 414 m). Ninety samples each for guar and tepary bean, and 50 mothbean samples were collected from field experiments conducted during the summers of 2017 and 2018. An additional 90 samples of both soybean and pigeon pea were obtained from two long-term experiments (2001–2008) conducted in the same location [3,19]. In all three experiments, aboveground biomass was collected from randomly clipped 0.5 m row lengths from experimental plots at 15-day intervals, starting at 45 days after planting. Apart from whole plant samples, a major proportion of collected biomass samples in these experiments were separated into leaf, stem, and pods fractions before laboratory analysis.

2.2. Laboratory and NIRS Analysis

All leaf, stem, pod, and whole plant samples were oven-dried at 60 °C until a constant weight. Dry samples were ground to pass a 2-mm filter using a Wiley grinding mill. Total nitrogen concentration in each sample was determined by the flash combustion method (Model Vario Macro, Elementar Americas, Inc., Mt. Laurel, NJ, USA) and then converted into CP by multiplying with a factor of 6.25 (Table 1). The IVTD was obtained for each sample by following the Daisy Digester procedures (ANKOM Technology, Macedon, NY, USA). The NDF and ADF concentrations were only determined in samples of guar and tepary bean, in accordance with the batch fiber analyzer techniques (ANKOM Technology, Macedon, NY, USA).

Table 1. Summary statistics of lab datasets used for calibration, cross-validation and external validation of crude protein (CP), neutral detergent fiber (NDF), acid detergent fiber (ADF), and in vitro true digestibility (IVTD) of four warm season legumes.

Species	Parameter	Calibration and Cross-Validation ($n = 70$)				External Validation ($n = 20$)			
		Min	Max	Mean	SD	Min	Max	Mean	SD
					(%)				
Guar	CP	3.94	34.87	17.66	8.66	3.69	33.56	15.07	9.50
	NDF	16.83	70.80	37.57	16.95	22.95	75.78	45.94	17.82
	ADF	8.90	58.39	27.19	15.29	12.79	62.93	34.70	16.57
	IVTD	40.35	95.22	79.27	14.11	42.96	94.37	73.38	16.08
Tepary bean	CP	4.50	31.12	15.76	7.78	5.94	30.25	19.35	8.13
	NDF	22.90	71.57	48.34	12.31	25.52	60.95	43.90	10.21
	ADF	15.32	59.16	34.92	11.99	17.08	48.16	30.36	10.39
	IVTD	55.88	93.16	75.50	10.83	60.23	92.56	81.34	8.56
Soybean	CP	4.15	39.75	21.16	11.03	6.31	36.12	19.73	8.94
	IVTD	42.45	99.30	78.25	16.28	57.66	98.31	80.21	12.38
Pigeon pea	CP	4.52	32.48	16.30	8.77	6.24	28.64	15.62	7.41
	IVTD	30.71	91.08	61.55	19.28	33.31	82.89	59.76	16.40

n, number of samples; Min, minimum value; Max, maximum value; SD, standard deviation.

Aliquots of ground samples were filled into ring cups to eliminate voids. Spectral reflectance (R) of monochromatic light, averaged over 10 spectra per sample, were collected by scanning spectrophotometer (Model SpectraStar 2600 XT-R, Unity Scientific, Columbia, MD, USA). Spectral data were obtained as the logarithm of the inverse of reflectance [log(1/R)] at 1-nm interval over the range of 680–2600 nm.

2.3. Calibration Techniques

Partial least squares (PLS) is an extensively used class of statistical methods, which includes regression, classification, and dimension reduction techniques. It uses latent variables, also called score vectors, to model the relationship between input and response variables. In the case of regression problems, PLS first generates the latent variables from the given data and uses them as new predictor variables. There are different types of PLS, based on techniques employed to extract the latent variables. Two approaches are used to extend PLS for modeling non-linear relations among data. The first approach is to reformulate the linear relationship between score vectors, u and v, by a non-linear model:

$$v = g(u) + h = g(X, w) + h \tag{1}$$

where g is the continuous function that models the existing non-linear relation. Generally, g is modeled using artificial neural networks, smoothing splines, polynomial, or radial basis functions. Remaining variables h and w denote a residual vector and a weight vector, respectively.

The second PLS approach is to apply kernel-based learning. The kernel PLS method transforms the input space data to higher dimensional feature space and linearly estimates PLS in that space. To avoid the mapping function Φ from projecting data to feature space, PLS applies the kernel trick which uses the fact that a value of the inner product of two vectors x and y in feature space can be calculated using a kernel function $k(x, y)$ [20]:

$$k(x, y) = \Phi(x)^T \Phi(y) \tag{2}$$

By using the kernel function, score vectors (u and v) can be identified and used to define the non-linear relationship. The kernel PLS approach is used to model complex non-linear relations easily in terms of implementation and computation.

Gaussian processes (GP) are kernel-based, probabilistic, non-parametric regression models. A Gaussian process involves a set of random variables such that every finite number of those variables possess joint Gaussian distributions. A Gaussian process, $f(x)$, can be described using a mean function $m(x)$ and a covariance function $k(x, x')$. The covariance function defines the smoothness of responses, and the basis function Φ projects the input space vector x to a higher dimension feature space vector $\Phi(x)$. A Gaussian process regression (GPR) model describes the response by using latent variables from a Gaussian process. A GPR model is represented as:

$$\Phi(x_i)^T w + f(x) \tag{3}$$

where $f(x) \sim GP(0, k(x, x'))$, and $f(x)$ are from a zero mean GP having a covariance function, $k(x, x')$ [21]. The covariance is specified by kernel parameters, which are also known as hyperparameters. GPR is a probabilistic model, and an instance of response y is:

$$P\left(y_i \middle| f(x_i), x_i\right) \sim N(y_i | \Phi(x_i)^T w + f(x_i), \sigma^2) \tag{4}$$

GPR is non-parametric as there is a latent variable $f(x_i)$ for each observation x_i. Noise variance σ^2, basis function coefficients w, and hyperparameters of the kernel can be estimated from the data while training the GPR model.

Support vector machine (SVM) is a popular machine-learning algorithm used for identifying linear as well as non-linear dependency between input vectors and outputs. SVMs are non-parametric models, which means parameters are selected, estimated, and tuned in such a way that the model capacity matches the data complexity [21]. Generally, SVM starts by observing the multivariate inputs X and outputs Y, estimates its parameters w, and then learns the performed mapping function $y = f(x, w)$, which approximates the underlying dependency between inputs and responses. The obtained function, also known as a hyperplane, must have a maximal margin (for classification) or the error of approximation (for regression) to predict the new data. In the case of SVM regression, Vapnik's error (loss) function is used with ε-insensitivity. It finds a regression function $f(x)$ that deviates from the actual responses (y) by values no more than ε and is considerably flat at the same time.

For non-linear regression problems, SVM maps the input space to feature space (a higher dimension space) using a mapping $\Phi(x)$ to find a linear regression hyperplane in that space. However, there is no need to know the mapping Φ, as the kernel function $k\left(x_i, x_j\right)$, which is the inner product of the vectors $\Phi(x_i)$ and $\Phi\left(x_j\right)$, can be used to find the optimal regression hyperplane in extended space. There are many kernel functions available to describe non-linear regressions, such as the polynomial kernel, RBF kernel, Gaussian Kernel, normalized polynomial kernel, etc. The learning problem in classification as well as in regression, leads to solving the quadratic programming (QP). The sequential minimal optimization (SMO) is considered as the most popular optimizer for solving SVM problems [22]. It divides the large QP problem into a set of small QP problems and analytically solves them.

2.4. Performance Evaluation

Apart from calibration, 10-fold cross-validations and external validations were conducted to assess the performance of the calibration techniques. The 10-fold cross-validation is a unique statistical way of performance evaluations of machine learning models in which ten repeated hold-out executions are obtained and averaged. In each execution, the model is trained with 90% of the data points and tested with the remaining 10%, and thus every data point is taken nine times for training and once for testing the model. For each species-based model, the original dataset of 90 samples for each species was split into two subsets (Figure 2). A subset of 70 samples was used for running calibration and 10-fold cross-validation. The other subset of 20 remaining samples was used only for external validation and neither used in calibration nor cross-validation of any model. For the global model, the original dataset consisted of 250 samples, involving 50 samples each of guar, tepary bean, soybean, pigeon pea, and mothbean. These samples were divided into six subsets (Figure 2). One random subset of 150 samples (30 samples per species) was employed for calibration as well as 10-fold cross-validation. Each of the remaining five subsets, comprising 20 samples of individual species, was used for external validation.

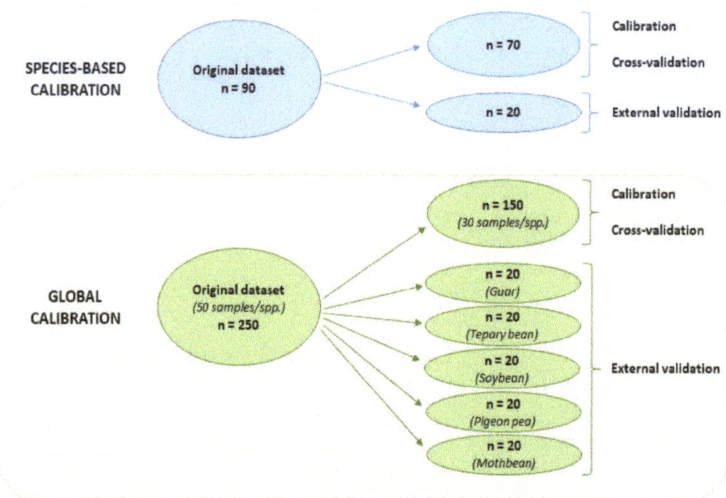

Figure 2. Diagram of the datasets, calibration, and different validation processes used in two calibration strategies.

Coefficients of determination, being upper-bounded by 1.0, are often adopted for meaningful comparisons across different models and therefore was used here as an estimate of prediction accuracy. To be precise, coefficient of determination in calibration (R^2_c), coefficient of determination in cross-validation (R^2_{cv}), and coefficient of determination in validation (R^2_v) were used for direct computation of the variance in the data captured at calibration, cross-validation, and external validation, respectively by each model. Additionally, root mean squared error estimation was also presented for comparing models, which were termed as $RMSE_c$, $RMSE_{cv}$, and $RMSE_v$ for calibration, cross-validation, and external validation, respectively.

2.5. Software

Regression models were calibrated, cross-validated, and externally validated using the *Weka* software, version 3.8 [23]. Weka is a suite of machine learning algorithms and is widely used for data mining. For implementing PLS, we used the PLS classifier package in Weka, which uses the prediction capabilities of PLSFilter. The PLSFilter runs the PLS regression on the given set of data and computes the beta matrix for prediction. By default, missing values are replaced, and the data are centered. For GP implementation, the Gaussian classifier for regression without hyperparameter-tuning was used. The kernel for the Gaussian classifier was configured as a polynomial. By default, missing values were replaced by the global mean. The SMOReg classifier was used to implement SVM in Weka. The classifier used the polynomial kernel and RegSMOImproved optimizer to learn SVM for regression. All remaining parameters, such as batch size, debugging, and filter type, which do not check capabilities, noise, etc., were kept as default.

3. Results and Discussion

The prediction accuracy of calibrated models is discussed by comparing their cross-validation (R^2_{cv}) and external validation (R^2_v) results to a scale proposed for NIRS calibrations [24]. According to the scale, the performance of a model is considered excellent if the R^2 of validations is greater than

0.95, and the resultant model can be used in any application. A model is assumed satisfactory with R^2 ranging from 0.9–0.95 and would be usable for most applications involving quality assurance. Models with R^2 ranging between 0.8–0.9 are considered moderately successful and can be used with caution for most applications, including research.

3.1. Guar

The chemical analysis of guar samples showed wide variability in parameters that define forage quality for different components (leaf, stem, or pod) of plant sampled at different growth stages (Table 1). The CP content for all 90 (70 + 20) guar samples ranged from 3.7% to 34.9%, while NDF concentrations ranged from 16.8% to 75.8%, ADF concentrations ranged from 8.9% to 62.9%, and IVTD from 40.3% to 95.2%.

Among the three techniques, the PLS technique performed best at calibrating each of the four forage quality parameters in guar with R^2_c of 0.98–0.99, though calibration results of SVM ($R^2_c = 0.94$–0.98) were also comparable (Table 2). While GP had a comparatively lower calibration accuracy with R^2_c ranging between 0.88–0.91 for IVTD, NDF and ADF, and R^2_c of 0.95 for CP of guar samples. Although PLS provided best calibrations out of the three, SVM gave better prediction accuracy in both cross-validation and external validation of all four indices of forage quality for guar. Thee GP approach generated the lowest R^2_{cv} for all four parameters and R^2_v for NDF and ADF.

Table 2. Calibration, cross-validation, and external validation statistics obtained for crude protein (CP), neutral detergent fiber (NDF), acid detergent fiber (ADF), and in vitro true digestibility (IVTD) in guar using three calibration techniques.

Parameter	Method	Calibration ($n = 70$)		Cross-Validation ($n = 70$)		External Validation ($n = 20$)	
		R^2_c	$RMSE_c$	R^2_{cv}	$RMSE_{cv}$	R^2_v	$RMSE_v$
CP	GP	0.95	1.84	0.93	2.20	0.96	2.12
	PLS	0.99	0.78	0.95	1.97	0.93	2.52
	SVM	0.98	1.23	0.97	1.56	0.98	1.27
NDF	GP	0.90	5.53	0.84	6.73	0.90	6.98
	PLS	0.98	2.17	0.85	6.66	0.93	5.52
	SVM	0.94	3.98	0.91	5.08	0.94	4.67
ADF	GP	0.91	4.79	0.86	5.77	0.92	6.02
	PLS	0.99	1.18	0.95	3.36	0.94	4.23
	SVM	0.97	2.46	0.95	3.51	0.96	3.78
IVTD	GP	0.88	4.92	0.81	6.10	0.93	5.63
	PLS	0.98	2.15	0.81	6.69	0.87	5.66
	SVM	0.94	3.51	0.83	5.88	0.94	4.19

GP, Gaussian processes; PLS; partial least square; SVM, support vector machine; R^2_c, determination coefficient in calibration; $RMSE_c$, root mean square error in calibration; R^2_{cv}, determination coefficient in cross-validation; $RMSE_{cv}$, root mean square error in cross-validation; R^2_v, determination coefficient in external validation; $RMSE_v$, root mean square error in external validation.

Among forage quality parameters, the greatest prediction accuracy was recorded for CP by all three techniques with R^2_{cv} of 0.93–0.97 and R^2_v of 0.93–0.98 (Table 2). In comparison, only the SVM technique resulted in a satisfactory prediction accuracy ($R^2_{cv} = 0.92$; $R^2_v = 0.94$) for NDF, based on the proposed scale [24]. Both the SVM and PLS techniques showed excellent accuracy at predicting ADF with R^2_{cv} and R^2_v between 0.94–0.96, while GP produced R^2_{cv} of 0.86. All three techniques resulted in relatively low prediction accuracy for IVTD, with R^2_{cv} ranging from 0.81–0.83. Overall, performances of SVM was most satisfactory among the three calibration methods, and it can be employed in NIRS-based prediction of CP, ADF, and NDF of guar. In contrast, use of IVTD predictions of guar would require caution, based on the type of application.

While currently a minor crop in the southern US, guar has a proven potential to serve as a multi-purpose legume and has potential for expansion in use. Guar is a common crop in regions of the Indian subcontinent, Africa, North and South America, and Australia [25]. Guar has been gaining attention as a forage resource in the southern US due to its capability of producing high N biomass under limited water conditions [3,5]. Therefore, this first report investigating the application of NIRS in guar would encourage the utilization of the technique in its research and forage management.

3.2. Tepary Bean

Results from the laboratory analysis of tepary bean samples showed high variability in all four of the quality indices, though the observed ranges were narrower than guar (Table 1). The concentration of CP varied from 4.5–31.1%, while NDF ranged from 22.9% to 71.6%. In contrast, ADF and IVTD ranged between 15.3–59.2% and 55.9–93.2%, respectively. Best calibration results for tepary bean were recorded using the PLS technique, with R^2_c of 0.98–0.99 (Table 3). Whereas, neither SVM nor PLS clearly resulted in better predictions for all quality indices when cross-validated and externally validated.

Table 3. Calibration, cross-validation, and external validation statistics obtained for crude protein (CP), neutral detergent fiber (NDF), acid detergent fiber (ADF), and in vitro true digestibility (IVTD) in tepary bean using three calibration techniques.

Parameter	Method	Calibration (n = 70)		Cross-Validation (n = 70)		External Validation (n = 20)	
		R^2_c	RMSE$_c$	R^2_{cv}	RMSE$_{cv}$	R^2_v	RMSE$_v$
CP	GP	0.94	1.89	0.90	2.42	0.94	2.20
	PLS	0.99	0.68	0.93	2.03	0.98	1.35
	SVM	0.97	1.35	0.95	1.74	0.94	1.94
NDF	GP	0.85	4.96	0.75	6.22	0.75	5.10
	PLS	0.98	1.64	0.84	5.09	0.75	5.53
	SVM	0.94	2.97	0.72	7.01	0.84	4.03
ADF	GP	0.87	4.60	0.78	5.62	0.86	3.90
	PLS	0.98	1.47	0.89	3.97	0.92	3.34
	SVM	0.96	2.45	0.86	4.52	0.95	2.23
IVTD	GP	0.87	4.02	0.75	5.39	0.75	4.25
	PLS	0.98	1.55	0.79	5.00	0.88	2.89
	SVM	0.93	2.86	0.75	5.70	0.82	3.82

GP, Gaussian processes; PLS; partial least square; SVM, support vector machine; R^2_c, determination coefficient in calibration; RMSE$_c$, root mean square error in calibration; R^2_{cv}, determination coefficient in cross-validation; RMSE$_{cv}$, root mean square error in cross-validation; R^2_v, determination coefficient in external validation; RMSE$_v$, root mean square error in external validation.

All calibration techniques showed best results at predicting CP in tepary bean samples with a R^2_{cv} or R^2_v above 0.90 among the forage quality characteristics (Table 3). The SVM technique resulted in the lowest RMSE$_{cv}$ value (1.74) for cross-validation of CP, whereas PLS had the lowest RMSE$_v$ of 1.35 for external validation among the three techniques. In contrast, PLS showed the lowest RMSE$_{cv}$ values of 5.09 and 3.97 and SVM had the lowest RMSE$_v$ of 4.03 and 2.23 for NDF and ADF, respectively. Both PLS and SVM produced satisfactory results at predicting ADF concentration in tepary bean with R^2_{cv} of 0.86–0.89 and R^2_v of 0.92–0.95 compared to GP, while all three techniques had comparatively low performance at predicting NDF in tepary bean with R^2_{cv} and R^2_v of 0.72–0.84 and 0.75–0.84, respectively.

In comparison to ADF, the NDF concentration of tepary samples were less accurately predicted by all three techniques (Table 3). Similar differences between prediction accuracy of ADF and NDF were also noticed for guar in this study, and also reported earlier in NIRS studies involving *Brassica napus* [26], *Lolium multiflorum* [11], and *Oryza sativa* [27]. Though PLS performed better at predicting IVTD in tepary bean compared to other two, all three techniques resulted in relatively low prediction accuracy with R^2_{cv} and R^2_v ranging between 0.75–0.79 and 0.75–0.88, respectively. Overall, both PLS and SVM could be considered as good among three tested techniques and hence can be employed for satisfactory predictions of CP and ADF in tepary bean. While prediction results of NDF and IVTD would need some caution if calibrations are developed with similar sample sizes (n = 70) as used in this study.

Tepary bean is a vining, warm-season legume species originated from the areas of the southwestern United States and northwestern Mexico, that may have value for multiple uses in dryland agricultural systems. Due to its spreading growth habit, and the ability to generate high N biomass with limited soil moisture, tepary bean could be an ideal summer forage for the Southern Great Plains [14]. This first study investigating the application of NIRS to attributes of forage quality in tepary bean showed that the technique could aid in quantifying its role in meeting animal nutrition needs.

3.3. Soybean

All three techniques (PLS, SVM, and GP) gave excellent accuracies at calibrating CP and IVTD of soybean samples with PLS again performing the best out of three with a R^2_c greater than 0.98 (Table 4). Among three techniques, SVM performed best at predicting CP with $RMSE_{cv}$ and $RMSE_v$ of 1.85 and 1.78, respectively, followed closely by PLS. All three calibration techniques produced better predictions of IVTD in soybean (R^2_{cv} > 0.84 and R^2_v > 0.89), compared to prediction accuracies obtained for guar and tepary bean. As observed for CP, SVM performed better than the other techniques in cross-validation (R^2_{cv} = 0.89) of IVTD, while the other two techniques performed better in external validation (R^2_v of 0.92–0.93). All three techniques can be employed for rapid NIR-based predictions of CP and IVTD in soybean forage samples, with SVM would be the best choice.

Table 4. Calibration, cross-validation, and external validation statistics obtained for crude protein (CP) and in vitro true digestibility (IVTD) in soybean using three calibration techniques.

Parameter	Method	Calibration (n = 70)		Cross-Validation (n = 70)		External Validation (n = 20)	
		R^2_c	$RMSE_c$	R^2_{cv}	$RMSE_{cv}$	R^2_v	$RMSE_v$
CP	GP	0.92	4.63	0.87	5.78	0.92	3.78
	PLS	0.98	2.16	0.84	6.92	0.93	3.46
	SVM	0.94	3.92	0.89	5.28	0.89	4.09
IVTD	GP	0.96	2.14	0.94	2.71	0.92	2.53
	PLS	0.99	0.80	0.96	2.05	0.94	2.24
	SVM	0.99	1.26	0.97	1.85	0.96	1.78

GP, Gaussian processes; PLS; partial least square; SVM, support vector machine; R^2_c, determination coefficient in calibration; $RMSE_c$, root mean square error in calibration; R^2_{cv}, determination coefficient in cross-validation; $RMSE_{cv}$, root mean square error in cross-validation; R^2_v, determination coefficient in external validation; $RMSE_v$, root mean square error in external validation.

Soybean was initially introduced as a forage into the US in the 19th Century, but is now one of the most widely grown grain legumes in the Southern Great Plains [14]. In the last two decades, there has been increased interest from researchers in utilizing soybean as a summer forage in the US [28–30]. Hence the need for rapid and low-cost techniques for estimating forage quality. The NIRS technique has not been exploited for forage quality predictions in soybean. A single report investigated modified PLS and multiple scatter correction methods for NIR predictions of CP, NDF, and ADF concentrations, using 353 soybean samples collected at one (R6) growth stage [31]. In comparison, calibrations developed in the present study, used data on IVTD and CP with just 70 soybean samples collected across a range of different growth stages. Thus, our observed ranges for CP (4.1–39.7) and IVTD (42.4–99.3%) were more diverse (Table 1). The accuracies (R^2_{cv} or R^2_v > 0.92) obtained in predicting CP in soybean forage by all three techniques were higher than the values reported [31], despite large differences in sample sizes (N = 70 vs. 353) used for developing calibrations. Therefore, this study showed machine learning algorithms could develop robust NIRS calibrations for precise analysis of forage quality of soybean with small sample sizes.

3.4. Pigeon Pea

Laboratory analyses for the current study showed wide variability in both CP (4.5–32.5%) and IVTD (30.7–91.1%) for forage samples of pigeon pea (Table 1). The CP concentration in pigeon pea was accurately calibrated ($R^2_c > 0.95$) by each of the three techniques (Table 5). All three techniques resulted in CP predictions with R^2_{cv} and R^2_v greater than 0.96. Both PLS and SVM also showed greater accuracies in predicting IVTD of pigeon pea with R^2_{cv} and R^2_v ranges of 0.91–0.92 and 0.96–0.97, respectively. Although lower than PLS and SVM, the performance ($R^2_{cv} = 0.86$) of GP-based calibrations were moderately satisfactory in IVTD predictions, following the proposed scale [24]. Overall, both PLS and SVM would provide excellent options for NIR predictions of CP and IVTD in pigeon pea.

Table 5. Calibration, cross-validation, and external validation statistics obtained for crude protein (CP) and in vitro true digestibility (IVTD) in pigeon pea using three calibration techniques.

Parameter	Method	Calibration ($n = 70$)		Cross-Validation ($n = 70$)		External Validation ($n = 20$)	
		R^2_c	$RMSE_c$	R^2_{cv}	$RMSE_{cv}$	R^2_v	$RMSE_v$
CP	GP	0.98	1.37	0.96	1.73	0.96	1.69
	PLS	1.00	0.43	0.97	1.46	0.98	1.02
	SVM	0.99	0.84	0.98	1.17	0.98	1.12
IVTD	GP	0.95	4.51	0.86	7.18	0.97	2.95
	PLS	0.99	1.93	0.92	5.49	0.96	3.09
	SVM	0.97	3.31	0.91	5.86	0.97	2.85

GP, Gaussian processes; PLS; partial least square; SVM, support vector machine; R^2_c, determination coefficient in calibration; $RMSE_c$, root mean square error in calibration; R^2_{cv}, determination coefficient in cross-validation; $RMSE_{cv}$, root mean square error in cross-validation; R^2_v, determination coefficient in external validation; $RMSE_v$, root mean square error in external validation.

Pigeon pea is another legume species that has seen the development of a range of cultivars for different uses in its home range, and areas of greater cultivation. This includes research on the value of cultivars of pigeon pea in the US for forage, grain, and pasture productivity [4,32]. Pigeon pea has a high degree of heat and drought tolerance, and the capacity for high levels of forage production in the US and other tropical and sub-tropical regions.

While pigeon pea is a broadly grown crop in much of the world, there was only one preliminary report that discussed the possible use of NIRS techniques to predict forage quality of pigeon pea [33]. That report used limited numbers of samples ($n = 48$), involving leaves and branches, that were mostly collected at one growth stage for calibrations of CP, NDF, and ADF concentrations; however, no validations were performed [33]. In contrast, the present study undertook both calibrations and validations using 90 (70 + 20) pigeon pea samples, involving leaves, stems, or seed pods, collected at different growth stages during a long-term experiment. Further, we investigated the NIR-based predictions of IVTD, which is assumed as an important quality parameter in pigeon pea forage [4]. Therefore, this study confirms that NIRS techniques could be effective tools for predicting forage quality of pigeon pea.

3.5. Global Calibrations

Global calibrations for CP and IVTD of warm-season legumes were developed with 150 samples, which included 30 samples each of guar, tepary bean, soybean, pigeon pea, and mothbean (Figure 2). As observed with the species-based calibrations for the four different legumes, the PLS technique performed best out of the three techniques for global calibrations both CP and IVTD (R^2_c of 0.97 and 0.94, respectively), while the GP technique was the least accurate (Table 6). In comparison, cross-validation of global models showed the SVM approach provided the greatest prediction accuracy for both CP (R^2_{cv} = 0.94) and IVTD (R^2_{cv} = 0.86), followed closely by PLS. Therefore, based on cross-validation results, the performance of global calibrations developed using SVM and PLS were satisfactory at predicting CP, and moderately satisfactory for IVTD.

When global calibrated models were validated using different external datasets for each of the five legume species, the predictions for CP by all three techniques resulted in sufficient accuracies with R^2_v ranging between 0.91–0.97 (Table 6). The SVM technique showed higher accuracy compared to the others in predicting CP, with the exception of guar, where the PLS approach provided slight improvements. Among species, the best CP predictions were noted for pigeon pea (R^2_v values of 0.98–0.99) for all three techniques. In contrast, IVTD predictions were not consistently accurate across all five species. The greatest accuracy was observed for IVTD predictions in pigeon pea with R^2_v of 0.97–0.98 under SVM and PLS. The lowest accuracy in predicting IVTD was noted for mothbean (R^2_v between 0.65–0.69 by SVM and PLS; 0.42 by GP). The best prediction accuracies for IVTD of soybean (R^2_v = 0.82–0.86 for all three techniques) and guar (PLS; R^2_v = 0.81) were moderately satisfactory. However, the performance of all three techniques was satisfactory at predicting IVTD in tepary bean (R^2_v of 0.91–92), which was better than the specific models developed for tepary bean (Table 3).

Overall, global-calibrated models for CP have the potential to offer sufficient prediction accuracies that are comparable to species-based calibration models. Diverse calibration sets that contain different legume species may allow the creation of robust, generalized models that provide predictions similar to species-based models. In some cases, global models may be capable of providing more accurate predictions, as was observed for IVTD predictions of tepary bean in this study.

The application of accurate globally calibrated models would be extremely useful for a broad range of end-users. They would reduce or eliminate the large amounts of time and other resources required to perform chemical analyses or the development and use of separate calibration sets for every species. However, adopting the global calibration approach for IVTD may not provide satisfactory predictions for all species. Some of the issues related to the low level of performance of calibrations for IVTD may be variability associated with using techniques that rely on rumen fluids in laboratory analyses [34]. Therefore, further investigations are required to compare the performance of global calibrations developed for IVTD of warm-season legumes derived using both rumen fluid and cellulose degradation methods.

Table 6. Calibration, cross-validation, and external (species) validation statistics of global models obtained for crude protein (CP) and in vitro true digestibility (IVTD) in warm-season legumes using three calibration techniques.

| Method | | Calibration (n = 150) | | Cross-Validation (n = 150) | | External Validation (n = 20) | | | | | | | | | |
| | | | | | | Guar | | Tepary Bean | | Soybean | | Pigeon Pea | | Mothbean | |
		R^2_c	$RMSE_c$	R^2_{cv}	$RMSE_{cv}$	R^2_v	$RMSE_v$	R^2_v	$RMSE_v$	R^2_v	$RMSE_v$	R^2_v	$RMSE_v$	R^2_v	$RMSE_v$
CP	GP	0.92	2.15	0.89	2.46	0.93	2.42	0.95	2.72	0.91	3.95	0.98	2.21	0.94	3.36
	PLS	0.97	1.15	0.92	2.02	0.94	2.36	0.94	2.49	0.94	2.47	0.98	2.03	0.94	3.10
	SVM	0.96	1.48	0.94	1.87	0.92	2.77	0.95	2.36	0.94	3.16	0.99	1.29	0.97	2.54
IVTD	GP	0.86	5.09	0.81	5.84	0.65	6.16	0.91	4.41	0.82	7.93	0.91	4.75	0.42	5.40
	PLS	0.94	3.28	0.85	5.28	0.81	5.00	0.90	5.53	0.88	5.19	0.98	2.21	0.69	4.50
	SVM	0.91	3.98	0.86	4.98	0.77	5.12	0.92	4.74	0.86	5.60	0.97	2.77	0.65	4.29

GP, Gaussian processes; PLS; partial least square; SVM, support vector machine; R^2_c, determination coefficient in calibration; $RMSE_c$, root mean square error in calibration; R^2_{cv}, determination coefficient in cross-validation; $RMSE_{cv}$, root mean square error in cross-validation; R^2_v, determination coefficient in external validation; $RMSE_v$, root mean square error in external validation.

4. Conclusions

The statistics obtained for calibration, cross-validation, and external validation in this study demonstrated that NIRS techniques could be effective for supplying rapid and accurate predictions of most attributes of forage quality (cell wall fractions, crude protein) for different warm-season legumes. Further, the applications of NIRS technique to guar, tepary bean, and mothbean represent the first reports of such tools to provide estimates of forage quality for these species. Though similar to PLS, the SVM technique performed consistently well in predicting quality parameters of five warm-season legumes under both species-based and global calibration strategies. The global calibration approach can be a useful approach for predicting CP in warm-season legumes, and reduce the time and resources required for traditional chemical analysis in the use of separate calibration equations for each species. However, the global model for IVTD was not accurate for all species. Further model development based on other analytical procedures may improve the consistency and reliability of the global approach. Machine learning algorithms like SVM could also allow the development of robust models with a relatively small number of samples. Additional research is required to refine the SVM approach for different NIRS applications.

Author Contributions: Conceptualization, G.S.B., H.K.B. and P.H.G.; methodology, G.S.B., H.K.B., P.H.G. and J.P.T.; formal analysis, G.S.B. and H.K.B.; data curation, G.S.B., B.K.N. and S.C.R.; writing—original draft preparation, G.S.B. and H.K.B.; writing—review and editing, B.K.N., P.H.G., H.S. and J.P.T.; visualization, G.S.B. and H.K.B.; supervision, P.H.G. and J.P.T.; project administration, P.H.G. All authors have read and agreed to the published version of the manuscript.

Funding: This work was partially funded by the Cooperative Agreement with USDA-ARS Grazinglands Research Laboratory and Oklahoma Agriculture Experiment Station Hatch project OKL03132.

Acknowledgments: The authors would like to acknowledge ARS technicians Cindy Coy, Delmar Shantz, Kory Bollinger, and Jeff Weik for their assistance in collecting, processing, and analyzing forage samples.

Conflicts of Interest: The authors declare no conflict of interest.

Disclaimer: Mention of trademarks, proprietary products, or vendors does not constitute guarantee or warranty of products by USDA and does not imply its approval to the exclusion of other products that may be suitable. All programs and services of the USDA are offered on a nondiscriminatory basis, without regard to race, color, national origin, religion, sex, age, marital status, or handicap.

Abbreviations

PLS, partial least square; SVM, support vector machine; GP, Gaussian processes; CP, crude protein; ADF, acid detergent fiber; NDF, neutral detergent fiber; IVTD, in vitro true digestibility; R^2_c, determination coefficient in calibration; R^2_{cv}, determination coefficient in cross-validation; R^2_v, determination coefficient in external validation; $RMSE_c$, root mean square error in calibration; $RMSE_{cv}$, root mean square error in cross-validation; $RMSE_v$, root mean square error in external validation.

References

1. Phillips, W.; Coleman, S. Productivity and economic return of three warm season grass stocker systems for the Southern Great Plains. *J. Prod. Agric.* **1995**, *8*, 334–339. [CrossRef]
2. Williams, M.; HaLmmond, A. Rotational vs. continuous intensive stocking management of bahiagrass pasture for cows and calves. *Agron. J.* **1999**, *91*, 11–16. [CrossRef]
3. Rao, S.C.; Northup, B.K. Capabilities of four novel warm-season legumes in the southern Great Plains: Biomass and forage quality. *Crop Sci.* **2009**, *49*, 1096–1102. [CrossRef]
4. Rao, S.C.; Northup, B.K. Pigeon pea potential for summer grazing in the southern Great Plains. *Agron. J.* **2012**, *104*, 199–203. [CrossRef]
5. Rao, S.C.; Northup, B.K. Biomass production and quality of indian-origin forage guar in Southern Great Plains. *Agron. J.* **2013**, *105*, 945–950. [CrossRef]
6. Foster, J.; Adesogan, A.; Carter, J.; Sollenberger, L.; Blount, A.; Myer, R.; Phatak, S.; Maddox, M. Annual legumes for forage systems in the United States Gulf Coast region. *Agron. J.* **2009**, *101*, 415–421. [CrossRef]
7. Baath, G.S.; Northup, B.K.; Gowda, P.H.; Turner, K.E.; Rocateli, A.C. Mothbean: A potential summer crop for the Southern Great Plains. *Am. J. Plant Sci.* **2018**, *9*, 1391. [CrossRef]

8. Rushing, J.B.; Saha, U.K.; Lemus, R.; Sonon, L.; Baldwin, B.S. Analysis of some important forage quality attributes of Southeastern Wildrye (Elymus glabriflorus) using near-infrared reflectance spectroscopy. *Am. J. Anal. Chem.* **2016**, *7*, 642. [CrossRef]

9. Brogna, N.; Pacchioli, M.T.; Immovilli, A.; Ruozzi, F.; Ward, R.; Formigoni, A. The use of near-infrared reflectance spectroscopy (NIRS) in the prediction of chemical composition and in vitro neutral detergent fiber (NDF) digestibility of Italian alfalfa hay. *Ital. J. Anim. Sci.* **2009**, *8*, 271–273. [CrossRef]

10. Volkers, K.; Wachendorf, M.; Loges, R.; Jovanovic, N.; Taube, F. Prediction of the quality of forage maize by near-infrared reflectance spectroscopy. *Anim. Feed Sci. Technol.* **2003**, *109*, 183–194. [CrossRef]

11. Yang, Z.; Nie, G.; Pan, L.; Zhang, Y.; Huang, L.; Ma, X.; Zhang, X. Development and validation of near-infrared spectroscopy for the prediction of forage quality parameters in Lolium multiflorum. *PeerJ* **2017**, *5*, e3867. [CrossRef] [PubMed]

12. Hill, N.; Cabrera, M.; Agee, C. Morphological and climatological predictors of forage quality in tall fescue. *Crop Sci.* **1995**, *35*, 541–549. [CrossRef]

13. Muir, J.P.; Pitman, W.D.; Dubeux Jr, J.C.; Foster, J.L. The future of warm-season, tropical and subtropical forage legumes in sustainable pastures and rangelands. *Afr. J. Range Forage Sci.* **2014**, *31*, 187–198. [CrossRef]

14. Baath, G.S.; Northup, B.K.; Rocateli, A.C.; Gowda, P.H.; Neel, J.P. Forage potential of summer annual grain legumes in the southern great plains. *Agron. J.* **2018**, *110*, 2198–2210. [CrossRef]

15. Agelet, L.E.; Hurburgh, C.R., Jr. A tutorial on near infrared spectroscopy and its calibration. *Crit. Rev. Anal. Chem.* **2010**, *40*, 246–260. [CrossRef]

16. Roggo, Y.; Chalus, P.; Maurer, L.; Lema-Martinez, C.; Edmond, A.; Jent, N. A review of near infrared spectroscopy and chemometrics in pharmaceutical technologies. *J. Pharm. Biomed. Anal.* **2007**, *44*, 683–700. [CrossRef]

17. Wang, K.; Chi, G.; Lau, R.; Chen, T. Multivariate calibration of near infrared spectroscopy in the presence of light scattering effect: A comparative study. *Anal. Lett.* **2011**, *44*, 824–836. [CrossRef]

18. Cui, C.; Fearn, T. Comparison of partial least squares regression, least squares support vector machines, and Gaussian process regression for a near infrared calibration. *J. Near Infrared Spectrosc.* **2017**, *25*, 5–14. [CrossRef]

19. Rao, S.; Mayeux, H.; Northup, B. Performance of forage soybean in the southern Great Plains. *Crop Sci.* **2005**, *45*, 1973–1977. [CrossRef]

20. Rosipal, R.; Kramer, N. Subspace, latent structure and feature selection techniques. *Lect. Notes Comput. Sci. Chap. Overv. Recent Adv. Part. Least Sq.* **2006**, *2940*, 34–51.

21. Williams, C.K.; Rasmussen, C.E. *Gaussian Processes for Machine Learning*; MIT Press: Cambridge, MA, USA, 2006; Volume 2.

22. Huang, C.-L.; Wang, C.-J. A GA-based feature selection and parameters optimizationfor support vector machines. *Expert Syst. Appl.* **2006**, *31*, 231–240. [CrossRef]

23. Platt, J. Probabilistic outputs for support vector machines and comparisons to regularized likelihood methods. *Adv. Large Margin Classif.* **1999**, *10*, 61–74.

24. Frank, E.; Hall, M.A.; Witten, I.H. *The WEKA Workbench; Online Appendix for "Data Mining: Practical Machine Learning Tools and Techniques"*, 4th ed.; Morgan Kaufmann: Cambridge, MA, USA, 2016.

25. Malley, D.; Martin, P.; Ben-Dor, E. Application in analysis of soils. In *Near-Infrared Spectroscopy in Agriculture*, 1st ed.; Roberts, C.A., Workman, J., Jr., Reeves, J.B., III, Eds.; American Society of Agronomy; Crop Science Society of America; Soil Science Society of America: Madison, WI, USA, 2004; pp. 729–783.

26. Baath, G.S.; Kakani, V.G.; Gowda, P.H.; Rocateli, A.C.; Northup, B.K.; Singh, H.; Katta, J.R. Guar responses to temperature: Estimation of cardinal temperatures and photosynthetic parameters. *Ind. Crop. Prod.* **2019**. [CrossRef]

27. Wittkop, B.; Snowdon, R.J.; Friedt, W. New NIRS calibrations for fiber fractions reveal broad genetic variation in Brassica napus seed quality. *J. Agric. Food Chem.* **2012**, *60*, 2248–2256. [CrossRef]

28. Kong, X.; Xie, J.; Wu, X.; Huang, Y.; Bao, J. Rapid prediction of acid detergent fiber, neutral detergent fiber, and acid detergent lignin of rice materials by near-infrared spectroscopy. *J. Agric. Food Chem.* **2005**, *53*, 2843–2848. [CrossRef]

29. Nielsen, D.C. Forage soybean yield and quality response to water use. *Field Crop. Res.* **2011**, *124*, 400–407. [CrossRef]

30. Beck, P.; Hubbell, D., III; Hess, T.; Wilson, K.; Williamson, J.A. Effect of a forage-type soybean cover crop on wheat forage production and animal performance in a continuous wheat pasture system. *Prof. Anim. Sci.* **2017**, *33*, 659–667. [CrossRef]

31. Asekova, S.; Han, S.-I.; Choi, H.-J.; Park, S.-J.; Shin, D.-H.; Kwon, C.-H.; Shannon, J.G.; LEE, J.D. Determination of forage quality by near-infrared reflectance spectroscopy in soybean. *Turk. J. Agric. For.* **2016**, *40*, 45–52. [CrossRef]

32. Rao, S.; Coleman, S.; Mayeux, H. Forage production and nutritive value of selected pigeonpea ecotypes in the southern Great Plains. *Crop Sci.* **2002**, *42*, 1259–1263. [CrossRef]

33. Berardo, N.; Dzowela, B.; Hove, L.; Odoardi, M. Near infrared calibration of chemical constituents of Cajanus cajan (pigeon pea) used as forage. *Anim. Feed Sci. Technol.* **1997**, *69*, 201–206. [CrossRef]

34. Roberts, C.A.; Stuth, J.; Flinn, P. Analysis of forages and feedstuffs. In *Near-Infrared Spectroscopy in Agriculture*, 1st ed.; Roberts, C.A., Workman, J., Jr., Reeves, J.B., III, Eds.; American Society of Agronomy; Crop Science Society of America; Soil Science Society of America: Madison, WI, USA, 2004; pp. 231–267.

Article

Thermal Imaging Reliability for Estimating Grain Yield and Carbon Isotope Discrimination in Wheat Genotypes: Importance of the Environmental Conditions

Sebastián Romero-Bravo [1,2,*], Ana María Méndez-Espinoza [2], Miguel Garriga [2], Félix Estrada [2], Alejandro Escobar [2], Luis González-Martinez [2], Carlos Poblete-Echeverría [3], Daniel Sepulveda [4], Ivan Matus [5], Dalma Castillo [5], Alejandro del Pozo [2] and Gustavo A. Lobos [2,*]

[1] Department of Agricultural Sciences, Universidad Católica del Maule, Curicó P.O. Box 684, Chile
[2] Plant Breeding and Phenomic Center, Faculty of Agricultural Sciences, Universidad de Talca, Talca P.O. Box 747, Chile; anmendez@utalca.cl (A.M.M.-E.); mgarriga@utalca.cl (M.G.); festrada@alumnos.utalca.cl (F.E.); escobar.opazo@gmail.com (A.E.); l.gonzalez.m@ieee.org (L.G.-M.); adelpozo@utalca.cl (A.d.P.)
[3] Department of Viticulture and Oenology, Stellenbosch University, Matieland 7602, South Africa; CPE@sun.ac.za
[4] Centro de Investigación y Transferencia en Riego y Agroclimatología (CITRA), Talca P.O. Box 747, Chile; dsepulveda18@gmail.com
[5] Centro Regional Investigación Quilamapu, Instituto de Investigaciones Agropecuarias, Chillán P.O. Box 426, Chile; imatus@inia.cl (I.M.); dalma.castillo@inia.cl (D.C.)
* Correspondence: sromero@ucm.cl (S.R.-B.); globosp@utalca.cl (G.A.L.)

Received: 10 April 2019; Accepted: 5 June 2019; Published: 13 June 2019

Abstract: Canopy temperature (Tc) by thermal imaging is a useful tool to study plant water status and estimate other crop traits. This work seeks to estimate grain yield (GY) and carbon discrimination ($\Delta^{13}C$) from stress degree day (SDD = Tc − air temperature, Ta), considering the effect of a number of environmental variables such as the averages of the maximum vapor pressure deficit (VPDmax) and the ambient temperature (Tmax), and the soil water content (SWC). For this, a set of 384 and a subset of 16 genotypes of spring bread wheat were evaluated in two Mediterranean-climate sites under water stress (WS) and full irrigation (FI) conditions, in 2011 and 2012, and 2014 and 2015, respectively. The relationship between the GY of the 384 wheat genotypes and SDD was negative and highly significant in 2011 (r^2 = 0.52 to 0.68), but not significant in 2012 (r^2 = 0.03 to 0.12). Under WS, the average GY, $\Delta^{13}C$, and SDD of wheat genotypes growing in ten environments were more associated with changes in VPDmax and Tmax than with the SWC. Therefore, the amount of water available to the plant is not enough information to assume that a particular genotype is experiencing a stress condition.

Keywords: remote sensing; phenotype; phenotyping; phenomics; *Triticum aestivum*; water deficit; stress; infrared

---

1. Introduction

Since the 1960s, crop temperature has been recognized as an indicator of water status [1]. When the plant is facing a water deficit, the stomata begin to close, reducing the transpiratory capacity (evaporative cooling) [2] and this results in increases in canopy temperature [3–7].

The development of infrared sensors/cameras has allowed a faster characterization of canopy temperatures [8]. At the same time, through computational analysis, it is possible to split the different parts of the image (e.g., soil, air, leaves, weeds) focusing only on the fraction(s) of interest [4,9,10]. Although thermal imaging does not directly measure stomatal conductance, the stomatal response is the main cause of changes in canopy temperature [10], so it is a useful tool to indirectly study spatial and temporal heterogeneity of leaf/canopy transpiration and the photosynthetic rate [10–12]. Indeed, in bread wheat grown in hot environments in Mexico under irrigation, a high correlation has been reported between temperature depression (TD = Ta − Tc) and leaf stomatal conductance ($r = 0.76 - 0.85$) and grain yield (GY; up to $r = 0.84$) [11,13]. Other researchers have used the concept of stress degree day (SDD), defined as the difference between leaf/canopy temperature (Tc) and air temperature (Ta) (SDD = Tc − Ta), which is equivalent to TD (but with positive values), mostly because canopy temperature in rainfed environments is lower than air temperature.

The main problem with the use of thermal assessments to estimate physiological and agronomic traits is that Tc is influenced by several environmental factors, such as air temperature and humidity, wind speed, net radiation, and soil water content [14–17]. Therefore, without detailed information about environmental factors, measurements of Tc are not sufficient to properly perform agronomic or physiological trait estimations.

Unlike irrigated conditions, a good correlation between Tc and GY under water deficit is not always expected [18]. However, it would be very useful for breeding programs to find such associations in stressful environments because the focus is on developing drought-tolerant cultivars with higher GY under water-limiting conditions.

It has been established that measurements of carbon isotope discrimination ($\Delta^{13}C$) in wheat are crucial for the selection of individuals with efficient water-use, mainly because this parameter is positively correlated with GY and negatively correlated with water-use efficiency (WUE) in moderately water-stressed to non-water-stressed environments [19–25]. The determination of $\Delta^{13}C$ is simple and relatively fast but needs expensive equipment or engagement of a paid analysis service; attempts have also been made to estimate $\Delta^{13}C$ by modeling the canopy spectral reflectance [24,26]. Under non-stressed conditions, the stomata remain open and the substomatal cavity is enriched with ^{12}C relative to the air; the heavier isotopic $^{13}CO_2$ has a lower diffusion speed than the lighter $^{12}CO_2$ [20]. Additionally, the ribulose bisphosphate carboxylase/oxygenase (RUBISCO) carboxylation enzyme in C3 plants has a higher affinity to $^{12}CO_2$. On the other hand, when stress forces the stomata to close, the proportion of $^{12}CO_2$ in the substomatal cavity is reduced, thus increasing the amount of fixed $^{13}CO_2$ [20]. Thus, daily conditions throughout the season will be summarized in the $\Delta^{13}C$ of leaves and kernels (calculation details in Section 2.2.1). In this sense, under the expected climate change scenarios predicted for the coming decades [27], the estimation of $\Delta^{13}C$ should be relevant in plant breeding programs oriented to environmental constraints [28–30].

Like all species, the phenotype of wheat plants is controlled by a large number of genes, and the expression of these is modulated, predominantly, in response to the environmental conditions (GxE) [31–33]. Consequently, it was hypothesized that the environmental conditions during and between seasons could interfere with the ability of canopy thermal imaging to estimate GY and $\Delta^{13}C$; in particular, the vapor pressure deficit (VPDmax) and soil water content (SWC), which can have a strong influence on canopy temperatures [34]. Therefore, the aim of this work was to study the reliability for estimating grain yield and carbon isotope discrimination in wheat genotypes growing under water stress (WS) and full irrigation (FI) conditions using thermal images, considering the relevance of the prevailing environmental conditions in estimation of the results.

2. Materials and Methods

2.1. Plant Material and Experimental Conditions

During four growing seasons (2011, 2012, 2014, and 2015), two sets of plant material were evaluated in two Mediterranean environments: (1) Cauquenes (c) (35°58′ S, 72°17′ W; 177 m.a.s.l.) under WS (rainfed) conditions during seasons 2011, 2012, and 2015, and under FI in 2015; and (2) Santa Rosa (sr)

(36°32′ S, 71°55′ W; 217 m.a.s.l.) under WS in 2011 and 2015, and FI conditions in 2011, 2012, 2014, and 2015. Each combination of season (year), location (c or sr), and water condition (WS or FI) was considered as an environment.

A collection of 384 advanced lines and cultivars of spring bread wheat (*Triticum aestivum* L.), were evaluated during 2011 and 2012. Plant material originated from three breeding programs: the Instituto de Investigaciones Agropecuarias in Chile (INIA-Chile) (153 genotypes), INIA-Uruguay (178 genotypes), and the International Wheat and Maize Improvement Centre CIMMYT (53 genotypes). In 2014 and 2105, a subset of 16 genotypes with contrasting tolerance to water deficit was studied.

Each genotype was established in plots of five rows (2.0 × 0.2 m) with a seeding rate of 20 g m^{-2}. Plots were fertilized with 260 kg ha^{-1} of ammonium phosphate (46% P_2O_5 and 18% N), 90 kg ha^{-1} of potassium chloride (60% K_2O), 200 kg ha^{-1} of Sul-Po-Mag (22% K_2O, 18% MgO, and 22% S), 10 kg ha^{-1} of Boronatrocalcita (11% B), and 3 kg ha^{-1} of zinc sulfate (35% Zn). Fertilizers were incorporated with a cultivator before sowing. During tillering, an extra 153 kg ha^{-1} of N was applied. Weeds were controlled with the application of flufenacet + flurtamone + diflufenican (96 g a.i.) as pre-emergence and a further application of MCPA (525 g a.i.) + metsulfuron-methyl (5 g a.i.) as post-emergent [35]; dates of sowing and the main phenological stages are shown in Table 1. Furrow irrigation was used at Santa Rosa, with the WS trials including one irrigation at the end of tillering (Zadocks stage 21–Z21; [36]) and FI comprising irrigations at the end of tillering (Z21), the flag leaf stage (Z37), heading (Z50), and early grain filling (Z71). At Cauquenes, the WS trials were purely rainfed and the FI trial during 2015 received sprinkler irrigation at Z37, Z50, and Z71. Approximately 50 mm was applied during each furrow/sprinkler irrigation application.

At each location, soil volumetric content (m^3 m^{-3}) was monitored periodically using 10HS sensors (Decagon Devices, Pullman, WA, USA), scanning the first 50 cm depth every 4 h. In order to generate the soil water content (SWC; mm), the volumetric values were multiplied by the soil depth (500 mm). Precipitation (mm), ambient temperature (°C), and relative humidity (%) were monitored hourly by autonomous weather stations (AWSs) belonging to the Red Agroclimática Nacional (National Agroclimatic Network, available at: www.agromet.inia.cl). Vapor pressure deficit (VPD; kPa) was determined hourly by the use of ambient temperature and relative humidity, according to Reference [37]. For analysis purposes, each environmental variable was studied as follows from sowing to harvest: (1) precipitation: daily summation; (2) ambient temperature: average of the daily maximum temperatures (Tmax); and (3) VPDmax: estimated at the highest ambient temperature and the corresponding relative humidity of each day, and then the average of the daily maximum VPDs (VPDmax) was calculated. Because water deficit in Mediterranean environments is present, primarily, between anthesis to grain filling, SWC was considered as the average of the daily mean values between anthesis and grain maturity.

Tmax, VPDmax, and SWC are summarized in Table 1 and Figure S1, and rainfall in Figure S2.

2.2. Evaluations

2.2.1. Grain Yield and Carbon Isotope Discrimination

Grain yield was evaluated by harvesting the whole plot (2 m^2) and was expressed as t ha^{-1}. Carbon isotope composition (δ^{13}C) was determined in mature kernels using an elemental analyzer (ANCA-SL, PDZ Europa, UK) coupled with an isotope ratio mass spectrometer, at the Laboratory of Applied Physical Chemistry at Ghent University (Belgium): δ^{13}C (‰) = (^{13}C/^{12}C)$_{sample}$/(^{13}C/^{12}C)$_{standard}$ − 1 [20], where the ^{13}C/^{12}C ratio of the sample refers to plant material and the ^{13}C/^{12}C ratio of the standard is calibrated against the international standards from Iso-Analytical (Crewe, Cheshire, UK). The carbon isotope discrimination (Δ^{13}C) of kernels was calculated as: Δ^{13}C (‰) = (δ^{13}C$_a$ − δ^{13}C$_p$)/[1 + (δ^{13}C$_p$)/1000], where a and p refer to air and the plant, respectively [20]. δ^{13}C$_a$ from the air was taken as −8.0‰.

2.2.2. Thermography

Thermal infrared images were taken using a portable infrared camera (i40, FLIR Systems, Sweden), at the soft dough (Z85) phenological stage. This camera provides images of 120 × 120 pixels (every pixel shows a temperature value) and has an uncooled infrared detector (microbolometer) in the spectral range from 7.5 to 13 microns. Infrared images were taken at ±2 h from the zenith (12:00 to 16:00 h), at a position of 45° from the horizontal, 0.5 m above the plant canopy, and a 3 m distance from the plot. Images were filtered using a process of interactive segmentation to exclude foreign matter from the picture (i.e., soil, weeds, neighboring plots, and air) using a custom MATLAB code [38]. To avoid surrounding plot noise, only the center of the image (30 × 30 pixels) was analyzed with a temperature frequency histogram (percentile level). The hottest and coldest pixels were eliminated, taking as a threshold the percentiles 1 downwards and 97.5 upwards, respectively. The remaining pixels were used to calculate the average canopy temperature (Tc), while the air temperature (Ta) was recorded from the AWS at the precise time the image was taken. Finally, Tc and Ta were used to calculate the SDD (°C) [39,40].

2.3. Statistical Design and Data Analysis

The experimental design for the trials at Cauquenes and Santa Rosa in seasons 2011 and 2012 was an alpha-lattice with two replicates; for this study, just one replicate ($n = 384$ genotypes) was assessed by thermography in each trial. For seasons 2014 and 2015, the experimental design was a random block with four replicates (16 genotypes; $n = 64$).

Correlations (x versus y) were performed through regression analysis: (1) genotype values: SDD versus GY and $\Delta^{13}C$; (2) environmental values: SDD, Tmax, VPDmax, and SWC versus GY and $\Delta^{13}C$; and (3) environmental values: VPDmax, Tmax, and SWC versus SDD.

Using the environmental (Tmax, VPDmax, and SWC), phenological (days between stages), physiological ($\Delta^{13}C$ and SDD) and productive (GY) information (Table 1), a clustering analysis was performed to verify whether the two water regimes evaluated (i.e., FI and WS) were grouped together, within and between seasons and locations, which is important in modeling and validation procedures. This consisted of a series of steps necessary to achieve a correct execution of the analysis methodology. For this study, clustering and hierarchical clustering were used, with the purpose of grouping the different environments studied. A group was defined as the set of elements that have a greater degree of similarity between the objects that belong to the same set [41]. The steps performed in the analysis were the following: obtaining the data, eliminating the columns that do not provide information to the grouping model, normalizing the data, then applying a method of hierarchical clustering using the "ward.D2" method [42] as a grouping form, and plotting the Euclidean distance between elements as a dendrogram. For clustering of groupings, a tree cluster was considered, which uses the Euclidean distance to identify the closeness of the nodes (environmental data points). In addition, this algorithm applies the principal component analysis (PCA) method to show the results with greater clarity [41]. The "ward.D2" was set to find two and three main data groups.

All statistical analysis was performed using R 3.0.0 [43].

3. Results

3.1. Environmental Conditions, Grain Yield, Carbon Isotope Discrimination, and Stress Degree Days

In general terms, the environmental conditions (Tmax, VPDmax, and SWC) varied according to the seasons, both within and between FI and WS conditions (Table 1, Figures S1 and S2). Under each water supply condition, minimum and maximum values from sowing to harvest were (Table 1): Tmax: 19.1 and 23.5 °C (FI) and 19.1 and 25.4 °C (WS); VPDmax: 1.35 and 1.92 kPa (FI) and 1.35 and 2.39 kPa (WS); and SWC: 198.3–542.7 mm (FI) and 180.4–418.8 mm (WS).

Table 1. Dates of sowing, anthesis, grain filling, and harvest, and mean values of grain yield (GY), carbon isotope discrimination in kernels ($\Delta^{13}C$), and stress degree day (SDD = Tc − Ta), determined at the soft dough stage (Z85), for wheat genotypes grown under full irrigation (FI) and water stress (WS) conditions, at Cauquenes (c) and Santa Rosa (sr) in 2011, 2012, 2014, and 2015. Each trial code is a combination of water regime, site, and season. Also, mean values (from sowing to harvest) of daily maximum temperature (Tmax), maximum vapor pressure deficit (VPDmax), and in the case of soil water content (SWC), the average of the daily mean values (from anthesis to mature grain) between 0 and 50 cm depth, are presented.

Trial Code	n	Dates					Means					
		Sowing	Anthesis	Grain filling	Harvesting		GY (t ha^{-1})	$\Delta^{13}C$ (‰)	SDD (°C)	Tmax (°C)	VPDmax (kPa)	SWC (mm)
FIsr 2011	384	31 Aug. 2011	24 Nov. 2011	22 Dec. 2011	11 Jan. 2012		8.03	18.0	1.81	23.4	1.76	542.7
FIsr 2012	384	07 Aug. 2012	05 Nov. 2012	19 Nov. 2012	28 Jan. 2013		9.83	18. 8	−1.70	21.5	1.52	404.8
FIsr 2014	64	27 Aug. 2014	24 Nov. 2014	17 Dec. 2014	22 Jan. 2015		9.90	-	1.60	23.5	1.92	212.3
FIsr 2015	64	29 Jul. 2015	20 Nov. 2015	08 Dec. 2015	25 Jan. 2016		9.38	18.8	1.38	21.5	1.63	198.3
FIc 2015	64	18 May 2015	23 Oct. 2015	12 Nov. 2015	23 Dec. 2015		8.46	16.9	1.12	19.1	1.35	246.2
					FI	Average	9.12	18.1	0.84	21.8	1.64	320.9
					FI	SD	0.84	0.90	1.44	1.79	0.22	148.9
					FI	Min.	8.03	16.9	−1.70	19.1	1.35	198.3
					FI	Max.	9.90	18.8	1.81	23.5	1.92	542.7
WSsr 2011	384	31 Aug. 2011	24 Nov. 2011	22 Dec. 2011	11 Jan. 2012		4.81	16.5	6.44	23.4	1.76	418.8
WSc 2011	384	07 Sep. 2011	29 Nov. 2011	13 Dec. 2011	05 Jan. 2012		1.68	14.2	12.29	25.4	2.39	320.9
WSc 2012	384	23 May 2012	11 Sep. 2012	25 Oct. 2012	23 Dec. 2012		3.18	15.0	2.17	20.6	1.51	225.5
WSsr 2015	64	29 Jul. 2015	20 Nov. 2015	08 Dec. 2015	25 Jan. 2016		7.40	18.5	3.31	21.5	1.63	180.4
WSc 2015	64	18 May 2015	23 Oct. 2015	12 Nov. 2015	23 Dec. 2015		8.13	16.9	2.16	19.1	1.35	283.8
					WS	Average	5.04	16.2	5.27	22.0	1.74	285.9
					WS	SD	2.45	1.49	4.29	2.18	0.36	91.8
					WS	Min.	1.68	14.2	2.16	19.1	1.35	180.4
					WS	Max.	8.13	18.5	12.3	25.4	2.39	418.8

Grain yield under WS conditions was 45% lower than under FI (Table 1). Also, the range of variation among seasons was much greater under WS (1.68–8.13 t ha^{-1}) compared to FI (8.03–9.9 t ha^{-1}). The Δ^{13}C data showed lower values (10.5%) and higher variability under WS conditions compared to FI conditions (Table 1). The average SDD was much higher (5.3 fold) under WS and had greater variability compared to FI conditions (Table 1).

3.2. Relationships between Grain Yield and Canopy and Ambient Temperatures in 384 Wheat Genotypes

The relationship between GY and SDD of the 384 genotypes was negative and highly significant in 2011 (r^2 = 0.52–0.68; p < 0.001) (Figure 1A). However, when SSD was compared with Δ^{13}C, the determination coefficients (r^2) were significant only in FIsr and WSsr (0.22 and 0.32, respectively), but not in WSc (Figure 1C). During the second season, r^2 values were much lower and not significant for both GY (r^2 = 0.03–0.12; p > 0.05) (Figure 1B) and Δ^{13}C (r^2 = 0.0002–0.04; p > 0.05) (Figure 1D). In terms of environmental conditions, both seasons showed important differences; Tmax (°C) values were higher in 2011 (WSc = 25.4, FIsr and WSsr = 23.4) than in 2012 (WSc = 20.6 and FIsr = 21.5). Consequently, VPDmax (kPa) in 2011 (WSc = 2.4, FIsr and WSsr = 1.8) was higher than in 2012 (WSc and FIsr = 1.5). In the case of SWC (mm), the values in 2011 (WSc = 381.7, FIsr = 550.7 and WSsr = 507.8) exceeded the values recorded in 2012 (WSc = 256.9 and FIsr = 399.0). Also, SDD (°C) was different between seasons, being higher in 2011 (WSc = 12.3, FIsr = 1.8, and WSsr = 6.4) than in 2012 (WSc = 2.2 and FIsr = −1.7).

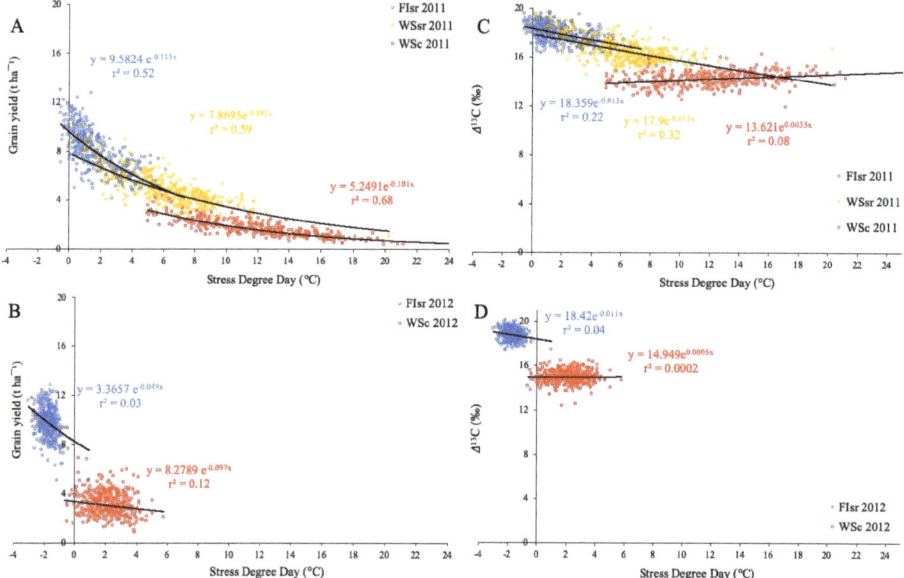

Figure 1. Relationship between stress degree day (SDD = Tc − Ta; where Tc is crop temperature and Ta air temperature, both measured at the soft dough stage (Z85) versus grain yield and carbon isotope discrimination in kernels for 384 spring bread wheat genotypes grown under two water regimes (full irrigation (FI) and water stress (WS)), in two locations (Santa Rosa (sr) and Cauquenes (c)), during the 2011 ((**A,C**) respectively) and 2012 seasons ((**B,D**), respectively). Regression lines and equations are presented for each water regime and location (determination coefficients are also included).

3.3. Environmental Effects on Grain Yield, Carbon Isotope Discrimination, and Stress Degree Day

The average GY of wheat genotypes under FI and WS conditions indicated different responses to environmental variables (Figure 2). Under WS conditions, GY decreased exponentially as SDD, Tmax, and VPDmax increased, which was not the case under FI conditions (Figure 2A–C). Similarly, Δ^{13}C also decreased incrementally in SDD, Tmax, and VPDmax (Figure 2E–G). No significant relationships were found between SWC and GY or $\Delta^{13}C$ (Figure 2D,H).

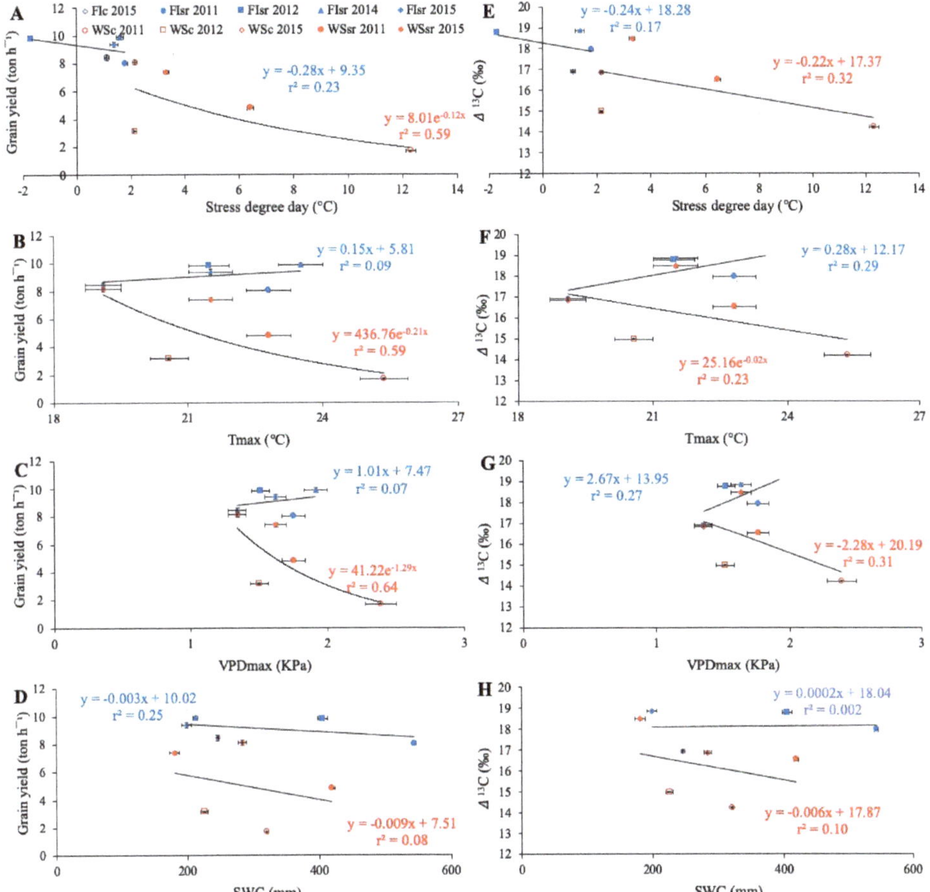

Figure 2. Average grain yield (**A–D**) and carbon isotope discrimination in kernels ($\Delta^{13}C$; **E–H**) of wheat genotypes growing in ten environments, in relation to the stress degree day (SDD = Tc − Ta; Tc is crop temperature and Ta air temperature, both measured at the soft dough stage Z85; **A** and **E**), the seasonal averages of daily maximum temperature (Tmax; **B** and **F**) and maximum vapor pressure deficit (VPDmax; **C** and **G**) and the soil water content between 0 and 50 cm depth (SWC; **D** and **H**). The environments corresponded to the water regime applied (full irrigation (FI) and water stress (WS)), the trial location (Santa Rosa (sr) and Cauquenes (c)), and growing seasons (2011, 2012, 2014, and 2015); the trial code is a combination of these factors. Regression lines and equations are presented for each water regime (determination coefficients are also included).

In relation to the environmental conditions during the study, when all the environments were combined (Table 2), SDD was only correlated with Tmax ($r = 0.64$; $p < 0.05$) and VPDmax ($r = 0.80$; $p < 0.01$). Under each water regime, close and significant relationships were found between SDD and Tmax and VPDmax, but only in plants growing in WS conditions (Figure 3). The relationship between SDD and SWC was not significant under either WS or FI conditions (Figure 3C).

Pearson correlation analysis (Table 2) showed that mean values of GY in the ten environments were highly correlated with SDD ($r = −0.81$; $p < 0.01$) and $\Delta^{13}C$ ($r = 0.92$; $p < 0.01$). Also, $\Delta^{13}C$ was negatively correlated with SDD ($r = -0.71$; $p < 0.05$). In concordance with this, GY and Δ^{13}C were primarily affected by Tmax and VPDmax but not by SWC (Figure 2B–F,C–G,D–H, respectively).

Table 2. Pearson's correlation matrix for stress degree day (SDD = Tc − Ta; Tc is crop temperature and Ta air temperature, both measured at the soft dough stage Z85), grain yield (GY), carbon isotope discrimination in kernels (Δ^{13}C), and seasonal averages of daily maximum temperature (Tmax), maximum vapor pressure deficit (VPDmax), and soil water content between 0 and 50 cm depth (SWC; between anthesis and mature grain). Data from the water regime applied (full irrigation and water stress), the trial location (Santa Rosa and Cauquenes), and the evaluated season (2011, 2012, 2014, and 2015).

	SDD	GY	Δ^{13}C	Tmax	VPDmax
GY	−0.81 **	-	-	-	-
Δ^{13}C	−0.71 *	0.92 **	-	-	-
Tmax	0.64 *	−0.38	−0.16	-	-
VPDmax	0.80 **	−0.52	−0.35	0.95 **	-
SWC	0.06	−0.06	−0.05	0.29	0.16

* Statistically significant ($p < 0.05$); ** Highly statistically significant ($p < 0.01$).

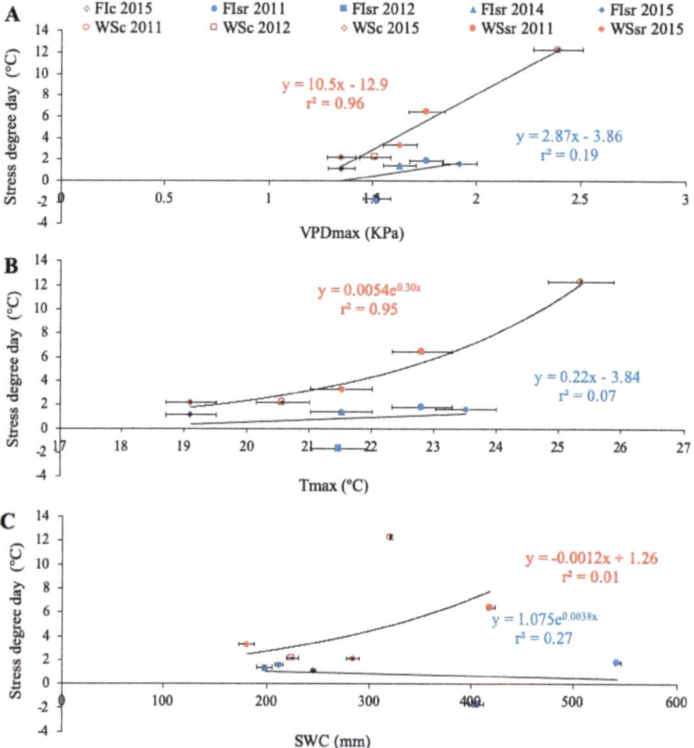

Figure 3. Relationships between the stress degree day (SDD = Tc − Ta; Tc is crop temperature and Ta air temperature, both measured at the soft dough stage Z85), and the seasonal averages of daily maximum vapor pressure deficit (VPDmax) (**A**), maximum temperature (Tmax; **B**) and soil water content between 0 and 50 cm depth (SWC; (**C**)). Mean values were the average of all genotypes growing in the particular environment according to the water regime (full irrigation (FI) and water stress (WS)), the trial location (Santa Rosa (sr) and Cauquenes (c)), and growing seasons (2011, 2012, 2014, and 2015). Regression lines and equations are presented for each water regime (determination coefficients are also included).

Finally, when the environmental, phenological, physiological, and productive information (Table 1) was included to generate the two- and three-group cluster dendrograms and plots (Figure 4A,B and Figure 4C,D, respectively), the main difference was found in the division of the first branch of the less stressful environment (green lines in Figure 4A) at the height of the first knot of the most stressful environment (origin of the blue and green lines in Figure 4B). Differences between the cluster plots were according to changes in the cluster dendrograms; the three groups were (Figure 4B): (i) lowest environmental limitations: FIsr 2012, FIsr 2014, FIsr 2015, and WSsr 2015; (ii) intermediate environmental limitations: WSc 2012, FIc 2015, and WSc 2015; and (iii) highest environmental limitations: WSc 2011, FIsr 2011, and WSsr 2011. The cluster plot that explains 61.3% of the variance (Figure 4C) shows a clear distance or separation between the groups with the lowest and the highest environmental constraints (green and red colors in Figure 4, respectively). In the three-group cluster plot, two of the groups overlap. However, even though WSc 2011, FIsr 2011, and WSsr 2011 had the highest SWC values, they also presented, on average, the greatest VPDmax, Tmax and SDD but the lowest GY (Table 1).

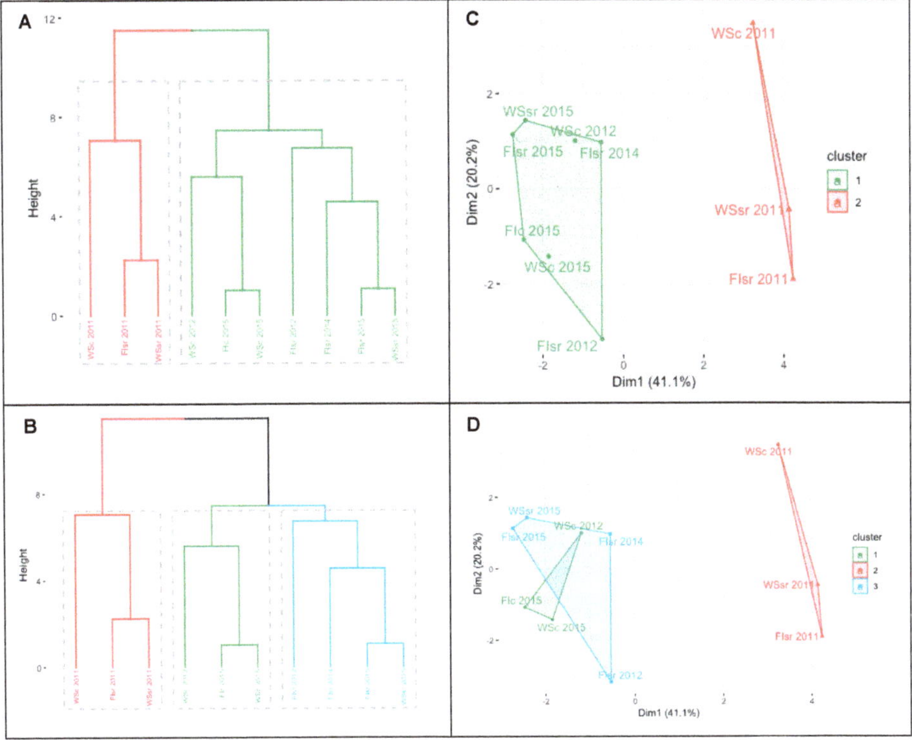

Figure 4. Cluster dendrogram ((**A**) two groups and (**B**) three groups) and plot ((**C**) two groups and (**D**) three groups) for the general characterization of the assessed environments according to the water regime applied (full irrigation (FI) and water stress (WS)), the trial location (Santa Rosa (sr) and Cauquenes (c)), and the evaluated season (2011, 2012, 2014, and 2015); the trial code is a combination of these factors. Data included the phenological (dates), productive (grain yield—GY), physiological (carbon isotope discrimination in kernels ($\Delta^{13}C$) and the stress degree day measured at the soft dough stage Z85 (SDD)), and environmental information (seasonal averages of daily maximum temperature and maximum vapor pressure deficit, and the soil water content between 0 and 50 cm depth). In the case of GY, $\Delta^{13}C$, and SDD, the mean values analyzed were the average of all genotypes growing in the particular environment.

4. Discussion

4.1. Environmental Effects on Grain Yield and Carbon Isotope Discrimination

Tolerance to WS usually implies some improvement or maintenance of metabolic processes that enables the plant to regulate cell water status and maintain leaf turgor under stressful conditions. One of the first mechanisms involved in reducing water loss by transpiration is stomatal control, which partially closes the stomata, thus affecting carbon assimilation and storage [44]. This gas exchange limitation between the atmosphere and the substomatal cavity is primarily driven by the surrounding environmental conditions (e.g., water availability, ambient temperature, relative humidity, wind speed, light intensity). To the extent that the diffusion of CO_2 through the stomata is more restrictive, the carbon isotope discrimination ($\Delta^{13}C$) between ^{12}C and ^{13}C will also be reduced, increasing the proportion of ^{13}C [24]. Therefore, in a particular environment, $\Delta^{13}C$ at the grain level provides an integrated assessment of the transpiration efficiency during the whole season [25,45]. As in other cereal studies [24,35,46–49], GY and $\Delta^{13}C$ in the current work had a strong and positive association ($r = 0.92$) (Table 2). Additionally, the evaluated environmental conditions generated similar responses in GY and $\Delta^{13}C$ (Figure 2), reaffirming the strong relationship that existed between these characters.

The SWC has also been used as an indicator of water stress in plants and is positively related to GY in wheat [25,50–52]. Working in the same species, Reference [53] evaluated the effect of water content in different soil profiles, concluding that a soil that was well irrigated throughout the first 50 cm of the profile obtained a greater yield and harvest index than a soil with dry upper layers. In the present work, even though FI environments always showed higher SWC than under WS in the same location and season (Figure S2A–D), the results did not show a significant correlation between SWC and GY (Table 2). Moreover, while the GY and $\Delta^{13}C$ were both higher under FI than WS (Figure 2D,H), neither GY nor $\Delta^{13}C$ were affected by increases in SWC (~200 to 550 mm) under either FI or WS.

Contrastingly, the environmental water demand (VPD; [54]), which is mainly driven by the ambient temperature and relative humidity, proved to influence both GY and $\Delta^{13}C$ under WS but not in FI (Figure 2C,G). The combination of high temperatures and low relative humidity, which is frequently encountered in the late stages of the growing season in Mediterranean climates (e.g., Santa Rosa and Cauquenes), caused an increase in the VPD. References [55,56] have assessed the effect of environmental variables on wheat physiology and GY, proposing that a high VPD environment should vary between 2.5 and 3.9 kPa. Therefore, the average values of VPDmax found in the present study (1.35 to 2.39 kPa; Figure 2C,G) could be considered moderately low to moderately high, although maximum values reached as high as 6.34 kPa in WSc in 2011 (Figure S1B).

Several studies have proven that growing cereals under non-limiting water conditions but with high VPD values leads to reduced GY and $\Delta^{13}C$ [21,55–57]. The present work, considering all measurements performed, shows a non-significant relationship between GY and VPDmax (Table 2). When FI and WS were analyzed separately, it was only the genotypes growing under WS that showed lower GY and $\Delta^{13}C$ as VPDmax increased (Figure 2C,G) and the VPD had a higher association with GY than $\Delta^{13}C$ (Table 2).

Likewise, the Tmax trends were similar to VPDmax (Figure 2B,F). However, despite the Tmax and VPDmax being relatively low (20.6 °C and 1.51 kPa, respectively) in the WSc 2012 trial, the lowest SWC (256.9 mm) was registered, especially after anthesis, and this generated a low GY (3.18 t ha^{-1}) (Table 1 and Figure S2B). On the other hand, the WSc 2011 trial had a relatively adequate SWC (320 mm) between anthesis and grain filling (Figure S2B), but due to the late sowing date (Table 1), the plants were exposed to higher Tmax and VPDmax (25.4 °C and 2.39 kPa, respectively) (Table 1), reaching ~40 °C and ~6 kPa for Tmax and VPDmax, respectively (Figure S1A,B), resulting in this trial having the lowest GY (1.68 t ha^{-1}).

Under high VPD, guard cell turgor may be decreased by direct evaporative losses from the guard cells and/or decreased water supply to the guard cells if the root or shoot hydraulic conductance is limiting [58], and this leads to a detrimental effect on plant production. Reference [56] tested the effect of VPD and ambient temperature on gas exchange and GY in wheat, finding that environments with high VPD (3.9 kPa) and high temperature (36 °C) increased respiration by up to 22% and decreased photosynthetic water-use efficiency by up to 64% compared to environments with high temperature and lower VPD (1.5 kPa). Indeed, environments with high VPD and temperature caused a reduction in leaf area and net assimilation of CO_2; however, in the case of plants under the same conditions but without water restriction, there was no decrease in GY. The same authors showed that GY was reduced by 7% in environments with water stress compared to no stress, which is concordant with the findings of the present study, where at similar VPDmax and Tmax (between 1.5 and 2 kPa, and between 21 and 24 °C, respectively), plants grown under WS showed lower GY than plants under FI (Figure 2B,C). In FI conditions, the plants had a GY that was higher than 8 t ha^{-1}, while in WS environments, the GY never exceeded that threshold; an exception was WSc 2015 (8.1 t ha^{-1}), which was influenced by an abnormally rainy season ("El Niño" phenomenon; Figure S2D).

4.2. The Potential of Stress Degree Day to Estimate Grain Yield

Stomatal closure causes a decrease in the transpiration rate, and as a consequence, there is a reduction in the cooling effect, which finally increases leaf/canopy temperature [10]. The reduction in the stomatal conductance could be a consequence of the limitations of the roots to absorb enough water to supply the atmospheric water demand [6]. Numerous studies have confirmed that the temperature of the canopy is associated with crop yield [59–61], as well as a series of physiological characteristics, including stomatal conductance [11], the hydric state of the plant [59], and the presence of deep roots.

In general, the present study establishes a negative and highly significant correlation ($r = -0.81$) between SDD and GY (Table 2). Analysis of the responses according to water regime (FI or WS) indicated that there was no significant relationship between GY and SDD in plants growing under FI conditions, but under WS conditions the correlation was moderately high ($r^2 = 0.59$) (Figure 2A). The $\Delta^{13}C$ followed the same pattern, but with lower determination coefficients (FI = 0.17 and WS = 0.34). When Reference [12] studied the relationship between GY and Tc in wheat genotypes grown with similar water regimes (FI and WS), they also found a stronger association under WS ($r^2 = 0.66$) than FI ($r^2 = 0.58$).

Despite similarities between the studies described above, there are also contradictory results for the relationship between SDD and GY. For example, References [62,63] found no significant relationships, whereas Reference [11] found a high and significant association in irrigated environments. These differences could be explained by the lower VPD registered in the studies of References [62,63] (~2.4 kPa) compared to that of Reference [11] (~5.5 kPa), with the latter case allowing a greater expression of the tolerance of each genotype to the environmental conditions. Although in the present work, there were trials that reached a VPDmax of 6.5 kPa (WSc 2011) (Figure S1), the seasonal averages were ~2.4 kPa.

Therefore, the low SDD values of plants growing under FI is likely due to the ability to meet the water demand of the air (VPD), thus maintaining a high transpiratory rate and allowing the plants to cool down their leaves; under this condition there is more CO_2 fixation, explaining the higher yields in FI. In this kind of environment where soil water availability is enough to compensate for VPD, the plants do not need to express their water deficit tolerance mechanisms, which in this case means there is a lower SDD versus GY data dispersion, implying lower coefficients of determination.

Similar to the GY and $\Delta^{13}C$, SDD was more sensitive to the VPDmax and Tmax than to SWC, with the WS environment having the most effect on plant temperature. Interestingly, an SDD of 2 °C seems to be the threshold between FI and WS environments (Figure 2A,E); SDD averages in FI were lower than 2 °C, while in WS they were greater than 2 °C.

When the relationship between SDD and GY was studied in individual genotypes under contrasting environments (seasons 2011 and 2012), the association (r^2 values) between these two variables depended on the environment. While in 2011 the relationships in FIsr, WSsr, and WSc were negative and moderately high ($r^2 = 0.52$, 0.59, and 0.68, respectively) (Figure 1A), the FIsr and WSc relationships in 2012 were low ($r^2 = 0.03$ and 0.12, respectively; WSsr was not sown in 2012) (Figure 1B). As seen before by the use of the average values per environment, the best determination coefficients were observed in more stressful conditions (FI < WSsr < WSc), likely associated with the higher trait-range during the first season; FIsr 2011, WSsr 2011, and WSc 2011 had a higher SDD data dispersion in relation to FIsr 2012 and WSc 2012 (Figure 1 and Figure S1E).

Except for the WSsr 2011 trial, the minimum SDD values of 2011 corresponded, approximately, to the maximums registered during 2012. As already explained, GY is influenced by SDD (Figure 2A,E, respectively), which in turn depends on the VPDmax and Tmax (Figure 3A,B). In this sense, although there was less SWC during the second season, Tmax and VPDmax were lower too, reaffirming that these last two variables would have a more significant impact on the transpiratory and cooling capacity than even the SWC.

Finally, because the main differences in the dendrograms between the two- and three-environment groups were found in the division of the first branch of the less stressful environment (green lines in Figure 4A) at the height of the first knot of the most stressful environment (origin of blue and green lines in Figure 4B), it is logical to think that there were at least three environmental conditions across the trials. In order to establish the possible differences between the superimposed groups (lower and intermediate environmental constraints; green and blue lines in Figure 4D), each trait average was contrasted (percentage of change) with the respective value in the higher environmental limitations group (2011 data; red line in Figure 4D). Using this as a form of normalization, it was possible to establish which traits had the largest and smallest differences between the superimposed groups; GY varied by 52.4% while SDD, $\Delta^{13}C$, Tmax, VPDmax, and SWC only varied within the range of 9.77 and 15.1% (data not shown), and these were traits that probably shared similar spatial coordinates in the cluster plot (Figure 4B).

For modeling purposes, it would then be desirable to cease grouping collected data according to text code treatments (e.g., FI and WS according to only the amount of applied water) and start associating them with the environmental conditions existing during the season (e.g., VPDmax). For an adequate estimation of GY and carbon isotope discrimination by thermal imaging, we will then need to use more complex models (e.g., tree-based neural networks) that allow us to identify the "type of environment" in which the collected data should be manipulated to generate and apply the appropriate model. Thus, a deeper environmental characterization would allow development of models with better fit and consistency between years.

5. Conclusions and Future Perspectives

The ability to predict GY through the use of thermal images is highly variable and will not only depend on the amount of water stored in the soil profile, but also on other environmental variables such as VPDmax and Tmax. To the extent that better environmental characterization can be achieved, an objective and integral classification of the assessed environment should then be possible. Because the environmental information usually originates from a standard AWS, characterization of the environment at the canopy level or in the first few centimeters above it would also be an important consideration. This would help to generate models with better predictive capacity, thus improving the consistency between seasons.

Supplementary Materials: The following are available online at http://www.mdpi.com/1424-8220/19/12/2676/s1, Figure S1: (**A**) Canopy temperature—air temperature (SDD; (**A**) at the early dough stage of grain filling (Zadoks Z83); (**B**) grain yield; (**C**) daily maximum temperature (from sowing to harvest; carbon isotope discrimination ($\Delta^{13}C$) in kernels; and (**D**) daily maximum VPD from sowing to harvest, for genotypes of wheat grown under full irrigation (FI) and water stress (WS) at Santa Rosa (sr) and Cauquenes (c) during four growing seasons (2011, 2012, 2014, and 2015). Box and whiskers show minimum, 25th percentile, median, mean, 75th percentile, and maximum values. Open symbols represent outlier data; Figure S2: Soil water content between 0 and 50 cm depth and precipitation according to the water regime applied (full irrigation (FI) and water stress (WS)), the trial location (Santa Rosa (sr) and Cauquenes (c)), and the evaluated seasons: 2011 (**A**), 2012 (**B**), 2014 (**C**), and 2015 (**D**); the trial code is a combination of these factors. Bars represent the precipitation and the arrows the phenological stages at Santa Rosa (orange) and Cauquenes (black). Solid arrows indicate anthesis and dashed arrows the early dough stage (Z85 from the Zadoks scale).

Author Contributions: S.R.-B., A.d.P., and G.A.L. contributed to the conception and design of the work. S.R.-B., A.M.M.-E., M.G., F.E., A.E., L.G.-M., C.P.-E., D.S., A.d.P. and G.A.L. performed acquisition, analysis, and interpretation of the data. D.C. and I.M. was in charge of the management of the field experiments and evaluation of agronomic traits. S.R.-B., A.d.P. and G.A.L. collaborated to generate and validate the version for publication.

Funding: This work was supported by the National Commission for Scientific and Technological Research CONICYT (Beca doctorado nacional 21130514, FONDECYT N°1110678, FONDEF IDEA 14I10106, and 14I20106, PAI 781413006) and the Universidad de Talca, Chile (Beca doctorado and the research programs "Adaptation of Agriculture to Climate Change-A2C2" and "Núcleo Científico Multidisciplinario").

Conflicts of Interest: The authors declare that the work and publication was conducted in the absence of any commercial or financial relationships that could be construed as a potential conflict of interest.

References

1. Tanner, C.B. Plant Temperatures 1. *Agron. J.* **1963**, *55*, 210–211. [CrossRef]
2. Zia, S.; Spohrer, K.; Merkt, N.; Wenyong, D.; He, X.; Müller, J. Non-invasive water status detection in grapevine (*Vitis vinifera* L.) by thermography. *Int. J. Agric. Biol. Eng.* **2009**, *2*, 46–54. [CrossRef]
3. Jones, H.G.; Stoll, M.; Santos, T.; Sousa, C.D.; Chaves, M.M.; Grant, O.M. Use of infrared thermography for monitoring stomatal closure in the field: Application to grapevine. *J. Exp. Bot.* **2002**, *53*, 2249–2260. [CrossRef] [PubMed]
4. Leinonen, I.; Jones, H.G. Combining thermal and visible imagery for estimating canopy temperature and identifying plant stress. *J. Exp. Bot.* **2004**, *55*, 1423–1431. [CrossRef] [PubMed]
5. Möller, M.; Alchanatis, V.; Cohen, Y.; Meron, M.; Tsipris, J.; Naor, A.; Cohen, S. Use of thermal and visible imagery for estimating crop water status of irrigated grapevine. *J. Exp. Bot.* **2007**, *58*, 827–838. [CrossRef] [PubMed]
6. Wang, W.S.; Pan, Y.J.; Zhao, X.Q.; Dwivedi, D.; Zhu, L.H.; Ali, J.; Li, Z.K. Drought-induced site-specific DNA methylation and its association with drought tolerance in rice (*Oryza sativa* L.). *J. Exp. Bot.* **2010**, *62*, 1951–1960. [CrossRef] [PubMed]
7. Zia, S.; Spohrer, K.; Wenyong, D.; Spreer, W.; Romano, G.; Xiongkui, H.; Joachim, M. Monitoring physiological responses to water stress in two maize varieties by infrared thermography. *Int. J. Agric. Biol. Eng.* **2011**, *4*, 7–15. [CrossRef]
8. Jackson, R.D.; Idso, S.B.; Reginato, R.J.; Pinter, P.J., Jr. Canopy temperature as a crop water stress indicator. *Water Resour. Res.* **1981**, *17*, 1133–1138. [CrossRef]
9. Luquet, D.; Bégué, A.; Vidal, A.; Clouvel, P.; Dauzat, J.; Olioso, A.; Tao, Y. Using multidirectional thermography to characterize water status of cotton. *Remote Sens. Environ.* **2003**, *84*, 411–421. [CrossRef]
10. Jones, H.G. Application of thermal imaging and infrared sensing in plant physiology and ecophysiology. *Adv. Bot. Res.* **2004**, *41*, 107–163. [CrossRef]
11. Amani, I.; Fischer, R.A.; Reynolds, M.P. Canopy temperature depression association with yield of irrigated spring wheat cultivars in a hot climate. *J. Agron. Crop Sci.* **1996**, *176*, 119–129. [CrossRef]
12. Cossani, C.M.; Pietragalla, J.; Reynolds, M. Temperatura del dosel vegetal y características de la relación planta-agua. In *Fitomejoramiento fisiológico I: Enfoques interdisciplinarios para mejorar la adaptación del cultivo*; Reynolds, M.P., Pask, A.J.D., Mullan, D.M., Chávez-Dulanto, P.N., Eds.; CIMMYT: Ciudad de Mexico, Mexico, 2013; pp. 60–68. ISBN 978-607-8263-18-9.
13. Fischer, R.A.; Rees, D.; Sayre, K.D.; Lu, Z.M.; Condon, A.G.; Saavedra, A.L. Wheat yield progress associated with higher stomatal conductance and photosynthetic rate, and cooler canopies. *Crop Sci.* **1998**, *38*, 1467–1475. [CrossRef]
14. Leinonen, I.; Grant, O.M.; Tagliavia, C.P.P.; Chaves, M.M.; Jones, H.G. Estimating stomatal conductance with thermal imagery. *Plant Cell Environ.* **2006**, *29*, 1508–1518. [CrossRef] [PubMed]
15. Grant, O.M.; Tronina, Ł.; Jones, H.G.; Chaves, M. Exploring thermal imaging variables for the detection of stress responses in grapevine under different irrigation regimes. *J. Exp. Bot.* **2007**, *58*, 815–825. [CrossRef] [PubMed]
16. Prashar, A.; Jones, H.G. Infra-red thermography as a high-throughput tool for field phenotyping. *Agronomy* **2014**, *4*, 397–417. [CrossRef]

17. Prashar, A.; Jones, H.G. Assessing drought responses using thermal infrared imaging. In *Environmental Responses in Plants. Methods and Protocols, Methods in Molecular Biology*; Duque, P., Ed.; Humana Press: New York, NY, USA, 2016; Volume 1398, pp. 209–219. [CrossRef]

18. Blum, A.; Shpiler, L.; Golan, G.; Mayer, J. Yield stability and canopy temperature of wheat genotypes under drought-stress. *Field Crops Res.* **1989**, *22*, 289–296. [CrossRef]

19. Condon, A.G.; Richards, R.A.; Farquhar, G.D. Carbon isotope discrimination is positively correlated with grain yield and dry matter production in field-grown wheat. *Crop Sci.* **1987**, *27*, 996–1001. [CrossRef]

20. Farquhar, G.D.; Hubick, K.T.; Condon, A.G.; Richards, R.A. Carbon isotope fractionation and plant water-use efficiency. In *Stable Isotopes in Ecological Research*; Rundel, P.W., Ehleringer, J.R., Nagy, K.A., Eds.; Springer: New York, NY, USA, 1989; pp. 21–40. [CrossRef]

21. Condon, A.G.; Richards, R.A.; Farquhar, G.D. The effect of variation in soil water availability, vapour pressure deficit and nitrogen nutrition on carbon isotope discrimination in wheat. *Aust. J. Agric. Res.* **1992**, *43*, 935–947. [CrossRef]

22. Araus, J.L.; Amaro, T.; Casadesus, J.; Asbati, A.; Nachit, M.M. Relationships between ash content, carbon isotope discrimination and yield in durum wheat. *Funct. Plant Biol.* **1998**, *25*, 835–842. [CrossRef]

23. Araus, J.L.; Villegas, D.; Aparicio, N.; Del Moral, L.F.; El Hani, S.; Rharrabti, Y.; Royo, C. Environmental factors determining carbon isotope discrimination and yield in durum wheat under Mediterranean conditions. *Crop Sci.* **2003**, *43*, 170–180. [CrossRef]

24. Lobos, G.A.; Matus, I.; Rodriguez, A.; Romero-Bravo, S.; Araus, J.L.; del Pozo, A. Wheat genotypic variability in grain yield and carbon isotope discrimination under mediterranean conditions assessed by spectral reflectance. *J. Integr. Plant Biol.* **2014**, *56*, 470–479. [CrossRef] [PubMed]

25. Cabrera-Bosquet, L.; Albrizio, R.; Nogués, S.; Araus, J.L. Dual $\Delta^{13}C/\delta^{18}O$ response to water and nitrogen availability and its relationship with yield in field-grown durum wheat. *Plant Cell Environ.* **2010**, *34*, 418–433. [CrossRef] [PubMed]

26. Garriga, M.; Romero-Bravo, S.; Estrada, F.; Escobar, A.; Matus, I.; del Pozo, A.; Astudillo, C.; Lobos, G. Assessing wheat traits by spectral reflectance: Do we really need to focus on predicted trait-values or directly identify the elite genotypes group? *Front. Plant Sci.* **2017**, *8*, 280. [CrossRef] [PubMed]

27. Moretti, C.L.; Mattos, L.M.; Calbo, A.G.; Sargent, S.A. Climate changes and potential impacts on postharvest quality of fruit and vegetable crops: A review. *Food Res. Int.* **2010**, *43*, 1824–1832. [CrossRef]

28. Lobos, G.A.; Hancock, J.F. Breeding blueberries for a changing global environment: A review. *Front. Plant Sci.* **2015**, *6*, 782. [CrossRef] [PubMed]

29. Camargo, A.V.; Lobos, G.A. Latin America: A development pole for phenomics. *Front. Plant Sci.* **2016**, *7*, 1729. [CrossRef] [PubMed]

30. Lobos, G.A.; Camargo, A.V.; del Pozo, A.; Araus, J.L.; Ortiz, R.; Doonan, J.H. Plant Phenotype and phenomics for plant breeding. *Front. Plant Sci.* **2017**, *8*, 2181. [CrossRef]

31. Sadras, V.O.; Reynolds, M.P.; De la Vega, A.J.; Petrie, P.R.; Robinson, R. Phenotypic plasticity of yield and phenology in wheat, sunflower and grapevine. *Field Crops Res.* **2009**, *110*, 242–250. [CrossRef]

32. Mora, F.; Castillo, D.; Lado, B.; Matus, I.; Poland, P.; Belzile, F.; von Zitzewitz, J.; del Pozo, A. Genome-wide association mapping of agronomic traits and carbon discrimination in a worldwide germplasm collection of spring wheat using SNP markers. *Mol. Breed.* **2015**, *35*. [CrossRef]

33. del Pozo, A.; Yáñez, A.; Matus, I.A.; Tapia, G.; Castillo, D.; Sanchez-Jardón, L.; Araus, J.L. Physiological traits associated with wheat yield potential and performance under water-stress in a mediterranean environment. *Front. Plant Sci.* **2016**, *7*, 987. [CrossRef]

34. Zhang, D.; Du, Q.; Zhang, Z.; Jiao, X.; Song, X.; Li, J. Vapour pressure deficit control in relation to water transport and water productivity in greenhouse tomato production during summer. *Sci. Rep.* **2017**, *7*, 43461. [CrossRef] [PubMed]

35. Hernández, J.; Lobos, G.A.; Matus, I.; del Pozo, A.; Silva, P.; Galleguillos, M. Using ridge regression models to estimate grain yield from field spectral data in bread wheat (*Triticum aestivum* L.) grown under three water regimes. *Remote Sens.* **2015**, *7*, 2109–2126. [CrossRef]

36. Zadoks, J.C.; Chang, T.T.; Konzak, C.F. A decimal code for the growth stages of cereals. *Weed Res.* **1974**, *14*, 415–421. [CrossRef]

37. Allen, R.G.; Pereira, L.S.; Raes, D.; Smith, M. Crop Evapotranspiration: Guidelines for Computing Crop Water Requirements. In *FAO Irrigation and Drainage Paper N° 56*; FAO: Rome, Italy, 1998; 300p.

38. Fuentes, S.; De Bei, R.; Pech, J.; Tyerman, S. Computational water stress indices obtained from thermal image analysis of grapevine canopies. *Irrig. Sci.* **2012**, *30*, 523–536. [CrossRef]

39. Idso, S.B.; Jackson, R.D.; Reginato, R.J. Remote sensing of crop yields. *Science* **1977**, *196*, 19–25. [CrossRef]

40. Jackson, R.D.; Reginato, R.J.; Idso, S.B. Wheat canopy temperature: A practical tool for evaluating water requirements. *Water Resour. Res.* **1977**, *13*, 651–656. [CrossRef]

41. Nerurkar, P.; Shirke, A.; Chandane, M.; Bhirud, S. Empirical analysis of data clustering algorithms. *Procedia Comput. Sci.* **2018**, *125*, 770–779. [CrossRef]

42. Murtagh, F.; Legendre, P. Ward's hierarchical agglomerative clustering method: Which algorithms implement ward's criterion? *J. Classif.* **2014**, *31*, 274–295. [CrossRef]

43. R Development Core Team. *R: A Language and Environment for Statistical Computing*; R Foundation for Statistical Computing: Vienna, Austria, 2011.

44. Ryan, A.C.; Dodd, I.C.; Rothwell, S.A.; Jones, R.; Tardieu, F.; Draye, X.; Davies, W.J. Gravimetric phenotyping of whole plant transpiration responses to atmospheric vapour pressure deficit identifies genotypic variation in water use efficiency. *Plant Sci.* **2016**, *251*, 101–109. [CrossRef]

45. Farquhar, G.D.; O'Leary, M.H.; Berry, J.A. On the relationship between carbon isotope discrimination and the intercellular carbon dioxide concentration in leaves. *Funct. Plant Biol.* **1982**, *9*, 121–137. [CrossRef]

46. Acevedo, E.H.; Baginsky, C.G.; Solar, B.R.; Ceccarelli, S. Discriminación isotópica de C13 y su relación con el rendimiento y la eficiencia de transpiracion de genotipos locales y mejorados de cebada bajo diferentes condiciones hídricas. *Inv. Agric.* **1997**, *17*, 41–54.

47. Rebetzke, G.J.; Condon, A.G.; Farquhar, G.D.; Appels, R.; Richards, R.A. Quantitative trait loci for carbon isotope discrimination are repeatable across environments and wheat mapping populations. *Theor. Appl. Genet.* **2008**, *118*, 123–137. [CrossRef] [PubMed]

48. del Pozo, A.; Castillo, D.; Inostroza, L.; Matus, I.; Méndez, A.M.; Morcuende, R. Physiological and yield responses of recombinant chromosome substitution lines of barley to terminal drought in a mediterranean type environment. *Ann. Appl. Biol.* **2012**, *160*, 157–167. [CrossRef]

49. Araus, J.L.; Cabrera-Bosquet, L.; Serret, M.D.; Bort, J.; Nieto-Taladriz, M.T. Comparative performance of δ^{13}C, δ^{18}O and δ^{15}N for phenotyping durum wheat adaptation to a dryland environment. *Funct. Plant Biol.* **2013**, *40*, 595–608. [CrossRef]

50. Passioura, J.B. Grain yield, harvest index, and water use of wheat. *J. Aust. Inst. Agric. Sci.* **1977**, *43*, 117–120.

51. French, R.J.; Schultz, J.E. Water use efficiency of wheat in a Mediterranean-type environment. I. The relation between yield, water use and climate. *Aust. J. Agric. Res.* **1984**, *35*, 743–764. [CrossRef]

52. Sun, H.Y.; Liu, C.M.; Zhang, X.Y.; Shen, Y.J.; Zhang, Y.Q. Effects of irrigation on water balance, yield and WUE of winter wheat in the north China plain. *Agric. Water Manag.* **2006**, *85*, 211–218. [CrossRef]

53. Li, F.M.; Liu, X.L.; Li, S.Q. Effects of early soil water distribution on the dry matter partition between roots and shoots of winter wheat. *Agric. Water Manag.* **2001**, *49*, 163–171. [CrossRef]

54. Anderson, D.B. Relative humidity or vapor pressure deficit. *Ecology* **1936**, *17*, 277–282. [CrossRef]

55. Dreccer, M.F.; Fainges, J.; Whish, J.; Ogbonnaya, F.C.; Sadras, V.O. Comparison of sensitive stages of wheat, barley, canola, chickpea and field pea to temperature and water stress across Australia. *Agric. For. Meteorol.* **2018**, *248*, 275–294. [CrossRef]

56. Rashid, M.A.; Andersen, M.N.; Wollenweber, B.; Zhang, X.; Olesen, J.E. Acclimation to higher VPD and temperature minimized negative effects on assimilation and grain yield of wheat. *Agric. For. Meteorol.* **2018**, *248*, 119–129. [CrossRef]

57. Sharifi, M.R.; Rundel, P.W. The effect of vapour pressure deficit on carbon isotope discrimination in the desert shrub *Larrea tridentata* (creosote bush). *J. Exp. Bot.* **1993**, *44*, 481–487. [CrossRef]

58. Franks, P.J. Stomatal control and hydraulic conductance, with special reference to tall trees. *Tree Physiol.* **2004**, *24*, 865–878. [CrossRef] [PubMed]

59. Blum, A.; Mayer, J.; Gozlan, G. Infrared thermal sensing of plant canopies as a screening technique for dehydration avoidance in wheat. *Field Crops Res.* **1982**, *5*, 137–146. [CrossRef]

60. Reynolds, M.P.; Balota, M.; Delgado, M.I.B.; Amani, I.; Fischer, R.A. Physiological and morphological traits associated with spring wheat yield under hot, irrigated conditions. *Funct. Plant Biol.* **1994**, *21*, 717–730. [CrossRef]

61. Olivares-Villegas, J.J.; Reynolds, M.P.; McDonald, G.K. Drought-adaptive attributes in the Seri/Babax hexaploid wheat population. *Funct. Plant Biol.* **2007**, *34*, 189–203. [CrossRef]

62. Idso, S.B.; Reginato, R.J.; Clawson, K.L.; Anderson, M.G. On the stability of non-water-stressed baselines. *Agric. For. Meteorol.* **1984**, *32*, 177–182. [CrossRef]

63. Hatfield, J.L. Measuring plant stress with an infrared thermometer. *HortScience* **1990**, *25*, 1535–1538.

Article

UAV-Borne Dual-Band Sensor Method for Monitoring Physiological Crop Status

Lili Yao, Qing Wang, Jinbo Yang, Yu Zhang, Yan Zhu, Weixing Cao * and Jun Ni *

National Engineering and Technology Center for Information Agriculture, Key Laboratory for Crop System Analysis and Decision Making, Ministry of Agriculture, Jiangsu Key Laboratory for Information Agriculture, Nanjing Agricultural University, Nanjing 210095, Jiangsu, China; 2017201083@njau.edu.cn (L.Y.); 2016101037@njau.edu.cn (Q.W.); 2018101166@njau.edu.cn (J.Y.); zhangyu@njau.edu.cn (Y.Z.); yanzhu@njau.edu.cn (Y.Z.)
* Correspondence: caow@njau.edu.cn (W.C.); nijun@njau.edu.cn (J.N.); Tel./Fax: +86-25-8439-6593 (J.N.)

Received: 29 December 2018; Accepted: 14 February 2019; Published: 17 February 2019

Abstract: Unmanned aerial vehicles (UAVs) equipped with dual-band crop-growth sensors can achieve high-throughput acquisition of crop-growth information. However, the downwash airflow field of the UAV disturbs the crop canopy during sensor measurements. To resolve this issue, we used computational fluid dynamics (CFD), numerical simulation, and three-dimensional airflow field testers to study the UAV-borne multispectral-sensor method for monitoring crop growth. The results show that when the flying height of the UAV is 1 m from the crop canopy, the generated airflow field on the surface of the crop canopy is elliptical, with a long semiaxis length of about 0.45 m and a short semiaxis of about 0.4 m. The flow-field distribution results, combined with the sensor's field of view, indicated that the support length of the UAV-borne multispectral sensor should be 0.6 m. Wheat test results showed that the ratio vegetation index (RVI) output of the UAV-borne spectral sensor had a linear fit coefficient of determination (R^2) of 0.81, and a root mean square error (RMSE) of 0.38 compared with the ASD Fieldspec2 spectrometer. Our method improves the accuracy and stability of measurement results of the UAV-borne dual-band crop-growth sensor. Rice test results showed that the RVI value measured by the UAV-borne multispectral sensor had good linearity with leaf nitrogen accumulation (LNA), leaf area index (LAI), and leaf dry weight (LDW); R^2 was 0.62, 0.76, and 0.60, and RMSE was 2.28, 1.03, and 10.73, respectively. Our monitoring method could be well-applied to UAV-borne dual-band crop growth sensors.

Keywords: CFD; airflow field test; monitoring method; spectral sensor; crop growth

1. Introduction

Accurate management of crop water and fertilizer in crop fields is an important prerequisite for ensuring high yield and quality of crops, sustainable use of cultivated land, and healthy development of the environment [1]. High-throughput, accurate, and real-time acquisition of crop-growth information is an important basis for the accurate management of crop water and fertilizer [2]. Monitoring technology based on the characteristics of the reflection spectrum has the advantages of being nondestructive, providing real-time information, and delivering high-efficiency analysis. Thus, it is widely used in crop-growth parameter acquisition. Various research institutions have developed spectral sensors to monitor crop growth, providing effective technical support for field-crop production management [3–5].

In 2004, Moya et al. [6] designed a chlorophyll-fluorescence test device, which uses sunlight as a light source. During the test, the leaf blade was required to be in a relatively static state, and the reflection=spectrum information of chlorophyll fluorescence in the 510 and 570 nm bands could be obtained from a short distance. Quantitative inversion of chlorophyll fluorescence could be achieved by the physiological reflectance index (PRI) calculated from the test results. In 2010, Ryu et al. [7] developed a normalized vegetation index spectroscopy sensor to achieve the inversion of vegetation-growth indicators. It has its own LED light source for illumination. When the test height was less than 3 m, the best test results could be achieved. Several commercial instruments are currently available for crop-growth monitoring. For example, the Greenseeker spectral sensor designed by Trimble USA can obtain the spectral information of reflection characteristics in crop canopy red and near-infrared bands and calculate the relevant vegetation index. For this device, the test distance should be kept at a height of 60–180 cm from the canopy [8,9]. The ASD FieldSpec4 spectrometer developed by American ASD Company can realize reflection-spectrum acquisition of the 350–2500 nm crop-canopy band, the data information is rich, and accuracy is high. Relying on sunlight as a light source, the test needs to be carried out at noon without wind or clouds, the test height should be kept between 30–120 cm, and the crop canopy needs to remain relatively static [10–12]. Holland Scientific designed and developed active light-source spectral sensors for monitoring crop growth, such as Crop Circle and Rapidscan. These instruments can emit light and receive reflection-spectrum information of a crop canopy in real time through their own light-source system. The test height should be kept within 3 m from the canopy, and the canopy structure needs to maintain a steady state [13–17]. These spectral crop-growth monitoring devices are simple in operation, easy to carry, high in test accuracy, intuitive in results, and can provide nondestructive access to crop-growth information, but they also have shortcomings, such as a small monitoring range, high labor intensity, and discontinuous monitoring, which cannot provide high-throughput information and real-time decision-making for large-scale crop-production management in the field.

Unmanned-aerial-vehicle (UAV) operation is highly efficient, flexible, easy, and has strong terrain applicability. Thus, UAVs are widely used in agricultural information-acquisition platforms [12], but so far there are few studies on the use of UAV-borne spectral sensors for monitoring crop growth. Krienke et al. attempted to measure the normalized vegetation index of lawn using a MikroKopterOktoKopter XL UAV equipped with a RapidScan CS-45 spectroscopy sensor. Flight height was maintained at 0.5–1.5 m above the lawn. However, test results were poor because the disturbance of the turf canopy from the downwash airflow field was ignored [18]. Shafian et al. mounted an image sensor on a fixed-wing UAV and collected the image information of a sorghum planting area at an altitude of 120 m. Pix4D software was used to splice, correct, and extract vegetation indices from each acquired image. The leaf area index (LAI) value of sorghum was simultaneously sampled and tested. The results show that the Normalized Difference Vegetation Index (NDVI) value extracted from the acquired image information had a higher linear fit with the sorghum LAI value obtained with the sampling test [19]. Schirrmann et al. used a UAV to work at an altitude of 50 m and obtain an RGB image of the wheat growth period. The acquired image was calibrated by Agisoft lens software, the distortion correction was modeled by Brown distortion model, and the final image mosaic and surface model generation results were improved by radiation pretreatment, from which the information such as crop coverage and plant height were extracted [20]. Zheng et al. used an OktoXL UAV equipped with a Cuubert UHD 185 hyperspectral camera to acquire hyperspectral images of a rice canopy at an altitude of 50 m. Images in the acquisition results were corrected using ENVI software. By comparing the synchronous test results of a ground-object spectrometer and the agronomic parameters obtained from laboratory chemical analysis, they proved that the image information acquired by the UAV platform equipped with imaging instruments could be used for the quantitative inversion of agronomic crop parameters [21]. Stroppianaet al. acquired a large number of multispectral images at a height of 70 m from the ground by using a 3DRobotics SOLO quadrotor UAV equipped with a Parrot Sequoia multispectral camera. By screening the acquired images, and then correcting and extracting the vegetation index, they proposed an automatic classification method for weeds and crops, which can be used for the classification and management of specific weeds in the field [22]. However, these studies simply installed image sensors on UAVs to obtain crop images from a high altitude. Although the influence of the UAV downwash airflow field is small, captured images can only be stored in the memory. Scientific-research personnel are required to use special software for image correction, cropping, splicing, enhancement, and other offline processing, and then analyze the relationship between the images and crop-growth parameters. The process is complex, requires specialized knowledge, and cannot acquire information in real time, which is not conducive to popularization [23,24].

In this paper, we studied the dual-band crop-growth sensor independently developed by Nanjing Agricultural University [25]. First, we investigated the monitoring method of the UAV-borne dual-band crop-growth sensor based on its spectral-monitoring mechanism and structural-design features. Then, we analyzed the spatial-distribution characteristics of the airflow field under the low-altitude hovering operation of the UAV. Finally, we built a UAV-borne crop-growth monitoring system to achieve high throughput and real-time access to rice- and wheat-growth information.

2. Materials and Methods

2.1. Test Equipment

2.1.1. Dual-Band Crop-Growth Sensor

There is a certain relationship between crop growth and the spectral reflectance of the crop canopy. As shown in Figure 1, the reflectivity of a wheat canopy is relatively low when the band is under 710 nm. Reflectivity linearly rises in the 710–760 nm band, and reaches a comparatively high level in the 760–1210 nm band. Wheat-canopy reflectivity shows significant differences between different nitrogen levels for the 460–730 and 760–1210 nm bands. In the 460–730 nm band, spectral reflectivity is negatively correlated with the amount of nitrogen fertilizer, with reflectivity being the lowest at N5 and highest at N1. In the 760–1210 nm band, spectral reflectivity is positively correlated with the amount of nitrogen fertilizer, with reflectivity being the highest at N5 and lowest at N1. The 710–760 nm band is an apparent transition zone. The spectral reflectivity of the wheat canopy at above 1150 nm is not very susceptible to the amount of nitrogen fertilizer, and there is little difference between spectral reflectivity at N2–N5. According to currently available research, there is a good linear relationship between leaf nitrogen accumulation (LNA) and spectral-reflectivity changes near 550, and 600–700 and 720 nm; there is a good linear relationship between leaf dry weight (LDW), and spectral-reflectivity changes at 580–700 and 770–900 nm; the spectral-reflectivity change at 460–680 nm and near 810 nm is closely correlated the LAI. Considering the sensitive bands of the three agronomic parameters, crop growth can be well-inverted with the ratio vegetation index (RVI) index constructed using the 730 and 810 nm bands [26–28].

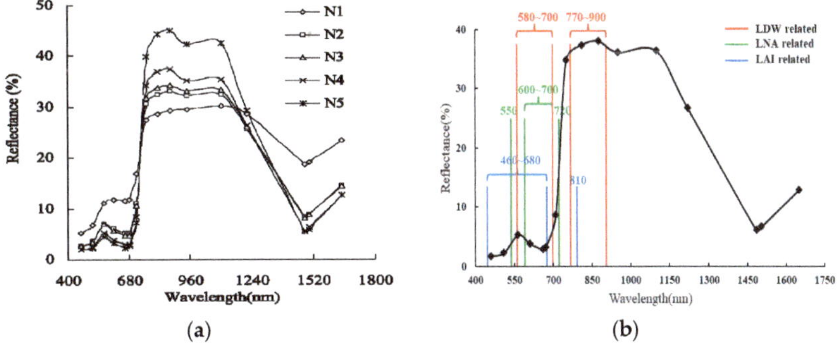

Figure 1. Characteristic curve of the spectral reflectivity of wheat canopy. (**a**) Multispectral reflectivity of wheat canopy under application of different amounts of nitrogen, and (**b**) characteristic spectral bands of agronomic wheat parameters.

The dual-band crop-growth sensor was designed by Nanjing Agricultural University, it is equipped with a dual-band detection lens, and its structure can be divided into an upward light sensor and a downward light sensor, as shown in Figure 2. The upward light sensor is used to acquire sunlight-radiation information at 730 and 815 nm wavelengths, and the downward light sensor was configured to receive crop-canopy-reflected light-radiation information of a corresponding wavelength. The structure is shown in Figure 2. It is packaged in a nylon case and weighs 11.34 g, with a test field of view of 27°. The crop-canopy RVI can be output in real time, and wireless transmission can achieve long-distance transmission and analysis.

With sunlight as the light source, dual-band crop-growth sensors require the testing object (i.e., wheat canopy) to remain relatively static so that the canopy presents the Lambertian reflection characteristics and the field of view of the sensor points vertically downward. During measurement, optical-radiation energy is converted to electric-energy signals by the sensor. Therefore, to ensure high sensitivity, measuring height should be maintained at 1.0–1.5 m above the canopy. The principle is shown in Figure 2.

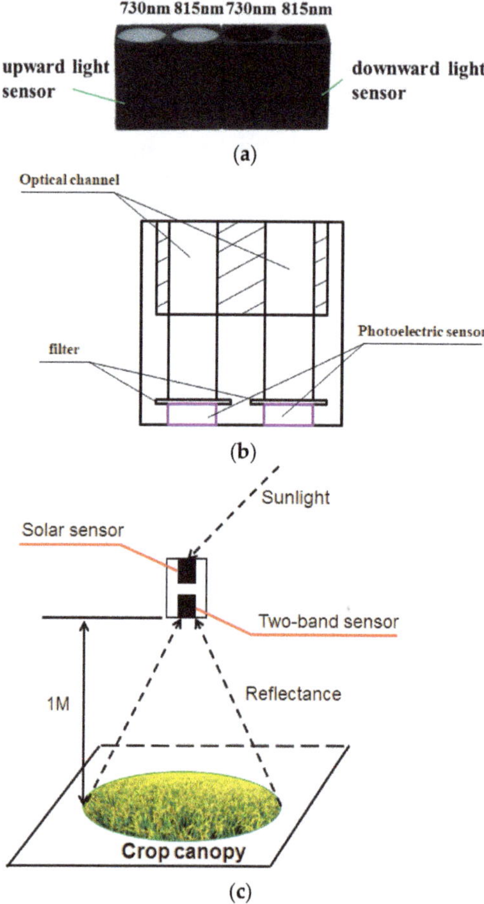

Figure 2. Introduction of dual-band crop-growth sensor. (**a**) Sensor; (**b**) sensor structure; (**c**) schematic diagram of dual-band crop-growth sensor test.

2.1.2. ASD FieldSpec HandHeld 2 Handheld Spectrometer

Developed by American Analytical Spectral Devices (AS, made by Advanced Systems Development, Inc., Alexandria, VA, America), the ASD FieldSpec HandHeld 2 handheld spectrometer can be used for the reflection-spectrum acquisition of different objects such as crops, marine organisms, and minerals. The device has the advantages of portability, simple operation, and accurate results. Test wavelength range is 325–1075 nm, wavelength accuracy is ±1 nm, spectral resolution is less than 3 nm, and test field of view is 25° [29,30].

2.1.3. LAI-2200C Vegetation Canopy Analyzer

The LAI-2200C (Made by LI-COR, Lincoln, NE, USA) is a vegetation-canopy analyzer manufactured by LI-COR, United States. The analyzer is light in weight, consumes little power, and is very suitable for outdoor measurements. In addition, the analyzer can work independently, perform unattended long-term continuous measurement, and automatically record data. The measurement principle is to measure the transmitted light at five angles above and below the vegetation canopy by using the "fisheye" optical sensor (which has a 148° vertical field of view and a 360° horizontal field of view), and calculate canopy-structure parameters such as LAI, average leaf dip angle, void ratio, and aggregation index by using the radiation-transfer model of a vegetation canopy [31–33].

2.1.4. Three-Dimensional Airflow Field Tester

The three-dimensional (3D) airflow field tester (South China Agricultural University, Guangzhou, China) uses three wind-speed sensors to test X-, Y-, and Z-axial airflow velocity. The test results are transmitted to a computer through the Zigbee module in real time. The power consumption of the sensor is little, and it can maintain continuous operation for a long time. The center point of the three test axes was fixed to a height of 60 cm from the ground, and the distance from each wind speed sensor to the center point was 15 cm. The three test axes were kept perpendicular to each other. The structure is shown in Figure 3.

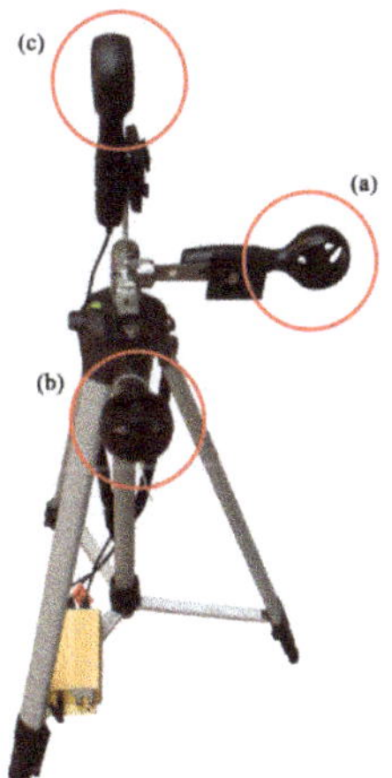

Figure 3. Structure of three-dimensional airflow field tester. (**a**) X-axial velocity test axis; (**b**) Y-axial velocity test axis; (**c**) Z-axial velocity test axis.

2.2. Research Methods

When the dual-band crop-growth sensor is measuring, test height is required to maintain a distance of 1–1.5 m from the crop canopy, and the crop canopy must remain relatively static, exhibiting Lambertian reflection characteristics. However, while the UAV is hovering at a low altitude, rotors rotate at a high speed, which causes the surrounding airflow to shrink and contract, forming an airflow field; this airflow field acts on the crop canopy, causing disturbance to it, destroying the Lambertian reflection characteristics of the crop canopy, affecting the test results, and even preventing completion of the test. Therefore, when a UAV is equipped with a dual-band crop-growth sensor for crop-growth monitoring, it is necessary to consider the disturbance effect of the downwash airflow field on the crop canopy.

To solve this problem, we analyzed the spatial distribution of the UAV rotor downwash airflow field by using computational fluid dynamics (CFD) and 3D airflow field testers. We then determined an acceptable deployment location for the dual-band crop-growth sensor based on the surface velocity and distribution range of the airflow field.

2.2.1. Airflow-Field Numerical Simulation

CFD is a branch of fluid mechanics. It uses computers as tools and applies various discrete mathematical methods to conduct numerical experiments, computer simulations, and analytical studies on various fluid-mechanics problems. The advantage of CFD lies in its ability to simulate the experimental process from basic physics theorems instead of expensive fluid-dynamics experiment equipment. According to the specific process of CFD numerical simulation analysis [25], we obtained 3D-sized data of the DJI phantom drone (a type of UAV) with a 3D scanning system, converted the 3D information of the UAV into a digital signal that could be processed by the computer, and used CFD to perform UAV mesh generation and numerical solution. The second-order upwind style was selected for calculation to improve accuracy [34–36]. The 3D scanning technology was only used to measure the profile data of the UAV fuselage and rotor, and construct a 3D UAV model. This is not a 3D model of the crop-canopy structure. The specific calculation process is shown in Figure 4.

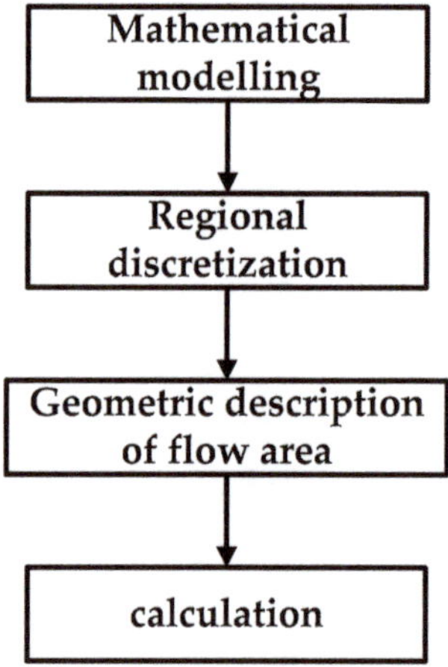

Figure 4. Computational fluid dynamics (CFD) mesh generation and numerical solution process.

Physical parameters such as UAV flight area, and rotor-rotation speed and direction, were set by CFD software, and UAV hovering-operation-state simulation analysis was carried out. In the grid simulation area with a diameter of 1.2 m and a height of 1.85 m, the velocity of each grid node is composed of the X-axle, Y-axle, and Z-axle velocity components. The flying height of the UAV was set to 1 m from the ground, and the ground-velocity nephogram distribution result was displayed by CFX's own post processing module, as shown in Figure 5.

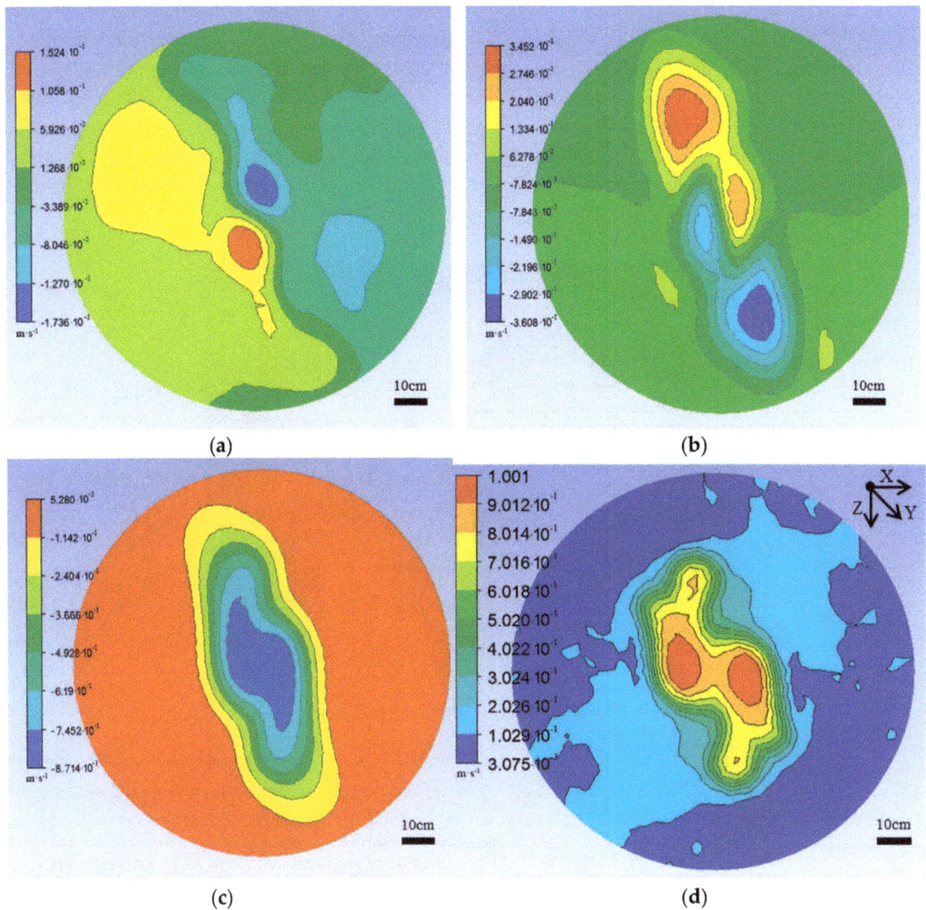

Figure 5. Ground-velocity nephogram distribution results. (**a**) X-axial velocity distribution nephogram; (**b**) Y-axial velocity distribution nephogram; (**c**) Z-axial velocity distribution nephogram; (**d**) combined velocity distribution nephogram.

Velocity intensity was analyzed, as shown in Figure 5. When the UAV was hovering at a height of 1 m from the ground, the Z-axial velocity component was much larger than the X-axial and Y-axial velocity components. The combined velocity depends mainly on the axial velocity component. The size of the axial velocity nephogram was elliptical, the central region had the highest wind speed, and it gradually decreased toward the periphery. The results of the combined velocity nephogram distribution show that, due to the opposite rotation of the two pairs of rotors of the quadrotor UAV, forward-rotating rotors generate a downwash flow, so the region with the highest wind speed is distributed directly below the two forward-rotating rotors, wind speed gradually decreases toward the periphery, and is also distributed in an elliptical shape. The long semiaxis of the region is about 0.35 m, and the short semiaxis is about 0.3 m.

2.2.2. Actual Test of Airflow Field

In order to verify the numerical simulation analysis results of the airflow field, the 3D airflow field testers were used to carry out a real-world test of the downwash airflow field under the hovering operation state of the UAV. Nine 3D airflow field testers were used to test the X-axial, Y-axial, and

Z-axial velocity components of the downwash airflow field. The testers were arranged in an array structure of three rows and three columns at equal intervals. The UAV was hovering and flying at a vertical position above the center of the array. Flight height was 1 m from the array plane. When the distance between adjacent testers was 0.6 m, the wind-speed data collected by the testers at the edge of the array were 0 m/s. Therefore, tester spacing was adjusted to 0.5 m, so testers at the edge of the array collected nonzero data, which met the test requirements. In the test, after the flight attitude of the UAV was stabilized, each three-dimensional airflow field tester stopped the test after collecting 100 datasets. The test process is shown in Figure 6.

Figure 6. Actual test of downwash flow field.

To reduce the error, the 10 maximum and 10 minimum values were removed from the 100 collected datasets before calculating the average value from the remaining data. The obtained result was solved by interpolation using the four-point spline-interpolation (V4) algorithm. When the adjacent tester spacing was 0.6 m, the edge-tester test result was 0 m/s. Therefore, the interpolation boundary was set to 0.6 m away from the center. The V4 interpolation algorithm is also called the interpolation algorithm based on biharmonic Green's functions. The difference surface is a linear combination of Green's functions centered on each sample. The surface is passed through various points by adjusting the weight of each point. Green's function of the spline satisfies the following biharmonic equation:

$$\frac{d^4\phi}{dx^4} = 6\delta(x) \tag{1}$$

The specific solution of Equation (1) is

$$\phi(x) = |x|^3 \tag{2}$$

When Green's function is used to interpolate N data points, the problem of w_i interpolation at x_i is

$$\frac{d^4w}{d^4} = \sum_{j=1}^{N} 6\alpha_j \delta(x - x_j) \tag{3}$$

$$w(x_i) = w_i \tag{4}$$

The specific solution to Equations (3) and (4) is that Green's function is linearly combined around each data point, eliminating the need for a uniform solution.

$$w(x) = \sum_{j=1}^{N} \alpha_j |x - x_j|^3 \tag{5}$$

Green's function of the plane space is shown in Table 1:

Table 1. Function in flat space.

Dimension m	Green's Function $\phi_m(x)$
1	$|x|^3$
2	$|x|^2(\ln|x|-1)$
3	$|x|$

Weight value α_j is obtained by using the x values and $w(x)$ of N points, and the interpolation results are obtained by substituting the weight value into Equation (5).

The data of each 3D airflow field tester were sorted, invalid data were eliminated, valid data were retained, and data results were analyzed and processed. According to the V4 interpolation principle and calculated by MATLAB software, the interpolation results of each 3D airflow field tester node data are displayed in Figure 7.

From the results of Figure 7, we can see that, when the UAV hovering flight height was 1 m from the test plane of the 3D airflow field testers, the Z-axial velocity component of the tester was much larger than the X-axial and Y-axial velocity components. The combined velocity mainly depends on the axial velocity component. In the distribution range, the axial velocity component is elliptical, and the center velocity is the largest and gradually decreases toward the periphery. The combined-velocity distribution results show that the influence range of the UAV downwash airflow field was also elliptical, the center-point velocity was the largest, and peripheral speed was gradually reduced. The long semiaxis of the affected area was about 0.45 m, and the short semiaxis was about 0.4 m. Test results are consistent with the CFD numerical simulation results.

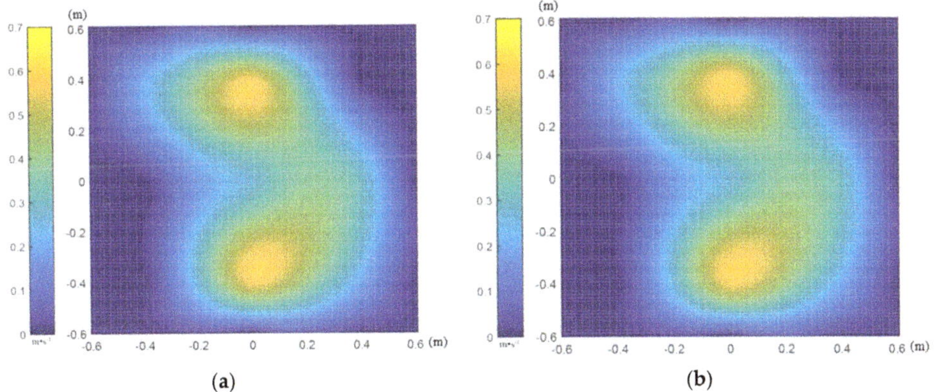

(a) (b)

Figure 7. *Cont.*

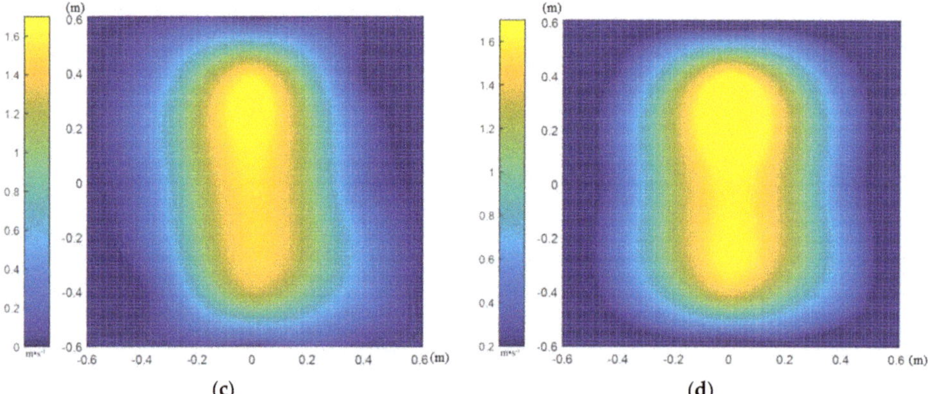

Figure 7. Three-dimensional tester interpolation results when the DJI phantom drone (unmanned aerial vehicle, UAV) hover-flight height is 1 m from 3D airflow field testers. (**a**) X-axial velocity profile; (**b**) Y-axial velocity profile; (**c**) Z-axial velocity profile; (**d**) combined velocity profile.

2.2.3. Dual-Band Crop-Growth-Sensor Deployment Method

According to CFD numerical simulation analysis and the actual test results of the 3D airflow field, combined with the test field of view of the dual-band crop-growth sensor, when the dual-band crop-growth sensor was deployed 60 cm away from the UAV rotors, the test area of the crop canopy retained Lambert characteristics, and measurement results were not affected by the airflow field, enabling normal testing. Therefore, we here designed a carbon-fiber sensor support with a length of 60 cm. One end of the support was fixed under the UAV spiral wing, and the other end extends outward along the UAV arm. The sensor was fixed to the rear end of the support through a mechanical structure. The support is connected to the UAV by a cantilever beam structure. We also designed supports of other lengths for experimental comparison.

In order to avoid the vibration impact of rotor high-speed rotation on the dual-band crop-growth sensor and the support during the flight, a damping rod was designed for shock absorption in order to maintain the stable state of the support and the dual-band crop-growth sensor. The support and damping rod were fixed with a triangular structure to improve the overall stability and shock resistance, as shown in Figure 8. The angle between support and damping rod is very important for the stability of the structure and the balance performance of the aircraft. Therefore, the optimal value of the angle was calculated by static equation analysis.

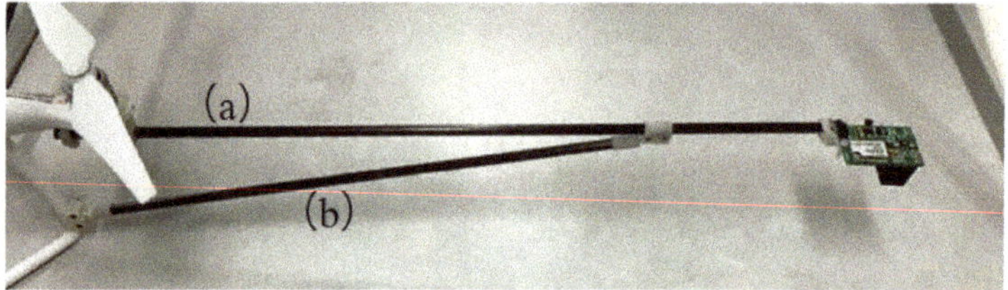

Figure 8. Design of UAV support and damping rod. (**a**) Support; (**b**) damping rod.

Taking the sensor support as the research object, the analysis diagram of the force sustained by the support is shown in Figure 9.

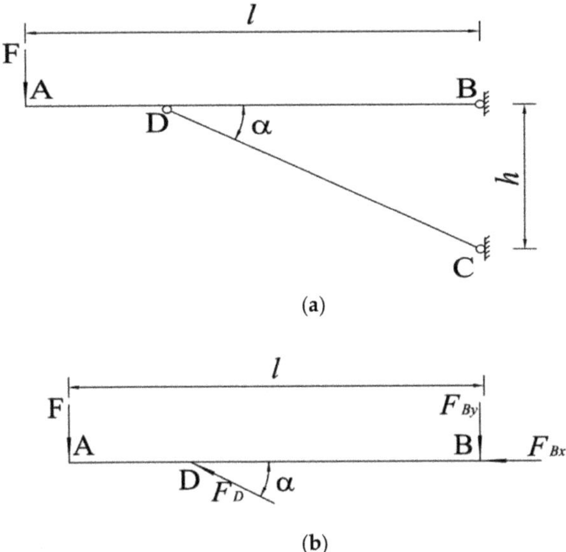

Figure 9. Analysis of the force sustained by UAV sensor support. (**a**) Schematic diagram of the structure of the support and damping rod; (**b**) analysis of the force sustained by each part of the support.

In Figure 9, AB is the sensor support, CD is the support damping rod, points C and D are the fixed points of the sensor support and the damping rod at the end of the UAV, and the force sustained by the sensor support is analyzed by the following static equations:

$$F = -F_{By} + F_D \sin \alpha \qquad (6)$$

$$F_D \cos \alpha = -F_{Bx} \qquad (7)$$

$$Fl = F_D \sin \alpha \cdot \frac{h}{\tan \alpha} \qquad (8)$$

Simultaneous calculations were performed on Equations (6)–(8) to obtain the following:

$$F_D = \frac{Fl}{h \cos \alpha} \qquad (9)$$

$$F_{Bx} = -\frac{Fl}{h} \qquad (10)$$

$$F_{By} = F\left(\frac{l}{h} \tan \alpha - 1\right) \qquad (11)$$

In Equations (6)–(11), F_D is the internal force of CD, F_{Bx} and F_{By} are the reaction forces in the horizontal and vertical directions of fixed-point B, respectively, α is the angle between AB and CD, h is the height difference between fixed points B and C, and l is the length of AB.

When the length of the UAV support is 60 cm, the value of α ranges from 9.5° to 90°. Since a certain length of the support needs to be used for fixing the UAV and dual-band crop-growth sensor, the actual value range of α is 13°–90°. From the derivation results of Equations (9)–(11), it can be seen that F_D and F_{By} decrease with the decrease of α, and the smaller the values of F_D and F_{By} are, the more stable the force sustained by the support structure is. Therefore, the optimal angle between sensor support and damping rod is 13°.

2.3. Field Trial

2.3.1. Test Design

Field Trial 1 was conducted at the Baipu Town (32°14′58.88 N 120°45′44.26 E) test base in Rugao City, Jiangsu Province, China, from February to May 2016. The test varieties were Ningmai 13 (V1) and Huaimai 33 (V2), three different gradients of nitrogen fertilizer treatment were set up, which were N_0 (0 kg/hm^2), N_1 (180 kg/hm^2), and N_2 (270 kg/hm^2), and each variety was repeated three times. Each planting area was 30 m^2 (5 × 6 m). In addition, the application rate of phosphate fertilizer was 120 kg/hm^2, and the application rate of potassium fertilizer was 135 kg/hm^2, which was applied once in the base fertilizer. Other cultivation-management measures were the same as those in general high-yield fields.

Field Trial 2 was conducted at the Lingqiao Township (33°35′53.27 N 118°51′11.01 E) Test Base in Huai'an City, Jiangsu Province, China, from July to October 2016. The test varieties were Nanjing 9108 (V1) and Lianjing 10 (V2), four different gradients of nitrogen fertilizer were applied, which were N_0 (0 kg/hm^2), N_1 (120 kg/hm^2), N_2 (240 kg/hm^2), and N_3 (360 kg/hm^2), each variety was repeated three times, and each planting area was 30 m^2 (5 × 6 m). In addition, the application rate of phosphate fertilizer was 105 kg/hm^2 and was applied once in the base fertilizer, the potassium fertilizer was 135 kg/hm^2, the base fertilizer was applied 50%, and, at the early boot stage, application was 50%. The other cultivation-management measures were the same as those in general high-yield fields.

2.3.2. Test Method

Field Trial 1 was used to test whether the proposed monitoring method can effectively avoid the disturbance range of the UAV downwash airflow field when acquiring data from the crop canopy. Additionally, Field Trial 1 was used to verify the accuracy and stability of the dual-band crop-growth sensor test results. The experiment was carried out in the middle of the wheat jointing stage. The test was carried out on a clear, windless, and cloudless day. Test time was between 10:00 and 14:00. The UAV was flown 1 m above the wheat canopy, and, as shown in Figure 10, the dual-band crop-growth sensor was deployed in three different horizontal distances from the UAV rotors: 0 (i.e., directly below the UAV), 30, and 60 cm. The sensors determined the RVI value of the wheat canopy by measuring three random points in each subarea and repeating the measurement of each point three times to obtain an average value. The ASD FieldSpec HandHeld 2 was used to measure the RVI value of the wheat canopy at the same time.

Figure 10. Comparison of field flights with the conditions of three support lengths.

Field Trial 2 was used to evaluate the applicability and accuracy of UAV-borne spectral sensors for crop-growth parameters. The test was carried out in the tillering, jointing, booting, and heading stages of the rice, test weather was sunny and windless, and test time was between 10:00 and 14:00. In the test, the UAV was made to hover at a height of 1 m above the rice canopy in different test plots, and the dual-band crop-growth sensor was deployed at a horizontal distance of 60 cm from the UAV rotors to obtain the RVI value at the 730 and 815 nm bands of the rice canopy. Three points were randomly measured in each subarea, and the measurement of each point was repeated three times to obtain an average value. The FieldSpec HandHeld 2 and LAI2200 testers were synchronously used to obtain the RVI and LAI values of the rice leaf layer. At the same time, in parallel with the test, the rice sample was destructively sampled, and the sample was placed at 105 °C for 30 min for fixing, then baked at 80 °C to constant weight, and weighed to obtain the LDW. After the sample was pulverized, the LNA was determined by the Kjeldahl method.

2.3.3. Analysis of Field-Test Data

The field-test datasets were statistically analyzed with Microsoft Excel 2010 software; the correlation of the model was evaluated by the coefficient of determination (R^2) and root mean square error (RMSE).

3. Results and Discussion

3.1. Field-Test Results

Figure 11 shows the test results of Field Trial 1, in which the dual-band crop-growth sensor was located at different positions relative to the UAV in the middle of the wheat-jointing stage. Parts a, b, and c of Figure 11 show the simple linear fitting results of the RVI test value of the UAV-borne spectral sensor when the length of the support was 0, 30, and 60 cm, respectively, and the RVI value of the handheld ASD FieldSpecHandHeld 2 spectrometer in the corresponding growth period. When the length of the support was 0 and 30 cm, the results were close, and R^2 values were 0.63 and 0.66, respectively. When the length of the support was extended to 60 cm, the curve fitting degree was obviously improved; R^2 was 0.81, and RMSE was 0.38.

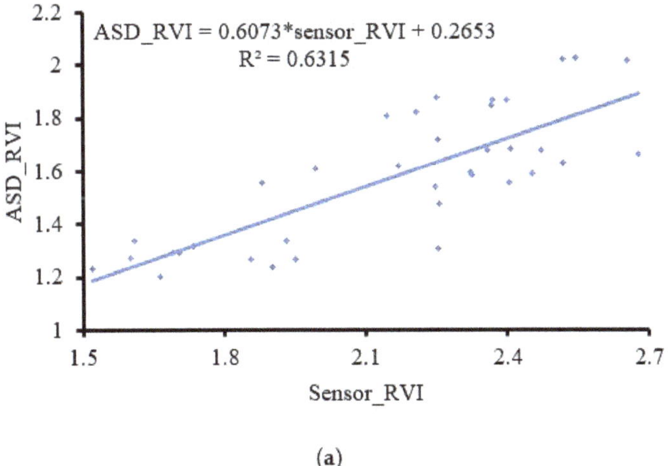

(a)

Figure 11. *Cont.*

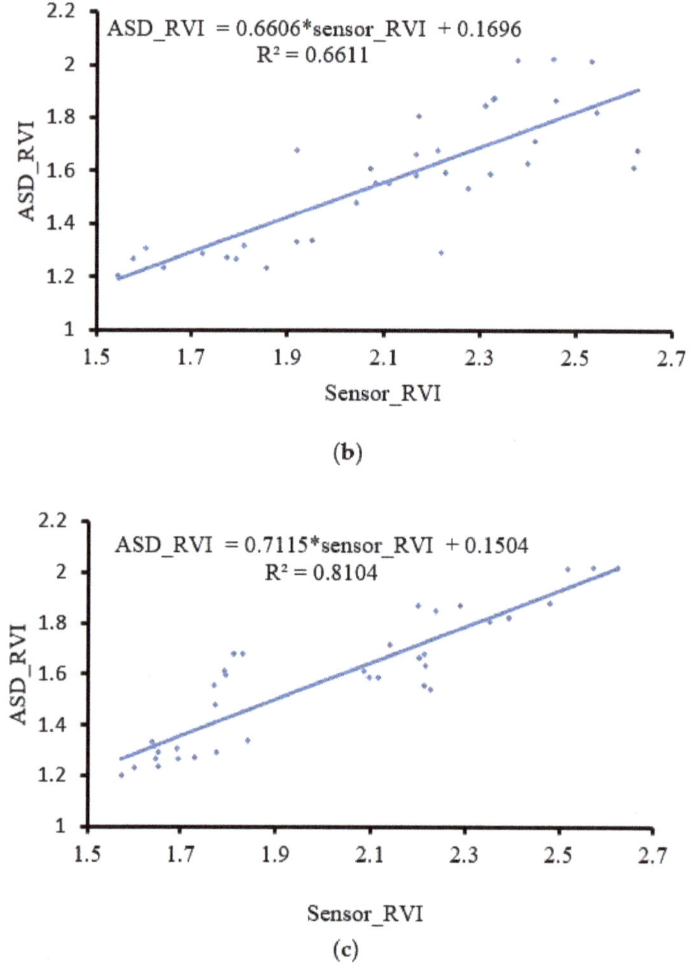

Figure 11. Fitting curves for the sensor and ASD data. (a) Sensor located under the UAV; (b) sensor located 30 cm from the rotor; (c) sensor located 60 cm from the rotor.

It can be seen from the above test results that the farther the dual-band crop growth sensor was deployed from the UAV rotors, the better the correlation between the test data and the ASD test results. According to CFD numerical simulation results and 3D airflow field-test results, when the dual-band crop-growth sensor was deployed directly below the UAV, the sensor-test field of view was completely within the disturbance range of UAV rotor downwash airflow field. When the dual-band crop-growth sensor was deployed 30 cm away from the UAV rotors, the sensor-test field of view included both the disturbance and nondisturbance zone of the rotor downwash airflow field. The correlation of the data results improved slightly, but it was still not ideal. When the dual-band crop-growth sensor was deployed 60 cm away from the rotors of the UAV, the sensor-test field was completely in the nondisturbance zone, and the correlation of the data results was significantly improved. In summary, the downwash airflow field generated by the rotation of the UAV rotors has a certain influence on the results of the dual-band crop-growth sensor. The proposed UAV-borne spectral sensor crop-growth monitoring method can effectively target areas of the crop canopy outside the disturbance range of the UAV downwash airflow field.

In Field Trial 2, the flying height of the UAV was 1 m from the rice canopy, the dual-band crop-growth sensor was deployed 60 cm away from the UAV rotors, and measurements were taken throughout the entire growth period of rice. Figure 12 shows the linear fitting results of the RVI values obtained by our method, and the LNA, LAI, and LDW obtained from the field test and the indoor chemical analysis test. R^2 values were 0.62, 0.76, and 0.60, and RMSE values were 2.28, 1.03, and 10.73, respectively. Using the proposed UAV-borne spectral sensor crop-growth monitoring method, rice-growth parameters in the entire growth period could accurately be obtained.

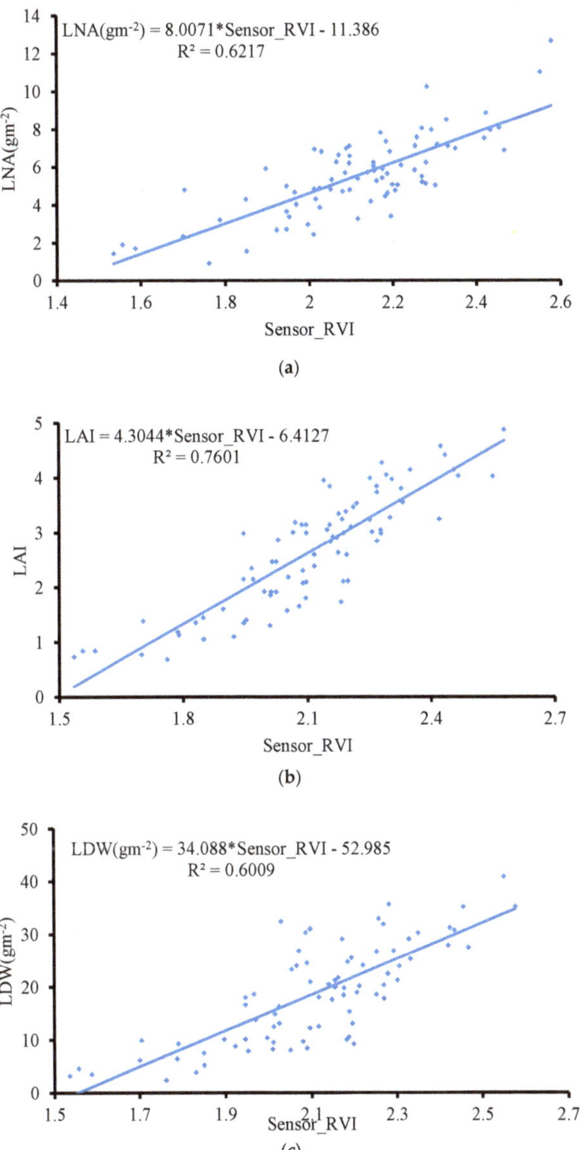

Figure 12. Spectral model for the UAV-borne crop-growth monitoring system. (**a**) LNA–RVI fitting curve; (**b**) LAI–RVI fitting curve; (**c**) LDW–RVI fitting curve.

3.2. Discussion

UAVs have the characteristics of simple operation and high efficiency. At present, there is little targeted research on monitoring crop growth by UAV-borne dual-band crop-growth sensors. Under a low-altitude hovering operation, the downwash airflow field generates strong disturbance on the crop canopy that disrupts the Lambertian reflection characteristics of the crop canopy and thus has a serious impact on the accuracy of the test, even causing the test to fail. Therefore, we used CFD numerical simulation and real-world 3D airflow field testing to analyze the spatial distribution of the downwash airflow field when the UAV is hovering at a height of 1 m above the crop canopy and determine the disturbance range on the crop canopy. Most of the current studies simply mounted the energy-type spectrum sensor on UAVs, lacking consideration for the disturbance influence of the crop canopy by the UAV downwash airflow field during the test [37]. Our study filled the gap in these studies. Our test showed that the influence range of the downwash airflow field of the UAV is elliptical, wind speed at the center point is the largest, and gradually decreases toward the periphery. The long semiaxis was about 0.45 m, and the short semiaxis was about 0.4 m. According to the distribution range of the downwash airflow field, we designed a support with a length of 0.6 m to deploy a dual-band crop-growth sensor so that the test field of view is extended beyond the distribution range of the downwash airflow field, and the disturbance effect of the downwash airflow field on the crop canopy was avoided. When the flight height of the UAV was kept at 1 m from the wheat canopy and the dual-band crop-growth sensor was deployed 0.6 m from the rotors, the linear fit R^2 of the test output RVI value and the handheld ASD Fieldspec2 spectrometer, the test RVI value was 0.81, and RMSE was 0.38. Therefore, the UAV-borne spectral sensor crop-growth monitoring method can effectively target areas of the crop canopy outside the disturbance range of the UAV downwash airflow field.

In the test process of the 3D airflow field testers, the wind-speed results of different dimensions measured by the testers were larger than the CFD numerical-simulation results. The main reason was that the test plane of the 3D airflow field testers was 60 cm away from the ground, and the arrangement was lattice. When the downwash airflow diffused downward through the lattice plane, the direction of the airflow changed. In the CFD numerical simulation analysis, the downwash airflow directly reached the ground plane, so the distribution states of the real-world test and numerical simulation differed. Although the analyzed UAV in this paper is a DJI Phantom drone, the proposed UAV-borne spectral sensor crop-growth monitoring method can be applied to various types of multirotor UAV structures.

4. Conclusions

1. We identified the UAV-borne spectral sensor crop-growth monitoring method, used CFD numerical simulation and an actual test of 3D airflow field to determine the distribution range of the UAV downwash airflow field above the surface of the crop canopy, and designed sensor supports to target areas of the crop canopy outside the disturbance range of the UAV downwash airflow field.

2. When the flying height of the UAV was 1 m from the crop canopy, the influence range of the downwash airflow field of the UAV was elliptical, central wind speed was the largest and gradually decreased toward the periphery, the long semiaxis was about 0.45 m, and the short semiaxis was about 0.4 m. When the designed sensor support length was 60 cm, the sensor-test field of view was completely outside the disturbance range of the UAV downwash airflow field.

3. The wheat test showed that, when the dual-band crop-growth sensor was deployed at 0, 30, and 60 cm from the UAV, the linear fit R^2 of the RVI value obtained by our method, and the RVI value measured by the ASD FieldSpec HandHeld 2 spectrometer, was 0.63, 0.66, and 0.81, respectively. When the length of the support was 60 cm, the fitting degree was obviously improved. The UAV-borne spectroscopy sensor crop-growth monitoring method can effectively avoid the disturbance range of the UAV downwash airflow field on the crop canopy.

4. The rice experiment showed that the RVI value measured by the UAV-borne spectral sensor had a good linear fitting relationship with the LNA, LAI, and LDW obtained from the field test and the indoor chemical analysis test. R^2 values were 0.62, 0.76, and 0.60, respectively, and RMSE values

were 2.28, 1.03, and 10.73, respectively. Using the UAV-borne spectral sensor crop-growth monitoring method, rice-growth parameters during the entire growth period could be accurately obtained.

Author Contributions: J.N., Y.Z. (Yan Zhu) and W.C. designed the research, J.N., Q.W., Y.Z. (Yu Zhang) and L.Y. performed the research, L.Y., J.Y. and J.N. analyzed the data, and L.Y. and Y.Z. (Yan Zhu) wrote the paper. All authors have read and approved the final manuscript submitted to the editor.

Funding: This research was funded by grants from the National Key Research and Development Program of China (2016YFD070030403), Primary Research and Development Plan of Jiangsu Province of China (Grant No. BE2016378), Jiangsu Agricultural Science and Technology Independent Innovation Fund Project (Grant No. CX(16)1006), the Priority Academic Program Development of Jiangsu Higher Education Institutions (PAPD), and the 111Project (B16026).

Acknowledgments: The authors thank all those who helped in the course of this research.

Conflicts of Interest: The authors declare no conflict of interest.

References

1. Pan, J.; Liu, Y.; Zhong, X.; Lampayan, R.M.; Singleton, G.R.; Huang, N.; Liang, K.; Peng, B.; Tian, K. Grain yield, water productivity and nitrogen use efficiency of rice under different water management and fertilizer-N inputs in South China. *Agric. Water Manag.* **2017**, *184*, 191–200. [CrossRef]
2. Cambouris, A.N.; Zebarth, B.J.; Ziadi, N.; Perron, I. Precision Agriculture in Potato Production. *Potato Res.* **2014**, *57*, 249–262. [CrossRef]
3. Guo, J.H.; Wang, X.; Meng, Z.J.; Zhao, C.J.; Zhen-Rong, Y.U.; Chen, L.P. Study on diagnosing nitrogen nutrition status of corn using Greenseeker and SPAD meter. *Plant Nutr. Fertil. Sci.* **2008**, *14*, 43–47.
4. Grohs, D.S.; Bredemeier, C.; Mundstock, C.M.; Poletto, N. Model for yield potential estimation in wheat and barley using the GreenSeeker sensor. *Eng. Agrícola* **2009**, *29*, 101–112. [CrossRef]
5. Ali, A.M.; Thind, H.S.; Sharma, S.; Varinderpal-Singh, V. Prediction of dry direct-seeded rice yields using chlorophyll meter, leaf color chart and GreenSeeker optical sensor in northwestern India. *Field Crop. Res.* **2014**, *161*, 11–15. [CrossRef]
6. Moya, I.; Camenen, L.; Evain, S.; Goulas, Y.; Cerovic, Z.G.; Latouche, G.; Flexas, J. A new instrument for passive remote sensing: 1. Measurements of sunlight-induced chlorophyll fluorescence. *Remote Sens. Environ.* **2004**, *91*, 186–197. [CrossRef]
7. Youngryel, R.; Dennisd, B.; Joseph, V.; Ma, S.; Matthias, F.; Ilse, R.M.; Ted, H. Testing the performance of a novel spectral reflectance sensor, built with light emitting diodes (LEDs), to monitor ecosystem metabolism, structure and function. *Agric. For. Meteorol.* **2011**, *150*, 1597–1606.
8. Martin, D.E.; Lan, Y.B. Laboratory evaluation of the GreenSeekerTM handheld optical sensor to variations in orientation and height above canopy. *Int. J. Agric. Biol. Eng.* **2012**, *5*, 43–47.
9. Ali, A.M.; Abouamer, I.; Ibrahim, S.M. Using GreenSeeker Active Optical Sensor for Optimizing Maize Nitrogen Fertilization in Calcareous Soils of Egypt. *Arch. Agron. Soil Sci.* **2018**, *64*, 1083–1093. [CrossRef]
10. Vohland, M.; Besold, J.; Hill, J.; Fründ, H.C. Comparing different multivariate calibration methods for the determination of soil organic carbon pools with visible to near infrared spectroscopy. *Geoderma* **2011**, *166*, 198–205. [CrossRef]
11. Kusumo, B.H.; Hedley, M.J.; Hedley, C.B.; Hueni, A.; Arnold, G.C. The use of Vis-NIR spectral reflectance for determining root density: Evaluation of ryegrass roots in a glasshouse trial. *Eur. J. Soil Sci.* **2010**, *60*, 22–32. [CrossRef]
12. Nawar, S.; Buddenbaum, H.; Hill, J.; Kozak, J.; Mouazen, A.M. Estimating the soil clay content and organic matter by means of different calibration methods of vis-NIR diffuse reflectance spectroscopy. *Soil Tillage Res.* **2016**, *155*, 510–522. [CrossRef]
13. Cao, Q.; Miao, Y.; Wang, H.; Huang, S.; Cheng, S.; Khosla, R. Non-destructive estimation of rice plant nitrogen status with Crop Circle multispectral active canopy sensor. *Field Crop. Res.* **2013**, *154*, 133–144. [CrossRef]
14. Bonfil, D.J. Wheat phenomics in the field by RapidScan: NDVI vs. NDRE. *Isr. J. Plant Sci.* **2016**, 1–14. [CrossRef]

15. Lu, J.; Miao, Y.; Wei, S.; Li, J.; Yuan, F. Evaluating different approaches to non-destructive nitrogen status diagnosis of rice using portable RapidSCAN active canopy sensor. *Sci. Rep.* **2017**, *7*, 14073. [CrossRef] [PubMed]

16. Miller, J.J.; Schepers, J.S.; Shapiro, C.A.; Arneson, N.J.; Eskridge, K.M.; Oliveira, M.C.; Giesler, L.J. Characterizing soybean vigor and productivity using multiple crop canopy sensor readings. *Field Crop. Res.* **2018**, *216*, 22–31. [CrossRef]

17. Zhou, Z.; Andersen, M.N.; Plauborg, F. Radiation interception and radiation use efficiency of potato affected by different N fertigation and irrigation regimes. *Eur. J. Agron.* **2016**, *81*, 129–137. [CrossRef]

18. Krienke, B.; Ferguson, R.B.; Schlemmer, M.; Holland, K.; Marx, D.; Eskridge, K. Using an unmanned aerial vehicle to evaluate nitrogen variability and height effect with an active crop canopy sensor. *Precis. Agric.* **2017**, *18*, 900–915. [CrossRef]

19. Shafian, S.; Rajan, N.; Schnell, R.; Bagavathiannan, M.; Valasek, J.; Shi, Y. Unmanned aerial systems-based remote sensing for monitoring sorghum growth and development. *PLoS ONE* **2018**, *13*, e0196605. [CrossRef]

20. Schirrmann, M.; Giebel, A.; Gleiniger, F.; Pflanz, M.; Lentschke, J. Monitoring Agronomic Parameters of Winter Wheat Crops with Low-Cost UAV Imagery. *Remote Sens.* **2016**, *8*, 706. [CrossRef]

21. Zheng, H.; Zhou, X.; Cheng, T.; Yao, X.; Tian, Y.; Cao, W.; Zhu, Y. Evaluation of a UAV-based hyperspectral frame camera for monitoring the leaf nitrogen concentration in rice. In Proceedings of the Geoscience and Remote Sensing Symposium, Beijing, China, 10–15 July 2016; pp. 7350–7353.

22. Stroppiana, D.; Villa, P.; Sona, G.; Ronchetti, G.; Candiani, G.; Pepe, M.; Busetto, L.; Migliazzi, M.; Boschetti, M. Early season weed mapping in rice crops using multi-spectral UAV data. *Int. J. Remote Sens.* **2018**, *39*, 5432–5452. [CrossRef]

23. Wu, M.; Huang, W.; Niu, Z.; Wang, Y.; Wang, C.; Li, W.; Hao, P.; Yu, B. Fine crop mapping by combining high spectral and high spatial resolution remote sensing data in complex heterogeneous areas. *Comput. Electron. Agric.* **2017**, *139*, 1–9. [CrossRef]

24. Sandwell, D.T. Biharmonic spline interpolation of GEOS-3 and SEASAT altimeter data. *Geophys. Res. Lett.* **2013**, *14*, 139–142. [CrossRef]

25. Ni, J.; Yao, L.; Zhang, J.; Cao, W.; Zhu, Y.; Tai, X. Development of an Unmanned Aerial Vehicle-Borne Crop-Growth Monitoring System. *Sensors* **2017**, *17*, 502. [CrossRef] [PubMed]

26. Feng, W.; Zhu, Y.; Tian, Y.; Cao, W.; Yao, X.; Li, Y. Monitoring leaf nitrogen accumulation in wheat with hyper-spectral remote sensing. *Eur. J. Agron.* **2008**, *28*, 23–32. [CrossRef]

27. Yan, Z.; Dongqin, Z.; Xia, Y.; Yongchao, T.; Weixing, C. Quantitative relationship between leaf nitrogen accumulation and canopy reflectance spectra in wheat. *Aust. J. Agric. Res.* **2007**, *58*, 1077–1085. [CrossRef]

28. Yan, Z.; Yingxue, L.; Wei, F.; Yongchao, T.; Xia, Y.; Weixing, C. Monitoring leaf nitrogen in rice using canopy reflectance spectra. *Can. J. Plant Sci.* **2006**, *86*, 1037–1046.

29. Szuvandzsiev, P.; Helyes, L.; Lugasi, A.; Szántó, C.; Baranowski, P.; Pék, Z. Estimation of antioxidant components of tomato using VIS-NIR reflectance data by handheld portable spectrometer. *Int. Agrophys.* **2014**, *28*, 521–527. [CrossRef]

30. Battay, A.E.; Mahmoudi, H. Linear spectral unmixing to monitor crop growth in typical organic and inorganic amended arid soil. *IOP Conf. Ser. Earth Environ. Sci.* **2016**, *37*, 012046. [CrossRef]

31. Fang, H.; Li, W.; Wei, S.; Jiang, C.J.A.; Meteorology, F. Seasonal variation of leaf area index (LAI) over paddy rice fields in NE China: Intercomparison of destructive sampling, LAI-2200, digital hemispherical photography (DHP), and AccuPAR methods. *Agric. For. Meteorol.* **2014**, *198–199*, 126–141. [CrossRef]

32. Sandmann, M.; Graefe, J.; Feller, C. Optical methods for the non-destructive estimation of leaf area index in kohlrabi and lettuce. *Sci. Hortic.* **2013**, *156*, 113–120. [CrossRef]

33. Kobayashi, H.; Ryu, Y.; Baldocchi, D.D.; Welles, J.M.; Norman, J.M. On the correct estimation of gap fraction: How to remove scattered radiation in gap fraction measurements? *Agric. For. Meteorol.* **2013**, *174–175*, 170–183. [CrossRef]

34. Jiradilok, V.; Gidaspow, D.; Damronglerd, S.; Koves, W.J.; Mostofi, R. Kinetic theory based CFD simulation of turbulent fluidization of FCC particles in a riser. *Chem. Eng. Sci.* **2006**, *61*, 5544–5559. [CrossRef]

35. Blocken, B.; Stathopoulos, T.; Carmeliet, J. CFD simulation of the atmospheric boundary layer: Wall function problems. *Atmos. Environ.* **2007**, *41*, 238–252. [CrossRef]

36. Nakata, T.; Liu, H.; Bomphrey, R.J. A CFD-informed quasi-steady model of flapping wing aerodynamics. *J. Fluid Mech.* **2015**, *783*, 323–343. [CrossRef] [PubMed]

37. Li, S.; Ding, X.; Kuang, Q.; Ata-UI-Karim, S.T.; Cheng, T.; Liu, X.; Tian, Y.; Yan, Z.; Cao, W.; Cao, Q. Potential of UAV-Based active sensing for monitoring rice leaf nitrogen satus. *Front. Plant Sci.* **2018**, *9*, 1834. [CrossRef]

Article

Multi-Pig Part Detection and Association with a Fully-Convolutional Network

Eric T. Psota [1,*], Mateusz Mittek [1], Lance C. Pérez [1], Ty Schmidt [2] and Benny Mote [2]

[1]   Department of Electrical and Computer Engineering, University of Nebraska–Lincoln, Lincoln, NE 68505, USA; mmittek@gmail.com (M.M.); lperez@unl.edu (L.C.P.)

[2]   Department of Animal Science, University of Nebraska–Lincoln, Lincoln, NE 68588, USA; tschmidt4@unl.edu (T.S.); benny.mote@unl.edu (B.M.)

*   Correspondence: epsota@unl.edu

Received: 27 November 2018; Accepted: 16 February 2019; Published: 19 February 2019

Abstract: Computer vision systems have the potential to provide automated, non-invasive monitoring of livestock animals, however, the lack of public datasets with well-defined targets and evaluation metrics presents a significant challenge for researchers. Consequently, existing solutions often focus on achieving task-specific objectives using relatively small, private datasets. This work introduces a new dataset and method for instance-level detection of multiple pigs in group-housed environments. The method uses a single fully-convolutional neural network to detect the location and orientation of each animal, where both body part locations and pairwise associations are represented in the image space. Accompanying this method is a new dataset containing 2000 annotated images with 24,842 individually annotated pigs from 17 different locations. The proposed method achieves over 99% precision and over 96% recall when detecting pigs in environments previously seen by the network during training. To evaluate the robustness of the trained network, it is also tested on environments and lighting conditions unseen in the training set, where it achieves 91% precision and 67% recall. The dataset is publicly available for download.

Keywords: computer vision; deep learning; image processing; pose estimation; animal detection; precision livestock

1. Introduction

Researchers have established that changes in animal behavior correlate with changes in health [1–3], however, the labor-intensive methods used to monitor behaviors do not lend themselves well to modern commercial swine facilities where just two seconds of daily observation per pig is recommended [4]. Because industry caretakers are typically responsible for thousands of pigs, it is nearly impossible for them to thoroughly assess the health and well-being of individuals using manual observation. This limitation is further compounded by the fact that the effectiveness of human visual assessments are inherently limited by both the attention span and subjectivity of observers [5].

By adapting modern technology to commercial operations, precision livestock farming aims to ease the burden on individual caregivers and provide a solution for continuous automated monitoring of individual animals [6–8]. Over the past two decades, researchers have approached this problem from a multitude of different angles [9,10]. These include 3D tracking via wearable ultra-wide band (UWB) devices [11,12], GPS motes [13,14], inertial measurement unit (IMU) activity trackers [15–18], RFID ear tags [19–21], and detection-free tracking using depth cameras [22,23].

This work aims to provide a generalizable vision-based solution for animal detection in group-housing environments. While animal tracking methods using radio identification devices directly provide data on individual animals, there are several disadvantages to using wearable methods when compared to video-based approaches [24,25]. Wearables must withstand harsh environments, they require costly installation and maintenance on a per animal basis, and the localization accuracy of both UWB and GPS systems is typically too low to detect animal orientation, activities, and social behaviors. In contrast, video already provides information-rich data that allows humans to identify precisely what each animal is doing at all times. However, converting digital video to meaningful data without human intervention is an ongoing pursuit in the research community. Existing solutions require sophisticated computer vision algorithms and they have begun to rely increasingly on advancements in machine learning and artificial intelligence [26,27].

Computer vision researchers experienced a dramatic shift in 2012 when Krizhevsky et al. demonstrated that a deep convolutional neural network, trained from scratch without any hand-crafted heuristics, could easily outperform all existing methods for image classification [28]. This shift was made possible by efficient implementations of convolutional neural network (CNN) training [29,30], improvements in computational resources via graphics processing units (GPUs) [31], and large (100K+ images) human-annotated datasets like PASCAL [32], MS-COCO [33], and CityScapes [34]. Modern computer vision applications that locate and classify objects in images often leverage established methods like YOLO [35,36], SSD [37], and Faster R-CNN [38,39] in their pipeline. While common tasks like pedestrian detection lend themselves well to pre-trained networks and existing datasets, there still exists unique challenges that require custom datasets and accompanying solutions.

This work introduces a fully convolutional neural network used to identify the location and orientation of multiple group-housed pigs. The target output of the network is an image-space representation of each pig's body part locations along with a method for associating them to form complete instances. To train the network, a new dataset is used that contains 2000 images with 24,842 uniquely labeled pig instances. The dataset is divided into a training set and a testing set, and the testing set is subdivided into two sets: one with images depicting the same environments as the training set, and another with images of new environments not represented in the training set. This dataset design allows the robustness of detection algorithms to be tested against novel animal presentations and environments. The three key contributions of this work are (1) a fully convolutional instance detection method, (2) a public dataset for training and evaluation, and (3) metrics that can be used to measure the performance of methods that detect both location and orientation.

The remainder of this paper is organized as follows. Section 2 discusses related work in vision-based animal detection and modern deep learning detection methods. Section 3 introduces the proposed method and the design of the fully-convolutional neural network. The dataset, evaluation methodology, and results are presented in Sections 4 and 4.6 discusses some key insights and additional practical considerations. Finally, concluding remarks are provided in Section 5.

2. Background

Visual detection of multiple moving targets with a static camera often begins with segmentation of foreground objects using background subtraction. If sufficient separation between targets exists, traditional computer vision methods like connected components can be used to easily identify unique instances [40]. However, this is hardly the case for group-housed animals that constantly engage each other socially and often prefer to lie in groups to preserve warmth.

Using Otsu's method [41] to adaptively threshold background from foreground and Kashiha's ellipse-fitting method [42], Nasirahmadi et al. [43] used computer vision to identify the lying behavior of group-housed pigs as a function of temperature. While their method achieved approximately 95% accuracy when locating pigs in the images, it was only tested on a single environment where the animals had consistent age and appearance, so it is difficult to predict how well the method generalizes. Ahrendt et al. [44] used a continuously updated model and a 5D Mahalanobis distance between foreground and background pixels to isolate and track the targets, however, the authors point out that the method struggles to identify separate instances when animals are moving quickly and/or overlapping one another. Nilsson et al. [45] used a region-based method to segment group-housed pigs using supervised learning, demonstrating accurate results when the pigs are fully visible (i.e., non-occluded) from a top-down perspective.

A popular approach to dealing with the difficulties of background subtraction using color images is to incorporate depth-imaging cameras [22,46–54]. However, while foreground extraction is made simple, depth image processing is still faced with the challenge of separating abutting pigs. To address this challenge, Ju et al. [27] introduced a top-down method that first detects bounding boxes around pigs using the YOLO object detector [36]. When YOLO inevitably detects multiple touching pigs with a single bounding box, their method proceeds by separating segmented images (obtained using a Gaussian mixture model) to establish boundaries between instances. While their overall method achieves 92% accuracy, it is limited to three touching pigs and results were obtained within a single environment over a ten minute window, so generalizability is difficult to assess.

Some approaches using the Kinect depth-sensing camera explicitly define the region of interest (pen) as a three-dimensional bounding box to separate background from foreground. Mittek et al. [23] use this approach to track pigs as ellipsoidal collections of points in the foreground. While their method achieves an average of 20 min of continuous tracking without errors, it requires manual initialization and does not introduce a detection method. The method introduced by Matthew et al. [55] uses regional grouping of surface normals to detect individual instances. They do not perform a formal assessment of the detection accuracy, but the resulting tracking algorithm maintains accurate tracking for an average of 22 s without requiring manual initialization.

Beginning in 2014 with the introduction of R-CNN by Girshick et al. [56], visual detection methods have predominantly used deep convolutional neural networks. Furthermore, they generally fall into one of two categories: (1) top-down approaches that define regions of interest before performing subsequent segmentation or keypoint annotation, or (2) bottom-up approaches that directly segment pixels or detect keypoints without explicitly detecting regions of interest. Mask R-CNN [57], for example, is a state-of-the-art top-down approach for performing instance-level object segmentation and keypoint detection. However, because it relies on a priori region proposal, it is inherently unable to separate objects with significant bounding box overlap—a common occurrence among group-housed animals. In contrast, bottom-up detection and classification is directly done per-pixel in the image space [58]. Two notable methods, OpenPose [59] and PersonLab [60], directly identify keypoints and affinity metrics that allow them to be clustered into whole instances.

Bottom-up keypoint detection was adapted to cow tracking by Ardö et al. [26]. They identified a set of landmarks visible from a top-down view to represent each cow's location and orientation, and trained a fully-convolutional neural network to detect them in the image space. A post-processing network was then used to convert the annotations to per-pixel orientation classification outputs, resulting in 95% accuracy in correctly labeling all cows in a given image. In a follow-up experiment, they applied the previously trained network to a new environment and observed that it only succeeded on 55% of images, indicating that the network was overfitting to a particular environment.

The proposed method introduces a new bottom-up strategy that identifies multiple pig instances in images as a collection of keypoints. Unlike the approach by Ardö et al. [26], this method preserves the association between keypoints and instances, making it possible to evaluate the performance of the method directly as a keypoint detection method. Furthermore, keypoints provide a precise representation of the pose of each animal, making it possible to identify activities and interactions between animals.

3. Proposed Method

The goal of the proposed method is to detect the location and orientation of all visible pigs in the pen environment. The first stage of the process aims to find the location of all pertinent body parts, while the second stage aims to associate them with one another to form whole instances. The following two sections (Sections 3.1 and 3.2) describe the method used to represent parts and associations within the image space. Section 3.3 then presents a method that interprets these image space representations and produces a set of unique animal instances. Finally, Section 3.4 introduces the fully-convolutional network that takes, as input, an image of the pen environment and attempts to produce the image space outputs described in Sections 3.1 and 3.2.

3.1. Representation of Body Part Location

The proposed method assumed that images of the pen environment were captured from a downward-facing camera mounted above the pen. When trying to detect and differentiate multiple animals in a group-housed setting, a top-down view has three distinct advantages over alternative visual perspectives. Firstly, animals are generally nonoccluded from the top-down perspective unless they are crawling over (or lying on top of) one another. Secondly, the size and appearance of animals is consistent from a top-down perspective, making it easier for a system to identify the animals. Thirdly, one can reliably approximate the 3D position of each animal from their projection onto the 2D image plane by assuming a constant height above the pen floor plane. Approximation is often necessary if 3D coordinates are desired and the single-camera system being used lacks the ability to measure depth.

From a top-down perspective, the part of the animal most likely to be visible was the surface of the back. Thus, in order to represent both the position and orientation of each pig, the proposed method used the image-space location of the tail and shoulder belonging to each animal. Assuming there are N animals in the pen, the tail and shoulder position of animal $n \in \{1, \ldots, N\}$ is denoted $\mathbf{t}_n = (x_{t_n}, y_{t_n})$ and $\mathbf{s}_n = (x_{s_n}, y_{s_n})$, respectively. More specifically, "tail" refers to a surface point along the center ridge of the back that is between the left and right ham. The term "shoulder" refers to a surface point along the center ridge of the back between the shoulder blades. The chosen representation also includes the 2D position of the left and right ears, denoted $\mathbf{l}_n = (x_{l_n}, y_{l_n})$ and $\mathbf{r}_n = (x_{r_n}, y_{r_n})$, respectively. While their visibility is not guaranteed, such as when the animal lies on its side or positions its head in a feeder, their locations can be used to approximate the pose of the head and/or assign animals with a unique visual marker in the form of an ear tag.

Figure 1a illustrates an example of an input image depicting a single pig. The location of the left ear, right ear, shoulder, and tail are represented by the red, green, blue, and yellow spots in the target mapping shown in Figure 1b, respectively. Finally, the superimposed visualization given in Figure 1c illustrates the locations of the four parts in reference to the original image. Note that the target mapping in Figure 1b has the same spatial dimensions (rows and columns) as the input image, but it contains four channels with each corresponding to a single body part type.

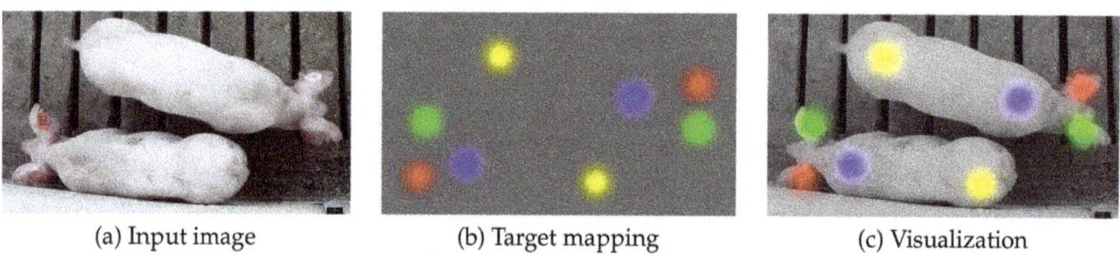

(a) Input image	(b) Target mapping	(c) Visualization

Figure 1. The original image (**a**) is mapped to a four-channel output (**b**), where the location of the left ear, right ear, shoulder, and tail are represented by Gaussian kernels in channels 1–4 of the output, respectively. Note that the four colors used in (**b**) are purely for visualization of the four separate channels of the target image. The overlay in (**c**) is provided to illustrate the locations relative to the original image.

To approximate the level of uncertainty inherent in the user annotations of each body part location, parts within the target mapping were each represented by 2D Gaussian kernels. While the distribution of the 2D Gaussian kernels is defined by a 2×2 covariance matrix, here we have only considered symmetric 2D Gaussian kernels that can be characterized by a single standard deviation multiplied by a 2×2 identity matrix. This standard deviation will be denoted σ_n for each animal $n \in \{1, \ldots, N\}$.

This mapping of uncertainty was meant to approximate the probability distribution of part locations annotated by a human given the original image **I**. The kernels were also scaled so that the magnitude at the center is 1.0. This allowed for a straightforward translation between kernels and 2D image coordinates via a simple thresholding operation. The first four channels of the target output, as defined in Table 1, were thus proportional to the probability that parts $\{l, r, s, t\}$ exist at each spatial location in the image.

Table 1. Channels 1–4 of the image-space representation used to represent pig locations and orientations. Each channel corresponds to a different body part. The locations of each part are marked with Gaussian kernels meant to represent the distribution of part locations provided by a human annotator.

Channel	1	2	3	4
Encoding	$\propto P(l\vert I)$	$\propto P(r\vert I)$	$\propto P(s\vert I)$	$\propto P(t\vert I)$

3.2. Representation of Body Part Association

Even if every body part location is detected correctly, parts must be associated with each other in order to identify individual whole-animal instances. A naive approach would be to associate each body part with its nearest neighbor in terms of Euclidian distance using an optimal bipartite assignment method, such as the Hungarian algorithm. However, due to the elongated shape of pigs, this approach was prone to failure in cluttered environments, as illustrated in Figure 2.

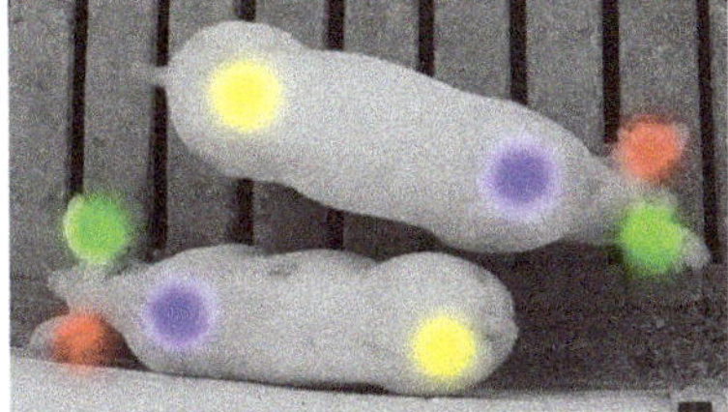
(a) Image with part detections

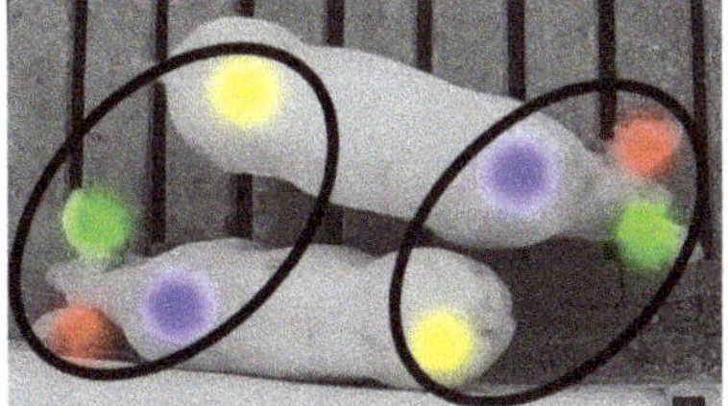
(b) Optimal grouping using Euclidean distance

Figure 2. Two nearby pigs with body parts properly annotated are depicted in (**a**). Whole instances must be formed by joining the parts together through part association. An optimal Euclidean nearest-neighbor part association is prone to failure when the animals are in close proximity, as illustrated in (**b**).

The proposed method used additional channels in the target output to encode body part locations with 2D vector offsets to other body parts belonging to the same animal. These offsets represented the direction and distance in pixels from one body part to another. While there were a total of $\binom{4}{2} = 6$ part pairs that exist between the four parts, the target output only represented three in order to reduce unnecessary redundancy (e.g., vectors joining tail-to-left-ear can be closely approximated by combining a vector joining the tail-to-shoulder and then the shoulder-to-left-ear). Specifically, 12 channels were used to represent three part pair associations, as listed in Table 2. The three part pairs and their associated channels are given below:

- Channels 1–4: Left Ear $\leftrightarrow$ Shoulder
- Channels 5–8: Right Ear $\leftrightarrow$ Shoulder
- Channels 9–12: Shoulder $\leftrightarrow$ Tail

Table 2. Channels 5–16 of the image-space representation used to represent pig locations and orientations. Pairs of neighboring channels corresponds to the x and y offset between neighboring parts. Overall, these 12 channels represent bidirectional vectors linking three pairs of body parts.

Channel	5	6	7	8	9	10
Encoding	$\propto (x_{l_n} - x_{s_n})$	$\propto (y_{l_n} - y_{s_n})$	$\propto (x_{s_n} - x_{l_n})$	$\propto (y_{s_n} - y_{l_n})$	$\propto (x_{r_n} - x_{s_n})$	$\propto (y_{r_n} - y_{s_n})$
Channel	11	12	13	14	15	16
Encoding	$\propto (x_{s_n} - x_{r_n})$	$\propto (y_{s_n} - y_{r_n})$	$\propto (x_{s_n} - x_{t_n})$	$\propto (y_{s_n} - y_{t_n})$	$\propto (x_{t_n} - x_{s_n})$	$\propto (y_{t_n} - y_{s_n})$

Each of the 12 channels encoded a real-valued offset from one point to another. Much like the part detection mappings, these vectors were encoded regionally into the spatial dimensions of the image. Figure 3 illustrates this encoding for a pair of side-by-side pigs. The diameter of the circular regions is denoted d_n for each pig n in the image, and it is proportional to the standard deviation used for the Gaussian kernel used in the previous section. For visualization purposes, each of the six images in Figure 3b,c represent the direction and distance between part pairs as a color, where the hue represents the direction and the saturation represents the magnitude (encoding provided in Figure 3d). Figure 3c further illustrates the lines joining the parts to one another.

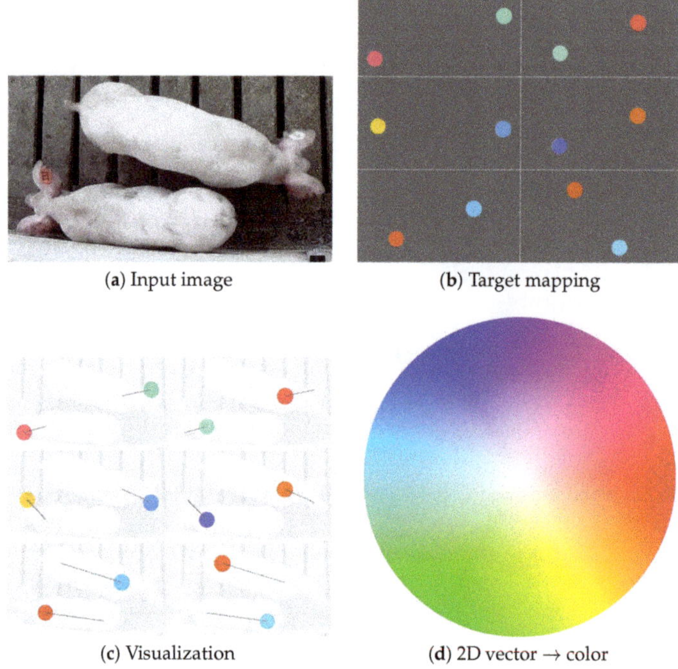

(**a**) Input image

(**b**) Target mapping

(**c**) Visualization

(**d**) 2D vector $\rightarrow$ color

Figure 3. The original image (**a**) is mapped to a 12-channel output (**b**), where vectors joining three pairs of body parts are encoded into circular regions in channels 5–16 of the output. Note that the four colors used in (**b**) are purely for visualization of the direction and magnitude of the vectors, where (**d**) provides a mapping between vectors and colors. The overlay in (**c**) is provided to illustrate the locations of the vector encodings and their magnitude and direction (illustrated by the gray line) in relation to the original image.

3.3. Instances from Part Detection and Association

The goal of the proposed method was to detect all visible parts and group them together in order to form whole-animal instances. Sections 3.1 and 3.2 provided a means to represent body part locations and vectors associated them to one another in the form of a 16-dimensional image-space mapping. To translate this image-space mapping to a set of discrete instance locations, one must follow a sequence of steps (illustrated in Figure 4).

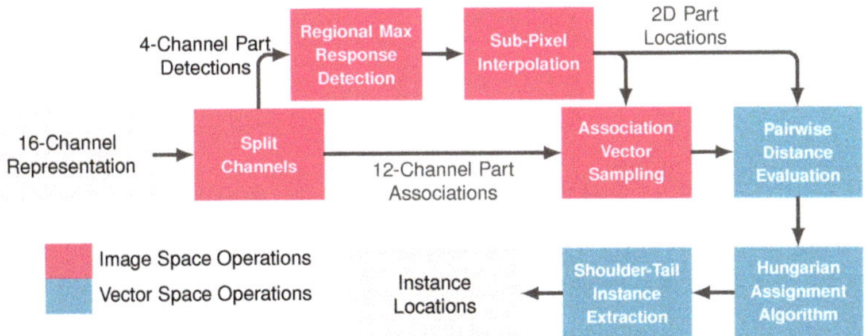

Figure 4. Flow diagram of the proposed method for converting the 16-channel image space representation to a set of 2D coordinates of each visible instance.

First, each 2D body part locations must have been extracted from the Gaussian kernels contained in channels 1–4 of the 16-channel representation image-space representation. Thus, the first step was to split channels into 4-channel part detections and 12-channel part associations. The precise 2D part locations were represented by the peaks of the Gaussian kernels in the image space mapping. Let $\mathbf{M_p}$ be the $R \times C$ image space map for body part $\mathbf{p} \in \{\mathbf{l}, \mathbf{r}, \mathbf{s}, \mathbf{t}\}$, which corresponds with the left ear, right ear, shoulder, and tail, respectively. Note that it was assumed that the number of rows and columns in the input image and output mappings are R and C. The part locations can be extracted from the image space using a form of regional max response detection defined by

$$\{\mathbf{p}\} = \left\{ (x,y) | \mathbf{M_p}(x,y) \geq \mathbf{M_p}(x',y') \text{ for all } (x',y') \in \mathcal{R}_{(x,y)} \right\} \quad \text{for } \mathbf{p} \in \{\mathbf{l}, \mathbf{r}, \mathbf{s}, \mathbf{t}\}, \tag{1}$$

where $\mathcal{R}_{(x,y)}$ was a region surrounding image space location (x,y). Part locations are therefore only detected if their value in the image space mapping was greater than that of its neighbors. This worked well for detecting the peak pixel coordinates of Gaussian kernels, but it can be further refined by using quadratic sub-pixel interpolation. Here, interpolation was performed by replacing the original integer coordinates (x,y) with real number coordinates using

$$(x,y) \leftarrow \left(x + \frac{\mathbf{M_p}(x-1,y) - \mathbf{M_p}(x+1,y)}{2\left(\mathbf{M_p}(x+1,y) + \mathbf{M_p}(x-1,y) - 2\mathbf{M_p}(x,y)\right)}, y + \frac{\mathbf{M_p}(x,y-1) - \mathbf{M_p}(x,y+1)}{2\left(\mathbf{M_p}(x,y+1) + \mathbf{M_p}(x,y-1) - 2\mathbf{M_p}(x,y)\right)} \right). \tag{2}$$

Given the complete set of detected body part locations

$$\{\mathbf{p}_1, \dots, \mathbf{p}_{N_p}\} = \{(x_{\mathbf{p}_1}, y_{\mathbf{p}_1}), \dots, (x_{\mathbf{p}_{N_p}}, y_{\mathbf{p}_{N_p}})\} \quad \text{for } \mathbf{p} \in \{\mathbf{l}, \mathbf{r}, \mathbf{s}, \mathbf{t}\}, \tag{3}$$

the next step was to estimate the locations of associated parts using an association vector sampling of the 12-channel part associations mapping. The 12 dimensions of the association mapping will be denoted $[\mathbf{M}^x_{l\to s} \mathbf{M}^y_{l\to s} \mathbf{M}^x_{s\to l} \mathbf{M}^y_{s\to l} \mathbf{M}^x_{r\to s} \mathbf{M}^y_{r\to s} \mathbf{M}^x_{s\to r} \mathbf{M}^y_{s\to r} \mathbf{M}^x_{s\to t} \mathbf{M}^y_{s\to t} \mathbf{M}^x_{t\to s} \mathbf{M}^y_{t\to s}]$ and the estimated location of an associated part $\mathbf{q}$ from location $\mathbf{p}_n$ can be obtained using

$$(\mathbf{p} \to \mathbf{q})_n = (x_{\mathbf{p}_n} - \mathbf{M}^x_{\mathbf{p}\to\mathbf{q}}(x_{\mathbf{p}_n}, y_{\mathbf{p}_n}), y_{\mathbf{p}_n} - \mathbf{M}^y_{\mathbf{p}\to\mathbf{q}}(x_{\mathbf{p}_n}, y_{\mathbf{p}_n})) \quad \text{for all} \quad n = 1, \dots, N_p. \tag{4}$$

To join parts together, the distance between the estimated part locations and the actual locations were first computed using a pairwise distance evaluation. Specifically, the association distance between two parts $\mathbf{p}_n$ and $\mathbf{q}_m$ is given by

$$d(\mathbf{p}_n, \mathbf{q}_m) = \frac{|(\mathbf{p} \to \mathbf{q})_n - \mathbf{q}_m| + |(\mathbf{q} \to \mathbf{p})_m - \mathbf{p}_n|}{2}, \tag{5}$$

where $|\mathbf{a}|$ denotes the L2-norm of vector $\mathbf{a}$. Overall, this collection of part-to-part distances forms a set of three unique distance matrices

$$\mathbf{D}_{\mathbf{p},\mathbf{q}} = \begin{bmatrix} d(\mathbf{p}_1, \mathbf{q}_1) & d(\mathbf{p}_1, \mathbf{q}_2) & \cdots & d(\mathbf{p}_1, \mathbf{q}_{N_q}) \\ d(\mathbf{p}_2, \mathbf{q}_1) & d(\mathbf{p}_2, \mathbf{q}_2) & \cdots & d(\mathbf{p}_2, \mathbf{q}_{N_q}) \\ \vdots & \vdots & \ddots & \vdots \\ d(\mathbf{p}_{N_p}, \mathbf{q}_1) & d(\mathbf{p}_{N_p}, \mathbf{q}_2) & \cdots & d(\mathbf{p}_{N_p}, \mathbf{q}_{N_q}) \end{bmatrix}, \tag{6}$$

where $(\mathbf{p} = \mathbf{l}, \mathbf{q} = \mathbf{s})$, $(\mathbf{p} = \mathbf{r}, \mathbf{q} = \mathbf{s})$, and $(\mathbf{p} = \mathbf{s}, \mathbf{q} = \mathbf{t})$. An optimal assignment between pairs of body parts that minimizes the sum of distances can be obtained by applying the Hungarian assignment algorithm to each distance matrix.

Finally, individual animals are identified as those that contain a joined shoulder and tail. The shoulder-tail instance extraction method began by identifying matches from the Hungarian assignment algorithm's output for $\mathbf{D}_{\mathbf{s},\mathbf{t}}$. Once all instances have been identified, the left and right ear detections can be joined to the shoulder locations of all instances via the Hungarian assignment algorithm's output for $\mathbf{D}_{\mathbf{l},\mathbf{s}}$ and $\mathbf{D}_{\mathbf{r},\mathbf{s}}$.

3.4. Fully-Convolution Network for Part Detection and Association Mapping

A fully-convolutional neural network was used to approximate the 16-channel target output, given a red-blue-green (RGB) image as input. Specifically, the hourglass network illustrated in Figure 5 was proposed. Hourglass networks with symmetry in the downsampling and upsampling stages have previously been used for pose estimation [61] and image segmentation [62]. Specifically, the proposed network was largely based on the popular SegNet architecture [62], which introduced a way to improve upsampling by sharing the indices of each max pooling layer with a corresponding max unpooling layer. This approach has been shown to achieve impressive performance in segmentation tasks by removing the burden of "learning to upsample" from the network.

The network architecture used by the proposed method also incorporated skip-connections in the form of depth concatenation immediately after max unpooling layers. Skip-connections have been demonstrated to encourage feature-reuse, thus improving performance with a fixed number of network coefficients [63]. They were also shown to drastically decrease the amount of training time required by the network. The U-net architecture, introduced by Ronneberger et al. [64], further demonstrated the power of skip-connections for hourglass-shaped networks.

During training, the objective function attempted to minimize the mean-squared error between the network output and the target ground truth. For the first four channels that correspond to part detections, gradients were back-propagated for all pixel locations regardless of their value. For the last 12 channels, gradients were back-propagated exclusively for pixel locations where the target output is assigned (non-zero). Therefore, a specific output was only encouraged in the regions surrounding the point locations. This type of selective training helped to ensure that the vector outputs did not tend toward zero in areas where part detections were uncertain. In essence, this approach was meant to separate the tasks of detection and association mapping.

Receptive Field

When designing neural networks for visual tasks, it is important that the network was able to "see" the entirety of the objects it is considering. This viewable area is often referred to as the receptive field. To derive the receptive field, the effective stride length between adjacent coordinates in the feature map is calculated using

$$s_{l_{\text{effective}}} = s_{l-1_{\text{effective}}} \times s_l,$$ (7)

where s_l is equal to the stride length at layer block l in the network and $s_0 = 1$. Note that, in the proposed network, all max pooling layers have $s_l = 2$ and all max unpooling layers have $s_l = 0.5$, while all other layers have $s_l = 1$. The overall stride length essentially relates the resolution of a downsampled feature map to the original input size. Given s_l for all l in the network, the receptive field size can be calculated using

$$r_l = r_{l-1} + (w_l - 1) \times s_{l-1},$$ (8)

where w_l is the convolutional kernel width at layer l and $r_0 = 1$. In the proposed network, each convolutional kernel had width $w_l = 3$. Because of the stochastic nature of max pooling and max unpooling operations, it was difficult to define their effective kernel size. Therefore, in this analysis, we have used the lower bound of $w_l = 1$ for all pooling operations.

Table 3 provides the receptive field of the network as a function of layer block for a subset of the 41 layer blocks featured in Figure 5. The receptive field represented the width of a square region in the original image space that affected a single pixel location at the output. While the receptive field of the proposed network was 363, the distance between any two image locations that can affect each others' outputs was 181 (the radius of the receptive field). In practice, it was recommended that the receptive field size should be considerably larger than the maximum object size due to a decaying effect that has been observed on the outer extremes of the square region [65]. It will be shown in Section 4 that the chosen image scale results in pigs that are typically much smaller than the receptive field radius.

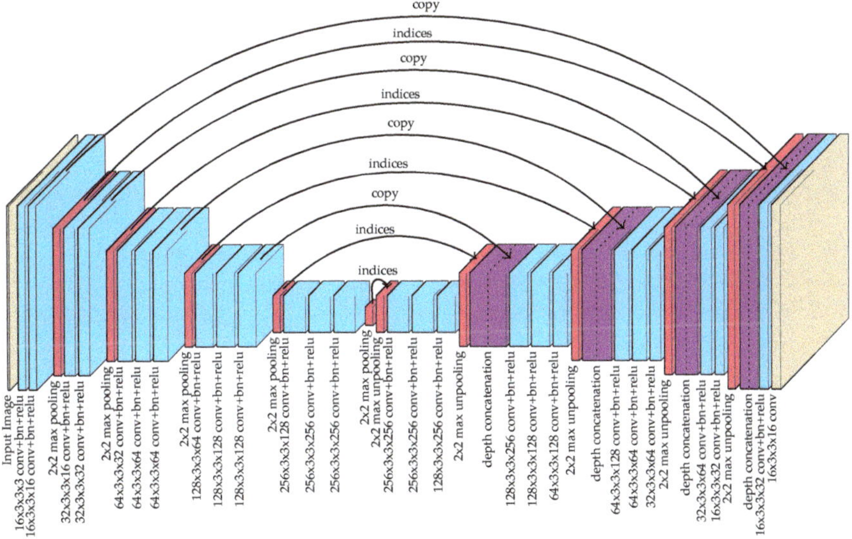

Figure 5. The hourglass-shaped network used by the proposed method to convert images to 16-channel image-space instance detection maps.

Table 3. Sampling of the receptive field calculations at the output of every layer of the proposed network. The different types of layers are abbreviated with the following notation: I: input image, C: convolution block, M: max pooling, U: max unpooling, D: depth concatenation, O: output image

Layer Type	I	C	C	M	C	⋯	C	M	U	C	⋯	C	U	D	C	O
l	1	2	3	4	5	⋯	18	19	20	21	⋯	37	38	39	40	41
s_l	1	1	1	2	1	⋯	1	2	0.5	1	⋯	1	0.5	1	1	1
$s_{l_{\text{effective}}}$	1	1	1	2	2	⋯	16	32	16	16	⋯	2	1	1	1	1
w_l	1	3	3	1	3	⋯	3	1	1	3	⋯	3	1	1	3	3
r_l	1	3	5	5	9	⋯	181	181	181	213	⋯	359	359	359	361	363

4. Experimental Results

4.1. Dataset

To the best of our knowledge, no open-source dataset exists for pig detection in group-housing environments. Therefore, to enable quantitative evaluation, a new dataset with 2000 annotated images of pigs was introduced. The dataset (http://psrg.unl.edu/Projects/Details/12-Animal-Tracking) depicts 17 different pen locations and includes pigs ranging in age from 1.5 to 5.5 months. Each unique image was randomly extracted from video recordings spanning multiple weeks in each location. More than two hours, on average, existed between samples at each location. Thus, a wide range of unique animal poses were represented in the dataset.

The dataset was divided into two subsets: 1600 images for training and 400 images for testing. Furthermore, the 400 testing images were subdivided into two additional subsets: 200 captured in the same environments seen in the training set (test:seen), and 200 images from environments previously unseen in the training set (test:unseen). The cameras used to capture the images included both a Microsoft Kinect v2 color camera with resolution 1080 × 1920 and Lorex LNE3162B and LNE4422 color/IR cameras with resolution 1520 × 2688. All of the environments were captured with the camera mounted above the pen looking down. The distance between the pen floor and the camera varied between 2.5 and 6.0 m, and the specific poses of the cameras ensured that the animal pen of interest was centered and entirely contained within the field of view. Variations in environment and image capture technology were used to ensure that the analysis emphasizes robustness.

Figure 6 shows sample images from the training set, depicting 13 different pen locations with color-coded annotations for each hand-labeled body part. Note that the last two images in Figure 6 depict the same environment, but one was captured with full color in the daytime and the other was captured with active IR at night. The first 200 images of the testing set (test:seen) were captured in the same environment as the training set, but at different times. Because more than two hours existed between subsequent randomly sampled images, it is likely that each test:seen image contained different animal poses than each training set image.

Figure 7 illustrates six sample images of the 200 images from the test:unseen set. Note that, not only was this environment previously unseen in the training set, but this set also included challenging lighting conditions that were also not represented among the training images. Twenty images from the training set were captured where the camera's IR night vision was activated, but all of the remaining 1580 training set images (and all of the test:seen images) were captured with overhead lights on. To achieve the challenging lighting conditions present in the test:unseen set, the lights were turned on at 6 am and off at 6 pm every day. For a short duration between approximately 6 pm and 8 pm, ambient lighting dimly illuminated the pens. After 8 pm, the cameras activated night-vision mode and captured IR images while actively illuminating the scene with built-in IR lights. Two of the four pens presented in the test:unseen set were also illuminated with IR flood lights. This had the effect of creating well lit scenes with harsh shadows and side-lighting.

In each of the images, a user manually annotated the location of the left ear (red), right ear (green), shoulder (blue), and tail (yellow) for each visible animal in that order. Annotations belonging to the same instance are connected with a continuous black line. If ears were not visible, they were not annotated, however, emphasis was placed on annotating both shoulders and tail for each instance even when these locations are occluded, i.e., both shoulder and tail were annotated as long as they are located in the pen of interest and their estimated positions were within the field of view of the camera.

It should be noted that, in some cases, pigs from adjacent pens were partially visible through bars that separate the pens. These partially visible pigs were not annotated. It was assumed that a camera placed above a pen is responsible for detecting only animals in that pen and, while some areas of the image belonging to the adjacent pen were masked out, it was difficult to remove partially visible pigs from the image without unintentionally masking out pigs within the pen of interest. In practice, areas of interests were defined by polygons for each image in the dataset and masking out was done by setting all pixels outside the area of interest to pure black. Examples of masked out regions can been seen in Figures 6 and 7, where the blacked out regions correspond to areas with pigs in adjacent pens.

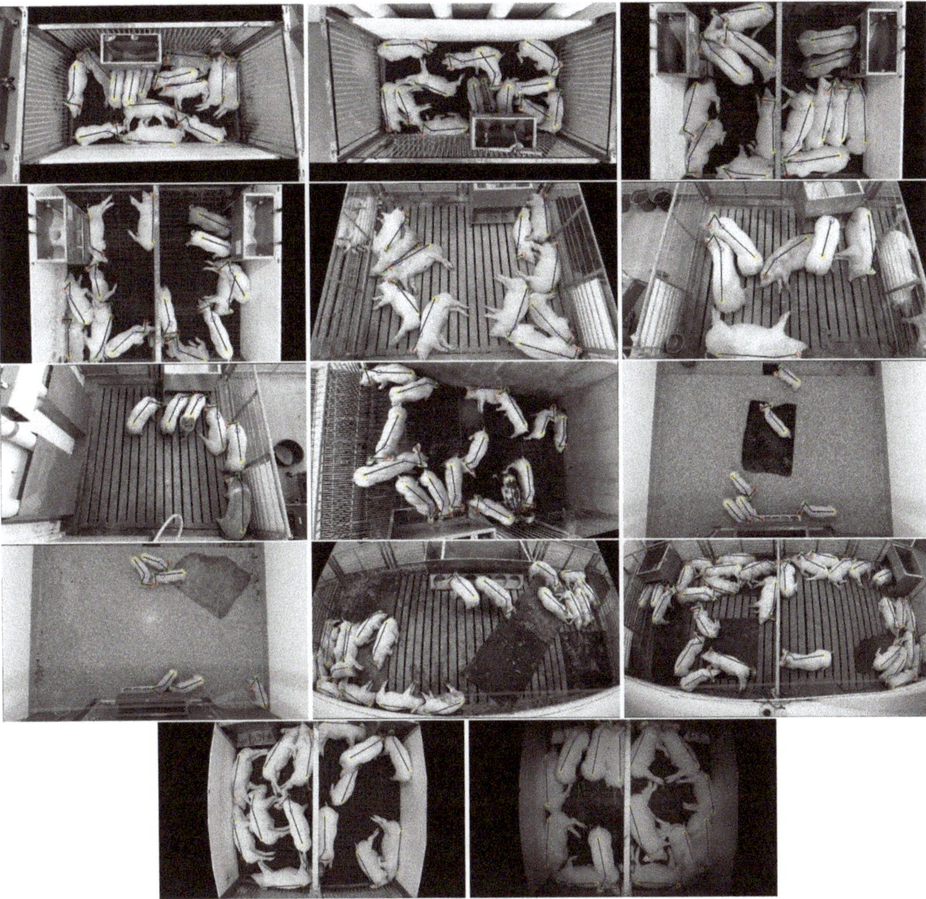

Figure 6. Sample images depicting different environments represented in the training set. The first 13 images (left-to-right, top-to-bottom) were captured during daylight hours with lights on. The last image (from the same environment as the 13th image) was captured using the infrared night vision mode used by the Lorex LNE3162B with active IR illumination.

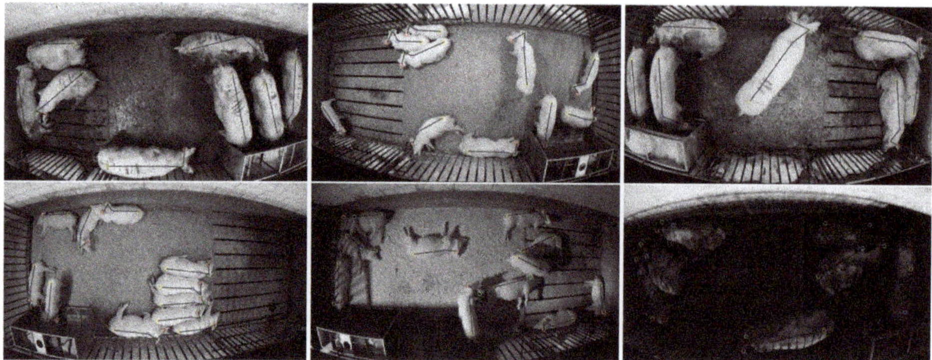

Figure 7. Sample images from the "unseen" portion of the testing set (test:unseen). These images depict environments and lighting conditions not represented in the training set.

4.2. Training Details

Prior to training the network, images were downsampled so that the number of columns was 480. This was empirically deemed to be a sufficient resolution for discerning the parts of interest while remaining small enough for the computing hardware to process multiple images per second. The average length of pigs in each image after downsampling is presented in the histogram of Figure 8. While the majority of the pigs had a body length of less than 100 pixels, there were some that exceed 140 pixels in length. For these pigs, it was important that the network was able to "see" the entirety of the pig as it estimated the body part locations and vector associations. In Section 3.4, the radius of the receptive field was found to be 181 using the proposed network. Therefore, the network was capable of observing the entire animal even in the most extreme images where the shoulder-to-tail distance approached 140 pixels.

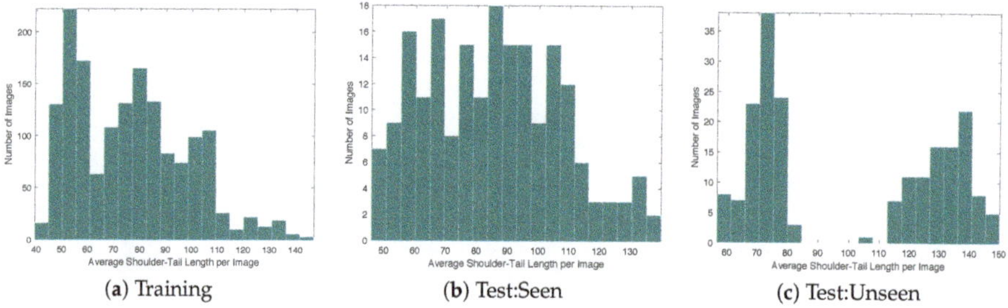

(a) Training   (b) Test:Seen   (c) Test:Unseen

Figure 8. Distribution of the average length from shoulder to tail in each partition of the dataset.

Target images for training the fully-convolutional network were created by adapting the size of the Gaussian kernels used to mark each part in channels 1–4 (Figure 1b) to the size of the animals. This adaptation encouraged continuity of image-space annotations between different environments and ages/sizes of pigs. Specifically, this was done by first computing the average distance between the shoulder and tail for all instances, denoted $\mu_{s\to t}$, to provide a numerical representation of the average size of pigs in the image space. Then, the shoulder-to-tail distance for instance n, given by $\delta_{s\to t}$, was combined with the average distance in order to compute the Gaussian kernel standard deviation, defined as $\sigma_n = 0.16 \times (\mu_{s\to t} + \delta_{s\to t})$. This combination was used to prevent unusual animal poses from shrinking the size of the kernels too much, while still allowing some adaptation to individual size variations. It should be noted that the scale factor of 0.16 was determined empirically to provide a suitable approximation to the variability of human annotations. If σ_n was too large, kernels belonging to nearby pigs interfered with each other and often resulted in a single part location being extracted by regional maximum detection. When σ_n was too small, the network training unfairly applied a penalty to kernels that were not exactly equal to the location provided by the human annotator, even if the kernel's location was within the natural variance of human annotations. Finally, the Gaussian kernels were then multiplied by a scalar value in order to set their maximum value to 1.0 and, in cases where two nearby Gaussian kernels for the same body part intersected, the output was simply assigned to the maximum of the two Gaussian kernel values. Scaling the kernels so that the peak is 1.0 helped to ensure that a fixed threshold can be used in peak detection regardless of σ_n.

The circular regions used to assign association vectors between parts in channels 5–16 (Figure 3b) should ideally have covered all possible pixel locations where the part might be detected. In practice, this area can be sufficiently covered by considering regions where the Gaussian kernel for each part had a magnitude greater than 0.2. In situations where one region intersected with another, the target output vector was composed of a weighted combination of the two intersecting vectors. The weights in these circumstances came from the corresponding Gaussian kernel magnitude at each pixel location.

The network was trained using heavy augmentation of both input and target images. Augmentations included random left-right flipping, random rotations sampled from a uniform distribution ranging from 0 to 360°, random scaling sampled uniformly between 0.5 and 1.5, and XY shifts uniformly sampled from the range of ±20 pixels along both dimensions. Careful consideration was needed for augmenting the 16-channel target image. Rotations and scaling were applied spatially to both the association vector regions and also the output values along pairs of channels that correspond to XY offsets between body parts. Left-right flips were handled by switching the labels for left and right ears.

4.3. Processing Details

After obtaining the 16-channel mapping from the trained network, each of the part location maps (channels 1–4) and the association maps (channels 5–16) were smoothened using a 5×5 averaging box filter. This step would not be necessary to extract information from ground truth mappings, but it was beneficial for reducing the effects of noise on regional maximum response detection. In practice, box filtering was done by adding an average pooling layer to the end of the neural network. The size of regions $\mathcal{R}_{(x,y)}$ used in (1) consisted of a 15×15 window surrounding each pixel (x,y).

The method was implemented in Matlab 2018b using the deep learning toolbox [66]. A desktop computer, equipped with an Intel i7-6700K CPU, 32 GB of RAM, and an NVIDIA GTX1070 GPU was used for training and inference. The approximate time required by the fully-convolutional neural network to perform forward inference is 0.24 s and it took an additional 0.01 s to find instance locations. Thus, the system was capable of fully processing four frames per second.

4.4. Instance Detection Performance Metric

The goal of the proposed method was to identify the location and orientation of each pig in a given image. Although the method generated detections and associations for four body parts, only the shoulder and tail location were used to identify a complete instance. This decision was based on two factors. Firstly, they are sufficient for approximating the center-of-mass location and orientation of each animal and, second, special emphasis was placed on ensuring their labeling by human annotators. Given a complete set of N ground truth shoulder-tail pairs $\{(\mathbf{s}_1, \mathbf{t}_1), \ldots, (\mathbf{s}_N, \mathbf{t}_N)\}$ and a set of M estimated shoulder-tail pairs $\{(\hat{\mathbf{s}}_1, \hat{\mathbf{t}}_1), \ldots, (\hat{\mathbf{s}}_N, \hat{\mathbf{t}}_M)\}$, an association method was needed to determine if an estimate corresponded to the ground truth, since both sets of pixel coordinates were unlikely to contain exactly the same values.

Bipartite matching problems are commonly solved with a Hungarian assignment, however, this can sometimes lead to matches between far-away pairs in order to minimize a global cost. For this particular matching problem where shoulder-tail pairs are associated with each other, there was likely to be very little distance between the ground truth and detected positions. Setting the maximum distance allowed between matching pairs can fix this issue, but it comes at the cost of introducing additional parameters that depended on image resolution and the relative sizes of animals. To avoid parameterization, the strict cross-check matching criteria was used here to assign estimates to the ground truth only when they were each others' minimum cost matches. More formally, two instances n and m matched if and only if

$$m = \operatorname*{argmax}_{m\in\{1,\ldots,M\}} |\mathbf{s}_n - \tilde{\mathbf{s}}_m| + |\mathbf{t}_n - \tilde{\mathbf{t}}_m| \tag{9}$$

and

$$n = \operatorname*{argmax}_{n\in\{1,\ldots,N\}} |\mathbf{s}_n - \tilde{\mathbf{s}}_m| + |\mathbf{t}_n - \tilde{\mathbf{t}}_m|, \tag{10}$$

where $\|$ denotes the L2 norm. Figure 9 illustrates the advantage of using the cross-check method instead of the unparameterized Hungarian algorithm.

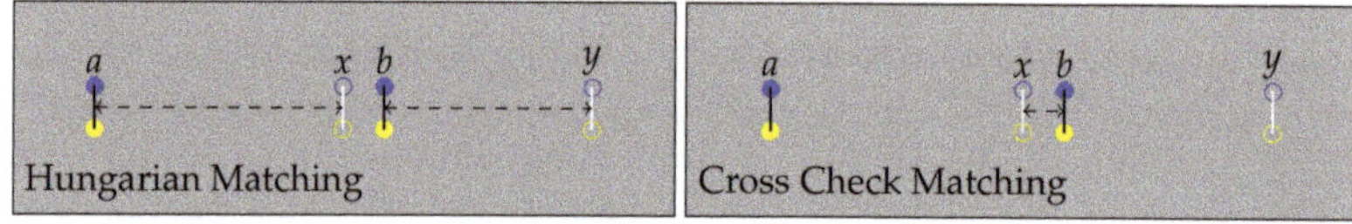

Figure 9. Example with two ground truth instance locations (solid circles connected with black lines) and two detected instances (empty circles connected with white lines) using matching results achieved with both the Hungarian algorithm and the proposed cross-check matching of Equations (9) and (10). While the detection x and ground truth location b in the middle are clearly nearest neighbors of one another, they are not matched by the Hungarian algorithm. Instead, in an effort to minimize the global matching cost, the Hungarian algorithm will assign a to x and b to y. In contrast, the cross-check matching method leaves the outer detection y and ground truth location a unmatched while assigning the two in the middle, x and b, together.

4.5. Instance Matching Results

In order to evaluate the effectiveness of the proposed vector part association method, it was compared to an alternative Euclidean part association method. The Euclidean part association method joins parts together by simply minimizing their Euclidean distance. This method, previously illustrated in Figure 2, removes the effects of part association vectors on detection performance and allows for a partial ablation study. Figure 10 presents the precision and recall for all three partitions of the dataset and Table 4 presents full numerical results over the training set. Each sample along the curve corresponds to a different threshold for part detection, where parts are detected only if the neural network output in channels 1–4 exceeds the threshold value.

The results show that the proposed vector part association method provided a significant boost to matching precision when compared to Euclidean part association. Less than 0.1% of detections were false positives compared to more than 5% when using Euclidean matching, regardless of threshold. Figure 10a,b illustrate nearly identical results across training and test:seen sets. This provides a strong indication that the method was not overfitting to the specific animal poses presented in the training set. Both Figure 10a,b demonstrate a minimum precision of ≈0.91 at a recall of ≈0.42 for the Euclidean matching method. This was because, at this threshold less than half of the animal parts were being detected but the ones that were detected are matched to their nearest neighbor. As a result, there was a relatively high likelihood that only a shoulder is detected, but not a tail, or vice versa. In an effort to form whole instances, the Euclidean method simply joined together nearest neighbors and many of these instances were not aligned with the ground truth. When the threshold was adjusted higher or lower, there was a higher likelihood that either both shoulder and tail were detected or neither was detected. In either case, this lead to improved precision, because the instances that were identified were more likely to be true positives.

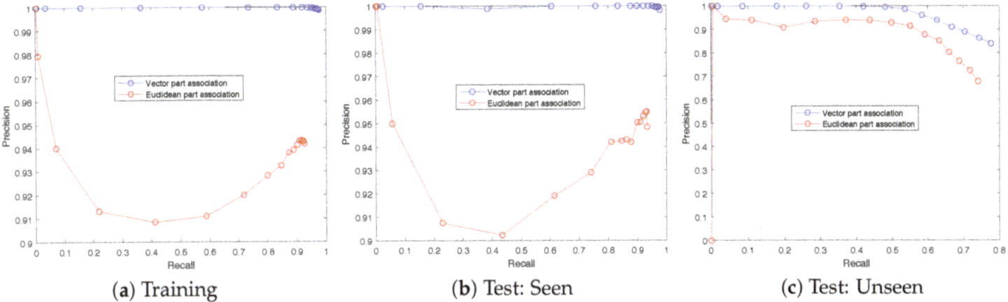

(a) Training   (b) Test: Seen   (c) Test: Unseen

Figure 10. Precision-recall curves for both the proposed method and an alternative association strategy that assigns parts to one another by minimizing Euclidean distance. The results are nearly identical across the training set (**a**) and test:seen set (**b**), and both illustrate a dramatic improvement achieved by joining parts together using the proposed vector mapping. The results on the test:unseen set (**c**) illustrate the limitations of the method when operating on different environments than those used for training.

Table 4. Detailed results obtained with the proposed method and an alternative association strategy that assigns parts to one another by minimizing Euclidean distance. The table includes true positives (TP), false positives (FP), false negatives (FN), precision, recall, and F-measure for different part detection thresholds.

Part Detection Threshold	Vector Matching						Euclidean Matching					
	TP	FP	FN	Recall	Precision	F-Measure	TP	FP	FN	Recall	Precision	F-Measure
0.10	20,217	27	525	0.975	0.999	0.987	19,170	1181	1572	0.924	0.942	0.933
0.15	20,160	20	582	0.972	0.999	0.985	19,127	1156	1615	0.922	0.943	0.932
0.20	20,092	17	650	0.969	0.999	0.984	19,058	1154	1684	0.919	0.943	0.931
0.25	19,999	13	743	0.964	0.999	0.981	18,971	1141	1771	0.915	0.943	0.929
0.30	19,865	10	877	0.958	0.999	0.978	18,851	1135	1891	0.909	0.943	0.926
0.35	19,675	7	1067	0.949	1.000	0.973	18,675	1162	2067	0.900	0.941	0.920
0.40	19,413	3	1329	0.936	1.000	0.967	18,418	1190	2324	0.888	0.939	0.913
0.45	19,029	2	1713	0.917	1.000	0.957	18,077	1195	2665	0.872	0.938	0.904
0.50	18,408	2	2334	0.887	1.000	0.940	17,526	1269	3216	0.845	0.932	0.887
0.55	17,287	2	3455	0.833	1.000	0.909	16,568	1281	4174	0.799	0.928	0.859
0.60	15,227	2	5515	0.734	1.000	0.847	14,871	1294	5871	0.717	0.920	0.806
0.65	11,929	0	8813	0.575	1.000	0.730	12,184	1189	8558	0.587	0.911	0.714
0.70	7565	0	13,177	0.365	1.000	0.534	8543	860	12,199	0.412	0.909	0.567
0.75	3261	0	17,481	0.157	1.000	0.272	4523	430	16,219	0.218	0.913	0.352
0.80	692	0	20,050	0.033	1.000	0.065	1414	90	19,328	0.068	0.940	0.127
0.85	53	0	20,689	0.003	1.000	0.005	143	3	20,599	0.007	0.979	0.014
0.90	1	0	20,741	0.000	1.000	0.000	5	0	20,737	0.000	1.000	0.000

In Table 5, the results are compared across all three partitions of the dataset with a fixed threshold of 0.25. While the F-measure was 0.981 at threshold 0.25, which was lower than the peak F-measure of 0.987 achieved at a threshold of 0.1, the decreased threshold produced more than twice the number of false positives. When F-measure values were nearly identical, the choice of threshold depended on how sensitive an application was to false positives and false negatives. The comparison at threshold 0.25 highlighted both the performance similarities across the training and test:seen sets and the discrepancies between both of those sets and the test:unseen set. One interpretation is that the discrepancy illuminates the importance of environment and lighting variations when training the neural network. The test:seen results were nearly identical to the training results, even though the specific poses of each animal were previously unseen. However, due to the use of heavy image augmentation, similar poses were likely represented during training. In contrast, the test:unseen results were much worse, likely due to the novel environments and challenging lighting conditions not present in the training set images.

Table 5. Instance detection results for the training set (1600 images), the test set with images of the same environments in the training set (200 images), and the test set with images of new environments and challenging lighting conditions (200 images). The part detection threshold was fixed at 0.25.

Evaluation Set	Vector Matching					
	TP	FP	FN	Recall	Precision	F-Measure
Training	19,999	13	743	0.964	0.999	0.981
Test: Seen	2273	1	94	0.960	1.000	0.980
Test: Unseen	1150	112	573	0.667	0.911	0.771

By digging deeper into the results and looking at specific examples, it is possible to learn more about the performance discrepancies. An example of 100% successful detections from both test:seen and test:unseen sets are shown in Figure 11. Here, the neural network output is illustrated for each of the 16 channels, and the final detections are shown below, where a green line connecting the shoulder to the tail defines a true positive detection. Note that, unlike the target part association maps illustrated in Figure 3, the outputs of the neural network do not clearly conform to the part locations. This is because the network was only trained to produce the correct vector (illustrated by color) at part locations and, at all other locations, the network essentially tried to minimize the cost with hypothetical part association vectors in case a part was present in that location. This attempt to minimize cost "in case" was most visible when comparing the part association maps of shoulder-to-tail and tail-to-shoulder (the bottom two part association maps in Figure 11). Even when the network was highly confident that a location belonged to a tail, it produced an association vector pointing from shoulder-to-tail at that location, just in case a shoulder at that location was mistaken for a tail.

Due to the similar lighting and overall appearance of test:seen image in Figure 11a, the method was able to identify every instance within the pen environment with high confidence (as indicated by the first four channel outputs of the neural network). However, in the test:unseen image (Figure 11b), the pig behind bars in the adjacent pen caused some confusion in the network. This was likely due to the fact that the network had never been exposed to this particular pen environment, and thus it had not been trained to ignore partial animals on the other side.

Alternatively, Figure 12 illustrates failure cases for both test:seen and test:unseen images. Each of the failures in the test:seen image occurred because of occlusions that made it difficult to discern the location of the shoulders and/or tail. In this case, it was even difficult for a human observer to confidently assign the ground truth locations. On the other hand, failures on the test:unseen image were not due to occlusions. They can instead by attributed to the unusual lighting conditions and the relatively large presentation of the animals in the image. Both of these properties were not represented in the training set, making it difficult for the neural network to interpret the image.

Figures 13 and 14 illustrate 24 failures from the test:seen and test:unseen set, respectively. In the test:seen sample set of Figure 13, 17 of the 23 false negatives can be attributed to occlusions or lack of visibility when the pig approached the edge of the image. Some other causes of error include unusual poses where the head was hidden, and situations where the pig had atypical markings. In contrast, only four false negatives out of 21 from the test:unseen sample set (Figure 14) can be attributed to occlusion. At least 10 can likely be attributed to lighting conditions. All three false positives occurred when a pig in an adjacent pen was laying next to the dividing bars. The outline of the bars on the pig's body appeared to confuse the network into interpreting this as a smaller body pointed in the orthogonal direction.

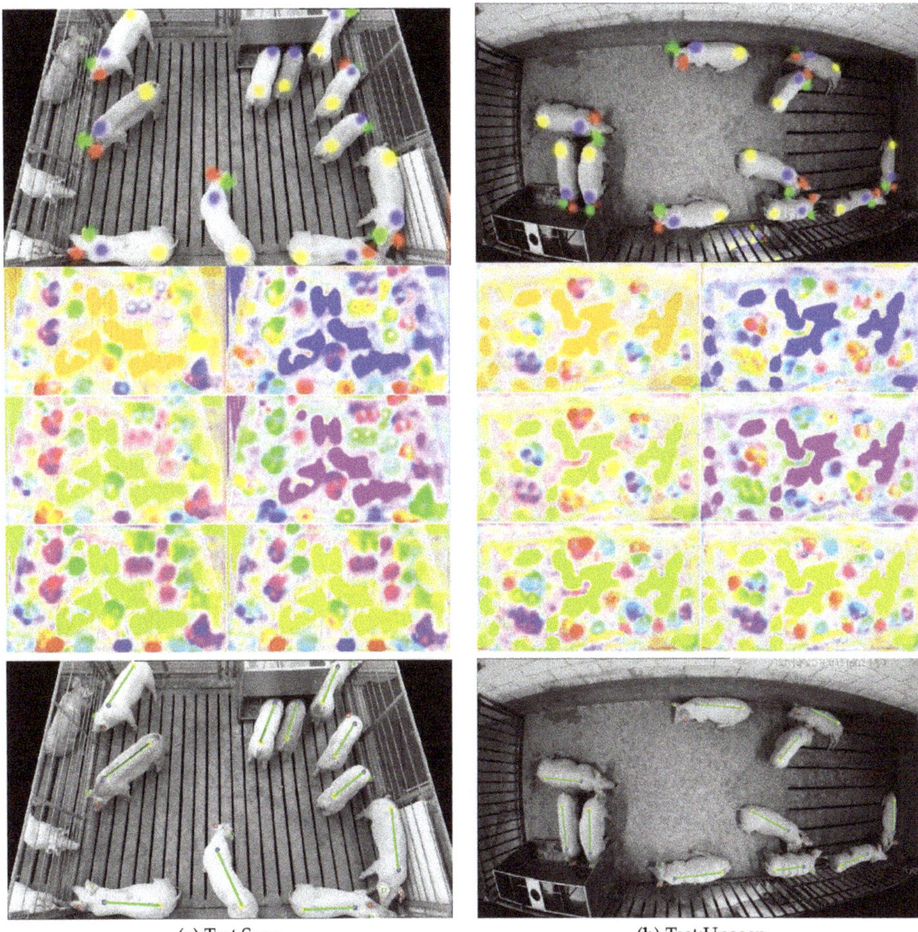

(a) Test:Seen   (b) Test:Unseen

Figure 11. Examples of successful instance detection from both the test:seen set (**a**) and the test:unseen set (**b**). The top images depict the first four channels of the neural network output. The middle image composed of six sub-images depicts the color-coded vector associations from the last 12 channels of the neural network output. The bottom images depict both ground truth locations and estimates using the following color coding: false negative (blue), and true positive (green). Note that these images depict 100% successful detections, so only true positives are present.

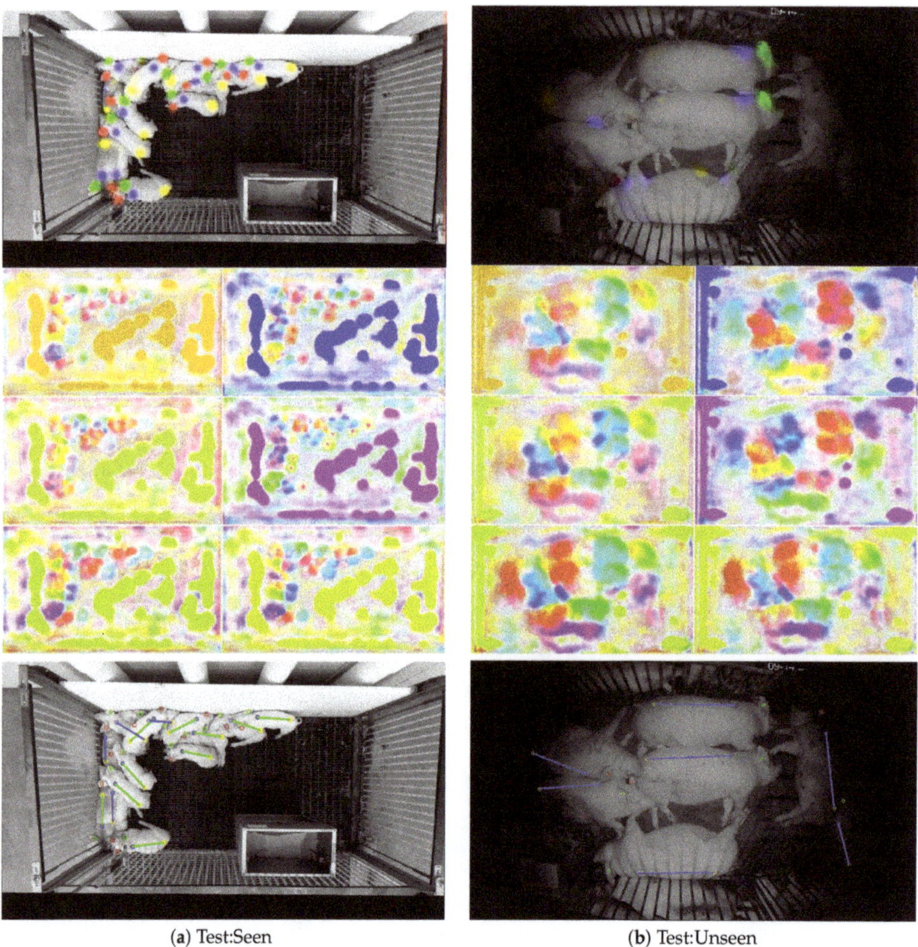

(a) Test:Seen   (b) Test:Unseen

Figure 12. Examples of unsuccessful instance detection from both the test:seen set (**a**) and the test:unseen set (**b**). The top images depict the first four channels of the neural network output. The middle image composed of six sub-images depicts the color-coded vector associations from the last 12 channels of the neural network output. The bottom images depict both ground truth locations and estimates using the following color coding: false negative (blue), and true positive (green).

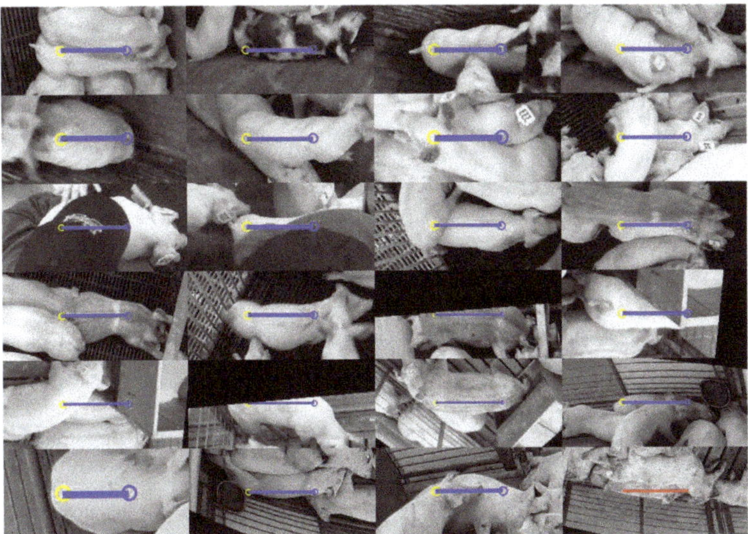

Figure 13. Twenty-four random samples of unsuccessful instance detections from the test:seen set using the following color coding: false negative (blue), and false positive (red).

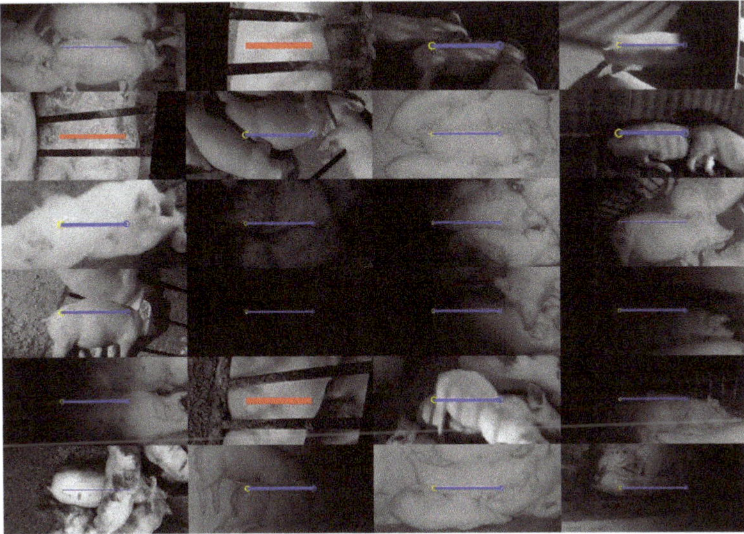

Figure 14. Twenty-four random samples of unsuccessful instance detections from the test:unseen set using the following color coding: false negative (blue), and false positive (red).

4.6. Discussion

The proposed method focuses on detecting the location and orientation of individual pigs in group-housing environments. Due to the lack of an available public dataset, it was necessary to create a new collection of annotated images. This presented a challenge in terms of capturing the appropriate level of variability and, while we believe that the chosen images sufficiently represent a variety of environments and ages of pigs, it would likely have been beneficial to include more camera angles and more than three types of cameras. Furthermore, the four body part locations were arbitrarily chosen as representatives of the location and orientation of each animal instance. A different set of body parts might have been equally effective.

Compared to datasets like ImageNet [67] and COCO [33], 2000 images may seem like an insufficient number of images to train a deep network. However, pig detection from an overhead camera is a much more specific task than classifying images into one of 1000 categories. With nearly 25,000 different animal poses captured, it is likely that any new pose presented to the network will strongly resemble one that already exists in the dataset. Augmentation is also critical to the success of network training. The chosen network contains nearly 4,000,000 coefficients, so it might be possible to overfit to 25,000 static animal poses, but it is much more difficult to overfit when the angle, size, and left-right orientation is randomized.

The fully-convolutional network introduced in Section 3.4 to estimate body part locations and association vectors was designed with sufficient complexity and a wide enough receptive field to achieve high performance levels in terms of precision and recall. However, the chosen hourglass architecture using max pooling/unpooling with skip connections and depth concatenations is almost certainly one of many network architectures capable of producing the desired 16-dimensional output representation. Atrous convolution, for example has been shown to be particular effective for creating a large depth of field [68] and spatial pyramid pooling has further been demonstrated to achieve excellent performance on multi-scale tasks [69]. A thorough investigation into the suitability of different network architectures might lead to a more accurate or efficient network for instance detection.

By inspecting the specific outputs of the network and the instance formation process, it becomes evident that errors are most commonly caused when the shoulder or tail of one pig occludes the same body part on another pig. Due to the network's inability to represent multiple part instances in the same image space location, it is only possible for one part instance to be detected in these situations. The vector associations inherently estimate the location of adjacent body parts, therefore, the occlusion can be inferred from the existing output of the network. Alternatively, it might also be possible to augment the dataset to explicitly label occlusions and build the network to detect such events.

In addition to shoulders and tails, the left and right ear were annotated in the dataset and explicitly detected by the network. While the results for instance-level detection do not evaluate the quality of these detections, we foresee them being integrated into future systems as a way to uniquely identify each instance. Ear tags are a common way for livestock to be identified in commercial facilities, and this may provide a convenient way to differentiate between individuals in an otherwise homogeneous population.

Perhaps the most likely application of this detection method would be within a larger tracking system, where detection serves as the first stage for video processing. To this end, the proposed per-frame detection method naturally lends itself to multi-object tracking (MOT). Specifically, a sub-category known as tracking-by-detection MOT methods directly process the outputs of per-frame detection methods, and their performance is often strongly tied to the quality of the detector [70]. For this reason, it is expected that high quality detection methods will eventually contribute to more reliable methods for multi-object tracking.

5. Conclusions

The proposed method and accompanying dataset introduced in this paper attempt to provide a robust solution to instance-level detection of multiple pigs in group-housing environments. A major contribution of this work is the introduction of an image space representation of each pig as a collection of body parts along with a method to join parts together to form full instances. The method for estimating the desired image space representation leverages the power of deep learning using a fully-convolutional neural network. Through gradual downsampling and upsampling, the network is able to consider large regions in the image space with a receptive field that covers even the largest pigs in the dataset.

Results demonstrate that the method is capable of achieving over 99% precision and over 95% recall at the task of instance detection when the network is tested and trained under the same environmental conditions. When testing on environments and lighting conditions that the network had not been trained to handle, the results drop significantly to 91% precision and 67% recall. These results can be interpreted in one of three ways: (1) networks should be fine-tuned to handle new environments, (2) a larger number and variety of images should be included in the dataset, or (3) the design and/or training methodology should be revised to improve the robustness to environmental variability. As the dataset and the number of environments grows, eventually there might be enough variety such that new environments add little to the network's ability to handle novel presentations. Regarding the third interpretation, while significant augmentations were applied to the input and output images during training, it is impossible for spatial transformations to mimic variations in lighting conditions. Therefore, a new set of non-uniform color-space transformations may provide a solution that improves the robustness of the trained network.

By introducing a new dataset and providing an accompanying method for instance-level detection, we hope that this work inspires other researchers to introduce competing methods and perform objective evaluations on the dataset.

Author Contributions: E.P., M.M., L.C.P., T.S. and B.M. conceived and designed the experiments and assisted in data acquisition. E.P. designed and implemented the neural network algorithm and performed the formal analysis. E.P., M.M., L.C.P., T.S. and B.M. wrote the paper.

Funding: This research was supported by the National Pork Board (NPB #16-122).

Acknowledgments: We would like to thank Lucas Fricke and Union Farms for their cooperation and allowing us to capture data in their swine facility.

Conflicts of Interest: The authors declare no conflict of interest.

References

1. Matthews, S.G.; Miller, A.L.; Clapp, J.; Plötz, T.; Kyriazakis, I. Early detection of health and welfare compromises through automated detection of behavioural changes in pigs. *Vet. J.* **2016**, *217*, 43–51. [CrossRef]

2. Wedin, M.; Baxter, E.M.; Jack, M.; Futro, A.; D'Eath, R.B. Early indicators of tail biting outbreaks in pigs. *Appl. Anim. Behav. Sci.* **2018**, *208*, 7–13. [CrossRef]

3. Burgunder, J.; Petrželková, K.J.; Modrý, D.; Kato, A.; MacIntosh, A.J. Fractal measures in activity patterns: Do gastrointestinal parasites affect the complexity of sheep behaviour? *Appl. Anim. Behav. Sci.* **2018**, *205*, 44–53. [CrossRef]

4. PIC North America. Standard Animal Care: Daily Routines. In *Wean to Finish Manual*; PIC: Hendersonville, TN, USA, 2014; pp. 23–24.

5. Tuyttens, F.; de Graaf, S.; Heerkens, J.L.; Jacobs, L.; Nalon, E.; Ott, S.; Stadig, L.; Van Laer, E.; Ampe, B. Observer bias in animal behaviour research: can we believe what we score, if we score what we believe? *Anim. Behav.* **2014**, *90*, 273–280. [CrossRef]

6. Wathes, C.M.; Kristensen, H.H.; Aerts, J.M.; Berckmans, D. Is precision livestock farming an engineer's daydream or nightmare, an animal's friend or foe, and a farmer's panacea or pitfall? *Comput. Electron. Agric.* **2008**, *64*, 2–10. [CrossRef]

7. Banhazi, T.M.; Lehr, H.; Black, J.; Crabtree, H.; Schofield, P.; Tscharke, M.; Berckmans, D. Precision livestock farming: An international review of scientific and commercial aspects. *Int. J. Agric. Biol. Eng.* **2012**, *5*, 1–9.

8. Tullo, E.; Fontana, I.; Guarino, M. Precision livestock farming: An overview of image and sound labelling. In *European Conference on Precision Livestock Farming 2013:(PLF) EC-PLF*; KU Leuven: Leuven, Belgium, 2013; pp. 30–38.

9. Kim, S.H.; Kim, D.H.; Park, H.D. Animal situation tracking service using RFID, GPS, and sensors. In Proceedings of the 2010 IEEE Second International Conference on Computer and Network Technology (ICCNT), Bangkok, Thailand, 23–25 April 2010; pp. 153–156.

10. Stukenborg, A.; Traulsen, I.; Puppe, B.; Presuhn, U.; Krieter, J. Agonistic behaviour after mixing in pigs under commercial farm conditions. *Appl. Anim. Behav. Sci.* **2011**, *129*, 28–35. [CrossRef]

11. Porto, S.; Arcidiacono, C.; Giummarra, A.; Anguzza, U.; Cascone, G. Localisation and identification performances of a real-time location system based on ultra wide band technology for monitoring and tracking dairy cow behaviour in a semi-open free-stall barn. *Comput. Electron. Agric.* **2014**, *108*, 221–229. [CrossRef]

12. Giancola, G.; Blazevic, L.; Bucaille, I.; De Nardis, L.; Di Benedetto, M.G.; Durand, Y.; Froc, G.; Cuezva, B.M.; Pierrot, J.B.; Pirinen, P.; et al. UWB MAC and network solutions for low data rate with location and tracking applications. In Proceedings of the 2005 IEEE International Conference on Ultra-Wideband, Zurich, Switzerland, 5–8 September 2005; pp. 758–763.

13. Clark, P.E.; Johnson, D.E.; Kniep, M.A.; Jermann, P.; Huttash, B.; Wood, A.; Johnson, M.; McGillivan, C.; Titus, K. An advanced, low-cost, GPS-based animal tracking system. *Rangel. Ecol. Manag.* **2006**, *59*, 334–340. [CrossRef]

14. Schwager, M.; Anderson, D.M.; Butler, Z.; Rus, D. Robust classification of animal tracking data. *Comput. Electron. Agric.* **2007**, *56*, 46–59. [CrossRef]

15. Taylor, K. Cattle health monitoring using wireless sensor networks. In Proceedings of the Communication and Computer Networks Conference, Cambridge, MA, USA, 8–10 November 2004.

16. Ruiz-Garcia, L.; Lunadei, L.; Barreiro, P.; Robla, I. A Review of Wireless Sensor Technologies and Applications in Agriculture and Food Industry: State of the Art and Current Trends. *Sensors* **2009**, *9*, 4728–4750. [CrossRef]

17. Escalante, H.J.; Rodriguez, S.V.; Cordero, J.; Kristensen, A.R.; Cornou, C. Sow-activity classification from acceleration patterns: A machine learning approach. *Comput. Electron. Agric.* **2013**, *93*, 17–26. [CrossRef]

18. Alvarenga, F.A.P.; Borges, I.; Palkovič, L.; Rodina, J.; Oddy, V.H.; Dobos, R.C. Using a three-axis accelerometer to identify and classify sheep behaviour at pasture. *Appl. Anim. Behav. Sci.* **2016**, *181*, 91–99. [CrossRef]

19. Voulodimos, A.S.; Patrikakis, C.Z.; Sideridis, A.B.; Ntafis, V.A.; Xylouri, E.M. A complete farm management system based on animal identification using RFID technology. *Comput. Electron. Agric.* **2010**, *70*, 380–388. [CrossRef]

20. Feng, J.; Fu, Z.; Wang, Z.; Xu, M.; Zhang, X. Development and evaluation on a RFID-based traceability system for cattle/beef quality safety in China. *Food Control* **2013**, *31*, 314–325. [CrossRef]

21. Floyd, R.E. RFID in animal-tracking applications. *IEEE Potentials* **2015**, *34*, 32–33. [CrossRef]

22. Mittek, M.; Psota, E.T.; Pérez, L.C.; Schmidt, T.; Mote, B. Health Monitoring of Group-Housed Pigs using Depth-Enabled Multi-Object Tracking. In Proceedings of the International Conference on Pattern Recognition, Workshop on Visual observation and analysis of Vertebrate and Insect Behavior, Cancun, Mexico, 4 December 2016; pp. 9–12.

23. Mittek, M.; Psota, E.T.; Carlson, J.D.; Pérez, L.C.; Schmidt, T.; Mote, B. Tracking of group-housed pigs using multi-ellipsoid expectation maximisation. *IET Comput. Vis.* **2017**, *12*, 121–128. [CrossRef]

24. Neethirajan, S. Recent advances in wearable sensors for animal health management. *Sens. Bio-Sens. Res.* **2017**, *12*, 15–29. [CrossRef]

25. Schleppe, J.B.; Lachapelle, G.; Booker, C.W.; Pittman, T. Challenges in the design of a GNSS ear tag for feedlot cattle. *Comput. Electron. Agric.* **2010**, *70*, 84–95. [CrossRef]

26. Ardö, H.; Guzhva, O.; Nilsson, M.; Herlin, A.H. Convolutional neural network-based cow interaction watchdog. *IET Comput. Vis.* **2017**, *12*, 171–177. [CrossRef]

27. Ju, M.; Choi, Y.; Seo, J.; Sa, J.; Lee, S.; Chung, Y.; Park, D. A Kinect-Based Segmentation of Touching-Pigs for Real-Time Monitoring. *Sensors* **2018**, *18*, 1746. [CrossRef]

28. Krizhevsky, A.; Sutskever, I.; Hinton, G.E. Imagenet classification with deep convolutional neural networks. In Proceedings of the Advances in Neural Information Processing Systems, Lake Tahoe, NV, USA, 3–6 December 2012; pp. 1097–1105.

29. LeCun, Y.; Bottou, L.; Bengio, Y.; Haffner, P. Gradient-based learning applied to document recognition. *Proc. IEEE* **1998**, *86*, 2278–2324. [CrossRef]

30. Jia, Y.; Shelhamer, E.; Donahue, J.; Karayev, S.; Long, J.; Girshick, R.; Guadarrama, S.; Darrell, T. Caffe: Convolutional architecture for fast feature embedding. In Proceedings of the 22nd ACM International Conference on Multimedia, Orlando, FL, USA, 3–7 November 2014; pp. 675–678.

31. Kirk, D. NVIDIA CUDA software and GPU parallel computing architecture. In Proceedings of the 6th international symposium on Memory management, Montreal, QC, Canada, 21–22 October, 2007; Volume 7, pp. 103–104.

32. Everingham, M.; Eslami, S.A.; Van Gool, L.; Williams, C.K.; Winn, J.; Zisserman, A. The pascal visual object classes challenge: A retrospective. *Int. J. Comput. Vis.* **2015**, *111*, 98–136. [CrossRef]

33. Lin, T.Y.; Maire, M.; Belongie, S.; Hays, J.; Perona, P.; Ramanan, D.; Dollár, P.; Zitnick, C.L. Microsoft coco: Common objects in context. In Proceedings of the European Conference on Computer Vision, Zurich, Switzerland, 6–12 September 2014; pp. 740–755.

34. Cordts, M.; Omran, M.; Ramos, S.; Rehfeld, T.; Enzweiler, M.; Benenson, R.; Franke, U.; Roth, S.; Schiele, B. The cityscapes dataset for semantic urban scene understanding. In Proceedings of the IEEE Conference on Computer Vision and Pattern Recognition, Las Vegas, NV, USA, 27–30 June 2016; pp. 3213–3223.

35. Redmon, J.; Divvala, S.; Girshick, R.; Farhadi, A. You only look once: Unified, real-time object detection. In Proceedings of the IEEE Conference on Computer Vision and Pattern Recognition, Las Vegas, NV, USA, 27–30 June 2016; pp. 779–788.

36. Redmon, J.; Farhadi, A. YOLO9000: Better, Faster, Stronger. In Proceedings of the IEEE Conference on Computer Vision and Pattern Recognition, Honolulu, HI, USA, 21–26 July 2017; pp. 7263–7271.

37. Liu, W.; Anguelov, D.; Erhan, D.; Szegedy, C.; Reed, S.; Fu, C.Y.; Berg, A.C. Ssd: Single shot multibox detector. In Proceedings of the European Conference on Computer Vision, Amsterdam, The Netherlands, 8–16 October 2016; Springer: Berlin/Heidelberg, Germany, 2016; pp. 21–37.

38. Girshick, R. Fast r-cnn. In Proceedings of the IEEE International Conference on Computer Vision, Santiago, Chile, 7–13 December 2015; pp. 1440–1448.

39. Ren, S.; He, K.; Girshick, R.; Sun, J. Faster r-cnn: Towards real-time object detection with region proposal networks. In Proceedings of the 28th International Conference on Neural Information Processing Systems, Montreal, QC, Canada, 7–12 December 2015; pp. 91–99.

40. Di Stefano, L.; Bulgarelli, A. A simple and efficient connected components labeling algorithm. In Proceedings of the 10th International Conference on Image Analysis and Processing, Venice, Italy, 27–29 September 1999; p. 322.

41. Otsu, N. A threshold selection method from gray-level histograms. *IEEE Trans. Syst. Man Cybern.* **1979**, *9*, 62–66. [CrossRef]

42. Kashiha, M.A.; Bahr, C.; Ott, S.; Moons, C.P.; Niewold, T.A.; Tuyttens, F.; Berckmans, D. Automatic monitoring of pig locomotion using image analysis. *Livest. Sci.* **2014**, *159*, 141–148. [CrossRef]

43. Nasirahmadi, A.; Richter, U.; Hensel, O.; Edwards, S.; Sturm, B. Using machine vision for investigation of changes in pig group lying patterns. *Comput. Electron. Agric.* **2015**, *119*, 184–190. [CrossRef]

44. Ahrendt, P.; Gregersen, T.; Karstoft, H. Development of a real-time computer vision system for tracking loose-housed pigs. *Comput. Electron. Agric.* **2011**, *76*, 169–174. [CrossRef]

45. Nilsson, M.; Ardö, H.; Åström, K.; Herlin, A.; Bergsten, C.; Guzhva, O. Learning based image segmentation of pigs in a pen. In Proceedings of the Visual Observation and Analysis of Vertebrate and Insect Behavior, Stockholm, Sweden, 24–28 August 2014; pp. 1–4.

46. Kongsro, J. Estimation of pig weight using a Microsoft Kinect prototype imaging system. *Comput. Electron. Agric.* **2014**, *109*, 32–35. [CrossRef]

47. Zhu, Q.; Ren, J.; Barclay, D.; McCormack, S.; Thomson, W. Automatic Animal Detection from Kinect Sensed Images for Livestock Monitoring and Assessment. In Proceedings of the 2015 IEEE International Conference on Computer and Information Technology, Ubiquitous Computing and Communications, Dependable, Autonomic and Secure Computing, Pervasive Intelligence and Computing, Liverpool, UK, 26–28 October 2015; pp. 1154–1157.

48. Stavrakakis, S.; Li, W.; Guy, J.H.; Morgan, G.; Ushaw, G.; Johnson, G.R.; Edwards, S.A. Validity of the Microsoft Kinect sensor for assessment of normal walking patterns in pigs. *Comput. Electron. Agric.* **2015**, *117*, 1–7. [CrossRef]

49. Lee, J.; Jin, L.; Park, D.; Chung, Y. Automatic Recognition of Aggressive Behavior in Pigs Using a Kinect Depth Sensor. *Sensors* **2016**, *16*, 631. [CrossRef]

50. Lao, F.; Brown-Brandl, T.; Stinn, J.; Liu, K.; Teng, G.; Xin, H. Automatic recognition of lactating sow behaviors through depth image processing. *Comput. Electron. Agric.* **2016**, *125*, 56–62. [CrossRef]

51. Choi, J.; Lee, L.; Chung, Y.; Park, D. Individual Pig Detection Using Kinect Depth Information. *Kips Trans. Comput. Commun. Syst.* **2016**, *5*, 319–326. [CrossRef]

52. Kim, J.; Chung, Y.; Choi, Y.; Sa, J.; Kim, H.; Chung, Y.; Park, D.; Kim, H. Depth-Based Detection of Standing-Pigs in Moving Noise Environments. *Sensors* **2017**, *17*, 2757. [CrossRef]

53. Pezzuolo, A.; Guarino, M.; Sartori, L.; González, L.A.; Marinello, F. On-barn pig weight estimation based on body measurements by a Kinect v1 depth camera. *Comput. Electron. Agric.* **2018**, *148*, 29–36. [CrossRef]

54. Fernandes, A.; Dórea, J.; Fitzgerald, R.; Herring, W.; Rosa, G. A novel automated system to acquire biometric and morphological measurements, and predict body weight of pigs via 3D computer vision. *J. Anim. Sci.* **2018**, *97*, 496–508. [CrossRef]

55. Matthews, S.G.; Miller, A.L.; Plötz, T.; Kyriazakis, I. Automated tracking to measure behavioural changes in pigs for health and welfare monitoring. *Sci. Rep.* **2017**, *7*, 17582. [CrossRef]

56. Girshick, R.; Donahue, J.; Darrell, T.; Malik, J. Rich feature hierarchies for accurate object detection and semantic segmentation. In Proceedings of the IEEE Conference on Computer Vision and Pattern Recognition, Portland, OR, USA, 23–28 June 2014; pp. 580–587.

57. He, K.; Gkioxari, G.; Dollár, P.; Girshick, R. Mask r-cnn. In Proceedings of the 2017 IEEE International Conference on Computer Vision (ICCV), Venice, Italy, 22–29 Ocotober 2017; pp. 2980–2988.

58. Long, J.; Shelhamer, E.; Darrell, T. Fully convolutional networks for semantic segmentation. In Proceedings of the IEEE Conference on Computer Vision and Pattern Recognition, Boston, MA, USA, 7–12 June 2015; pp. 3431–3440.

59. Cao, Z.; Simon, T.; Wei, S.E.; Sheikh, Y. Realtime Multi-Person 2D Pose Estimation using Part Affinity Fields. In Proceedings of the IEEE Conference on Computer Vision and Pattern Recognition, Honolulu, HI, USA, 21–26 July 2017.

60. Papandreou, G.; Zhu, T.; Chen, L.C.; Gidaris, S.; Tompson, J.; Murphy, K. PersonLab: Person Pose Estimation and Instance Segmentation with a Bottom-Up, Part-Based, Geometric Embedding Model. In Proceedings of the European Conference on Computer Vision, Munich, Germany, 8–14 September 2018.

61. Newell, A.; Yang, K.; Deng, J. Stacked hourglass networks for human pose estimation. In Proceedings of the European Conference on Computer Vision, Zurich, Switzerland, 6–12 September 2016; Springer: Berlin/Heidelberg, Germany, 2016; pp. 483–499.

62. Badrinarayanan, V.; Kendall, A.; Cipolla, R. SegNet: A Deep Convolutional Encoder-Decoder Architecture for Image Segmentation. *IEEE Trans. Pattern Anal. Mach. Intell.* **2017**, *39*, 2481–2495. [CrossRef]

63. Huang, G.; Liu, Z.; Van Der Maaten, L.; Weinberger, K.Q. Densely connected convolutional networks. In Proceedings of the CVPR, Honolulu, HI, USA, 21–26 July 2017; Volume 1, p. 3.

64. Ronneberger, O.; Fischer, P.; Brox, T. U-net: Convolutional networks for biomedical image segmentation. In Proceedings of the International Conference on Medical Image Computing and Computer-Assisted Intervention, Munich, Germany, 5–9 Ocotber 2015; Springer: Berlin/Heidelberg, Germany, 2015; pp. 234–241.

65. Luo, W.; Li, Y.; Urtasun, R.; Zemel, R. Understanding the effective receptive field in deep convolutional neural networks. In Proceedings of the 30th International Conference on Neural Information Processing Systems, Barcelona, Spain, 5–10 December 2016; pp. 4905–4913.

66. *MATLAB*, version 9.5.0 (R2018b); The MathWorks Inc.: Natick, MA, USA, 2018.

67. Deng, J.; Dong, W.; Socher, R.; Li, L.J.; Li, K.; Fei-Fei, L. Imagenet: A large-scale hierarchical image database. In Proceedings of the 2009 IEEE Conference on Computer Vision and Pattern Recognition (CVPR 2009), Miami, FL, USA, 20–25 June 2009; pp. 248–255.

68. Chen, L.C.; Papandreou, G.; Kokkinos, I.; Murphy, K.; Yuille, A.L. Deeplab: Semantic image segmentation with deep convolutional nets, atrous convolution, and fully connected crfs. *IEEE Trans. Pattern Anal. Mach. Intell.* **2018**, *40*, 834–848. [CrossRef]

69. Chen, L.C.; Zhu, Y.; Papandreou, G.; Schroff, F.; Adam, H. Encoder-Decoder with Atrous Separable Convolution for Semantic Image Segmentation. In Proceedings of the ECCV, Munich, Germany, 8–14 September 2018.

70. Luo, W.; Xing, J.; Milan, A.; Zhang, X.; Liu, W.; Zhao, X.; Kim, T.K. Multiple object tracking: A literature review. *arXiv* **2014**, arXiv:1409.7618.

Article

Monitoring of the Pesticide Droplet Deposition with a Novel Capacitance Sensor

Pei Wang [1,2,*,†], **Wei Yu** [2,†], **Mingxiong Ou** [1,2], **Chen Gong** [1,2] **and Weidong Jia** [1,2,*]

[1] Key Laboratory of Modern Agricultural Equipment and Technology, Ministry of Education of PRC, Jiangsu University, Zhenjiang 212300, China; myomx@ujs.edu.cn (M.O.); chengong@ujs.edu.cn (C.G.)
[2] Key Laboratory of Plant Protection Engineering, Ministry of Agriculture and Rural Affairs of PRC, Jiangsu University, Zhenjiang 212300, China; 2211616028@stmail.ujs.edu.cn
* Correspondence: wangpei@live.cn (P.W.); jiaweidong@ujs.edu.cn (W.J.); Tel.: +86-188-528-98152 (P.W.)
† The first two authors Pei Wang and Wei Yu contributed same to this publication as the co-first authors. Their role orders in this publication should be regarded equally as the first authors.

Received: 25 December 2018; Accepted: 24 January 2019; Published: 28 January 2019

Abstract: Rapid detection of spraying deposit can contribute to the precision application of plant protection products. In this study, a novel capacitor sensor system was implemented for measuring the spray deposit immediately after herbicide application. Herbicides with different formulations and nozzles in different mode types were included to test the impact on the capacitance of this system. The results showed that there was a linear relationship between the deposit mass and the digital voltage signals of the capacitance on the sensor surface with spray droplets. The linear models were similar for water and the spray mixtures with non-ionized herbicides usually in formulations of emulsifiable concentrates and suspension concentrates. However, the ionized herbicides in formulation of aqueous solutions presented a unique linear model. With this novel sensor, it is possible to monitor the deposit mass in real-time shortly after the pesticide application. This will contribute to the precision application of plant protection chemicals in the fields.

Keywords: capacitor sensor; deposit mass; pesticide droplets; formulations; ionization

---

1. Introduction

The usage of pesticides has increased significantly in the last two decades in China [1]. Meanwhile, its average application dose is 2.5 times of the global average level [2,3]. The over usage of pesticides will cause unnecessary invest for the farmers, leading to more residue of the pesticide ingredients in the food products and the soil, and damage the eco-system as well [4–6] Furthermore, it can also induce the resistant property of the weeds, insects, and diseases, which makes the pest management strategies more complicated [7,8]. Thus, the effective approaches to monitor the droplet deposition doses during the pesticide application are in urgent demands.

To enhance the assessment efficiency of the deposition quality for the spray, several approaches have been introduced based on the image processing technology [9]. Various systems have been open for users to identify the pesticide deposition effect by analyzing the images of the water sensitive paper after the application, for instance, the Swath Kit [10], the USDA Image Analyzer [11], the Droplet Scan [12], the Deposit Scan [13], the Image J [14], and the Drop Vision-Ag [15]. Most of the new technologies are applied for measuring the coverage, droplet density, and or the droplet sizes. Few methodologies have been reported to evaluate the droplet deposition doses after pesticide application.

Conventionally, the droplet deposition dose is measured based on the elution procedure [16]. Pigments like the poinsettia and the methylene blue are usually selected to replace the pesticide for the spray preparation. After the application, plant leaves or the Petri dishes will be collected and the pigment ingredients will be eluted with deionized water from the leaf surfaces or the Petri dishes. Then the deposition dose can be calculated from the concentration of the eluent by the measurement with a spectrophotometer [17]. Salyani and Serdynski [18] reported a prototype sensor for the spray deposit monitoring based on the measurement of the electric capacitance of the parallel copper conductors with spray deposited between the copper gaps. Thus, it could indicate the mass information of the droplet deposit with the voltage signals, which enables the real-time and rapid measurements of the pesticide deposition doses. Recently, Zhang et al. [19] applied a similar system for droplet deposition evaluation of the aerial spraying. Meanwhile, a fringing capacitive sensor with similar principles was also applied for water content measurement in the field [20]. However, the prototype equipment was tested with the electrolyte solution with high conductivity, such as NaCl and NaOH, which is not a common feature of the typical spray mixture. As is known, besides the water solvent and the emulsion in water with the ingredients as ionic compound, most pesticide formulations with the organic ingredients are hardly ionized in the water mixture. Thus, further studies are required to verify the fitness and the measurement accuracy of this novel sensor before its application in the agricultural practice.

The objectives of this study are, firstly, to test the sensors application capability on the deposition dose measurement of spray mixtures with pesticide in different formulations; secondly, to evaluate the measuring accuracy of droplet deposition doses of spray mixture with several common pesticide formulations including the emulsifiable concentrates (EC), suspension concentrates (SC), and aqueous solutions (AS).

2. Materials and Methods

2.1. Implementation of the Droplet Deposit Sensing System

A leaf like capacitor (Yingtai Tech., Tianjin, China) was adopted for the implementation of the droplet deposit sensing system in this study. The sensor was designed based on a capacitor with 84 parallel coppers (Figure 1). The coppers were separated into two groups and connected respectively as two electrode plates of the capacitor. The whole structure of the capacitor was packaged (painted) with insulation material of ceramic. Lim et al. has presented that, the capacitance varies according to the dielectric constant of the media composition, the air or the spray, inside the gap of the electrodes [21]. The dielectric constant changes when the ratio of each component in the media composition varies. The dielectric constant can be calculated due to the equation

$$C_s = \frac{\varepsilon_a S_a + \varepsilon_s S_s}{d} \tag{1}$$

where C_s is the capacitance of the capacitor with droplets depositing onside, ε_a is the dielectric constant of the air, ε_s is the dielectric constant of the spray mixture, S_a is the surface area of the electrode contacting to the air, S_s is the surface area of the electrode contacting to the spray droplet, and d is the distance between the copper electrodes.

When considering the total area of the electrodes (S), which should be the sum of S_a and S_S, the equation of the dielectric constant can be revised as

$$C_s = \frac{\varepsilon_a(S - S_s) + \varepsilon_s S_s}{d} = \frac{\varepsilon_a S}{d} + \frac{(\varepsilon_s - \varepsilon_a)S_s}{d} = C_0 + \frac{(\varepsilon_s - \varepsilon_a)S_s}{d} \tag{2}$$

where, C_0 is a constant for each sensor.

Theoretically, the surface area of the electrode contacting to the spray droplet (S_S) can be calculated from the droplets' mass (m_s) and density (ρ_s). Then, the deposit mass of the spray can be calculated following the equations

$$m = \rho_s S_s d = \rho_s \frac{(C_s - C_0)d}{\varepsilon_s - \varepsilon_a} d = C_s \frac{\rho_s d^2}{\varepsilon_s - \varepsilon_a} - \frac{\rho_s C_0 d^2}{\varepsilon_s - \varepsilon_a} \tag{3}$$

Thus, a linear model can be fitted to calculate the deposit mass of the spray according to the measurement of the capacitance of the capacitor with droplets depositing onside.

With the leaf like capacitor adopted in this study, it could provide the analog signals to indicate its capacitance. The sensor is implemented with an operational amplifier circuit as shown in Figure 1d. With a capacitor in fixed capacitance, the leaf like capacitance sensor will share different voltages when the dielectric constant inside it is changed. The capacitor was linked to an analog to digital converter (ADS1115, Texas Instruments, Dallas, TX, USA), which could convert the analog signal of the real time capacitance into the digital voltage signals. The digital signal was then processed by a 32-bit microcontroller (STM32F4 EXPLORER, ST Microelectronics, Geneva, Switzerland). The microcontroller could read the input digital signal and display the voltage information representing the analog signal on a thin film transistor-liquid crystal display (TFT screen). Linear models of deposit mass to the voltage signals were developed due to different herbicide spray in this research and installed in the microcontroller. Thus, the microcontroller could calculate the deposit results of each measurement and then display that on the TFT screen. All the measurement results were recorded with a trans-flash Card (SD card) which was mounted on the microcontroller processing board. Figure 1 presents the structure of the system.

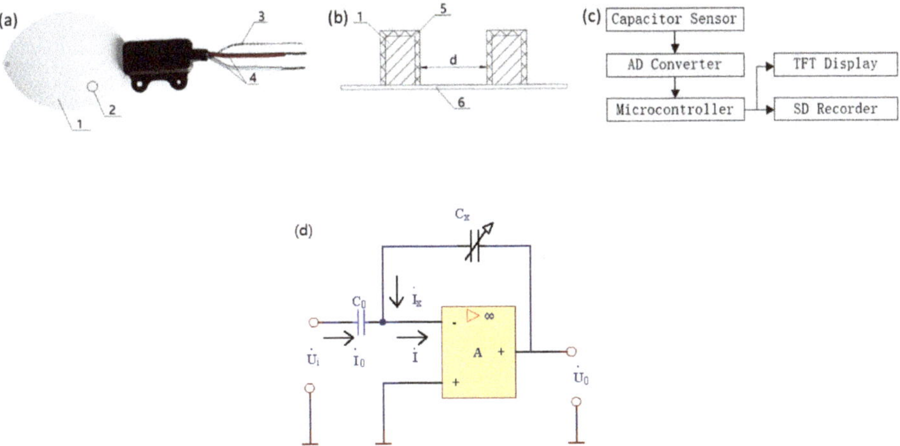

Figure 1. The structure of the leaf-like capacitor sensor and the deposit monitoring system. (**a**) Is the leaf-like sensor. (**b**) Is the electrode structure on the resin board. (**c**) Is the implementation diagram of the deposit monitoring system. (**d**) Is an electric schematic of the circuit of the leaf like sensor. 1 = electrodes, 2 = capacitor, 3 = data cable, 4 = power cable, 5 = insulating coating, 6 = resin board. The sensor is in 11.2 cm of length, 5.8 cm of width and 0.075 cm of thickness. The distance between each electrode is 1.58–1.78 mm. The width of the electrode is 0.59–0.79 mm and the thickness is 0.01–0.02 mm. The thickness of the insulating layer is about 2 μm. In (**d**), C_0 is a fixed capacitor, C_x is the leaf like sensor.

2.2. Experimental Setup and Processing

To develop the models of deposit mass to the voltage signals, experiments were conducted to measure the voltage signals of the leaf like sensor with different herbicide spray depositing onside. Herbicides in different formulation types were selected for the spray preparation with the suggested concentration on the product labels including the Yudasheng[®] (SC, 20% a.i. atrazine, Xinnong Guotai, Beijing, China), Lvlilai Butachlor[®] (EC, 50% a.i. butachlor, Lvlilai, Suzhou, China), Ruidefeng Lanhuoyan[®] (AS, 41% a.i. glyphosate isopropylamine salt, Noposion, Shenzhen, China), as well as the water for control. The sprays were prepared according to the details listed in Table 1. A hanging orbit sprayer (Figure 2) was used for the herbicide application. The application pressure was set at 0.3 MPa, while the application height was set at 50 cm above the sensor as is common in field conditions. Standard flat fan nozzles with different sizes were selected for the spraying (Lechler® ST 110-01/015/02/03, Lechler GmbH, Metzingen, Germany). Before spraying onto the sensors, the medium droplet sizes were measured for each nozzle with a laser particle size analyzer (Winner® 318B, Winner Particle Instrument, Ji'nan, China). During the testing experiment of the sensor, the moving velocity of the nozzles was fixed for three repeated measurements of each treatment and then varied randomly to obtain different deposit on the sensor surface. For each application, the signals from the sensor were recorded for 40 times over three seconds after the application. Mass of the sensor was measured with an electrical scale (YP102N, Sop-Top, Shanghai, China) before and after application. The deposit mass was calculated from the two mass measurements.

Table 1. Preparation of the sprays of the experiment

No	Formulation	Ingredient	Product Amount	Solution/Water Volume (mL)
1		water		5000
2	SC	atrazine	75 mL	5000
3	EC	butachlor	50 g	5000
4	AS	glyphosate isopropylamine salt	225 mL	5000

Figure 2. The hanging orbit sprayer for herbicide application in the experiment.

2.3. Data Analysis

The data analysis was processed with R Studio 0.98.490 [22]. The linear model was employed to fit the relationship between the voltage signals and the deposit mass of the spray. Analysis of variance (ANOVA) was carried out for the evaluation of differences between each treatment. Data were tested for normal distribution using the Shapiro-Wilk test ($p > 0.05$). Equality for heterogeneity of variances was tested using Levene's test for each treatment ($p > 0.05$).

3. Results and Discussion

3.1. Measurement Impact Analysis

To evaluate the impact of application factors as the nozzle types (droplet sizes) and the formulation types of the herbicides on the sensor's measurement results, experiments were conducted. The measuring results were presented in Figures 3 and 4.

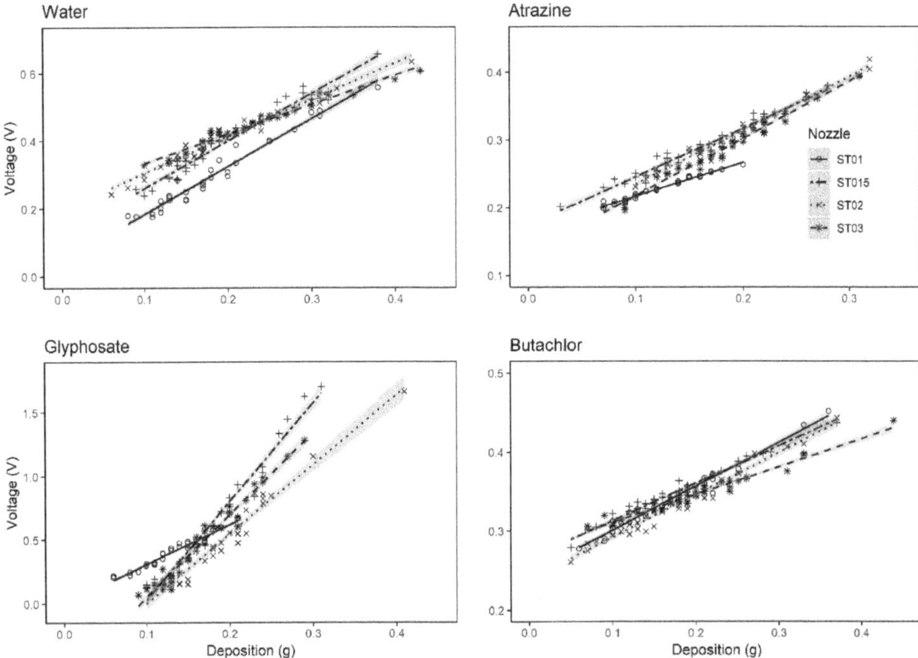

Figure 3. Linear regression models of the deposit data of sprays with herbicides in different formulations. The shadow area represents the standard error of the regressed linear models.

Linear models were employed for the regression of the signal voltage to deposit mass curves of each treatment. All the models were compared by ANOVA ($\alpha = 0.05$). It indicated that, when the nozzle type was fixed, there was no significant differences between the deposit to voltage curves of the spray mixture with EC (butachlor), SC (atrazine), and water. However, considering the AS (glyphosate), the voltage signal curve of the spray deposit was markedly different from the other groups. When the deposit is greater than 0.2 g on the leaf like sensor, the signal voltages were much higher of the sensor with AS droplets than the EC, SC, or water droplets. To explain this difference, the ionization property of the glyphosate isopropylamine salt should be concerned. The glyphosate is a weak acid herbicide. It usually exists as a salt compound and will divide into two ions with opposite charges (the cation and the anion) after being dissolved in the water, while the other herbicides with formulations like EC or SC usually contain organic ingredients which could not dissolve and ionize in the water. The concentration of ions would vary the dielectric constant of the spray mixture. As a result, the voltage signals of the electric capacity would be higher when the concentration of the glyphosate was increased.

In Figure 4, deposit curves of different nozzles were compared on each herbicide formulation. It presented that, for the spray mixture with same herbicide formulations, there was no significant differences of the deposit curves between the treatments with each nozzle type. Therefore, the spray mixture could be classified into two groups due to the ionization feature of the ingredients.

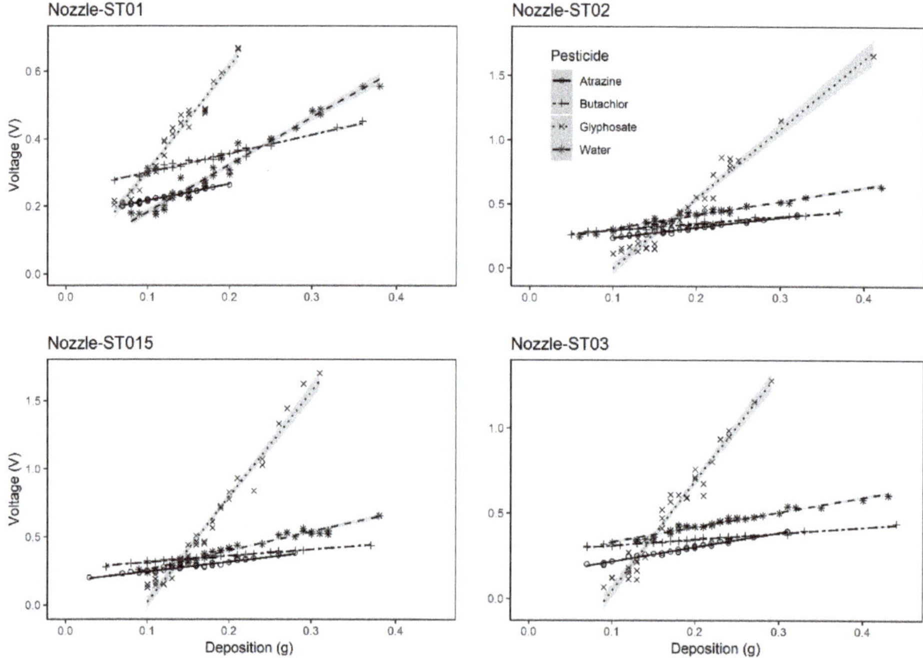

Figure 4. Linear regression models of the deposit data of sprays with droplet size spectrum generated from different nozzles. The shadow area represents the standard error of the regressed linear models.

3.2. Statistical Modeling

Considering the little differences of the deposit curves of water, SC and EC sprays, the data could be gathered for the modeling of the signal voltage to the deposit mass for all of the non-ionized herbicides, while the ionized herbicides should be tested for separated models. In this research, the glyphosate was taken as an example of the cases for ionized ingredients application. Since, the nozzle types did not show any impact on the signal voltages of the sensor measurement, the data of different herbicides but with the same nozzle type could be gathered for the modeling. Linear models were applied respectively for the regression of the two cases. The linear regression fitted well to the grouped data, which corresponding to the results of a former study by Zhang et al. [19].

With the linear model of the electric signals to the deposit of the non-ionized sprays, the algorithm could be programmed and installed in the microcontroller. Thus, the sensor was able to give precise information of the quantified spray deposit in the field. The dielectric constant of a dielectric material depends on the frequency of the applied electric field. The behavior of the dielectric constant is affected by three types of polarization including the orientation polarization, the ionic polarization and the electronic polarization. The ionized sprays may show all the three types of polarization, so that all polarization mechanisms are acting at lower frequencies till the water dipoles fail to follow the field alternations and so the orientation polar ceases to play, after which the polarization is dominated by ionic polarization. This continues until the ions' oscillations cannot follow the field alternation and it gets out the play at which the reaming will be the electronic polarization. When the frequency of the applied field equal to the resonance frequency of ionic movement, the dielectric constant show resonance behavior. The dielectric constant is proportional to density of the induced dipoles. Therefore, it is expected to increase with the salt concentration of the electrolyte [23]. Thus, the electric signal of ionized spray will be affected by the mount, concentration and other parameters of the electric charges in the mixture. Therefore, the model of one variable regression could not be used to simulate the application deposit with different ingredients. However, in this certain case of glyphosate application with the recommended doses on the product label, the linear regression model presented in Figure 5 could be adopted with the algorithm program similar to the one used for the deposit detection of the non-ionized sprays.

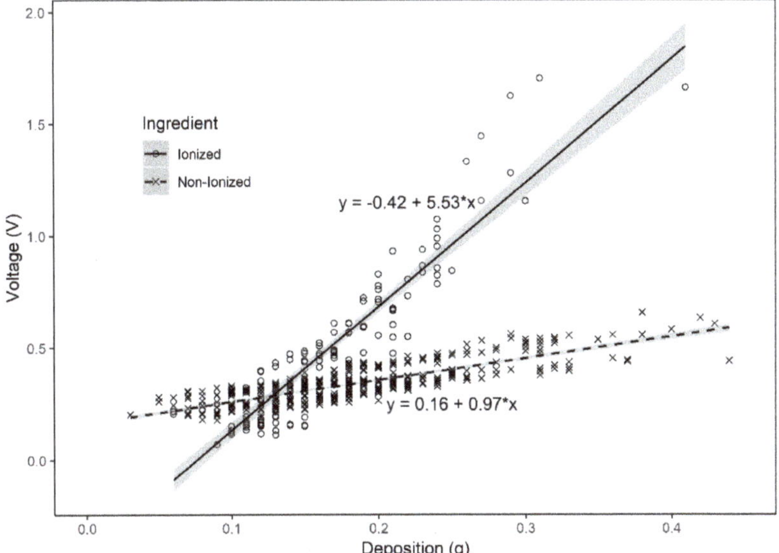

Figure 5. Linear regression models of the deposit data of sprays with ionized and non-ionized ingredients. The shadow area represents the standard error of the regressed linear models (ionized: $R^2 = 0.837$, non-ionized: $R^2 = 0.882$).

4. Conclusions

According to this study, we can conclude that the electric capacitor sensor has shown strong potential for its application on the real-time monitoring of the herbicide spraying deposit. The system and model implemented in this study can already be applied in the deposit measurement of the herbicide with non-ionized ingredients. For other water-soluble herbicides with ionized ingredients. However, further studies are required for the measuring model simulation, as the concentration and the valence of the ions will have impact on the dielectric constant of the spray mixture, which can directly affect the signal voltage representing the electric capacity of the sensor.

Author Contributions: Conceptualization, P.W.; Data curation, W.Y.; Formal analysis, P.W.; Funding acquisition, M.O. and W.J.; Investigation, W.Y.; Methodology, P.W., W.Y., M.O., and C.G.; Project administration, W.J.; Resources, C.G. and W.J.; Software, W.Y.; Supervision, P.W., M.O., and W.J.; Writing—original draft, P.W.; Writing—review & editing, P.W.

Funding: This research was funded by the National Key Research and Development Plan of China (grant number 2017YFD0700901), the National Natural Science Foundation of China (grant number 51475215, 31601676), and the Advanced Talent Research Funding of Jiangsu University (grant number 5501200004).

Acknowledgments: In addition, the authors would like to thank Huitao Zhou, Guangyang Mao, Shengnan Cao, Wanting Yang, and Shuai Zang for technical help in this study.

Conflicts of Interest: The authors declare no conflict of interest.

References

1.   National Bureau of Statistics of China. 2018. Available online: http://data.stats.gov.cn/easyquery.htm?cn=C01 (accessed on 25 December 2018).

2. Zhang, C.; Shi, G.; Shen, J.; Hu, R. Productivity effect and overuse of pesticide in crop production in China. *J. Integr. Agric.* **2015**, *14*, 1903–1910. [CrossRef]

3. Zhang, Y.; McCarl, B.A.; Luan, Y.; Kleinwechter, U. Climate change effects on pesticide usage reduction efforts: A case study in China. *Mitig. Adapt. Strateg. Glob. Chang.* **2018**, *23*, 685–701. [CrossRef]

4. Bonny, S. Herbicide-tolerant transgenic soybean over 15 years of cultivation: Pesticide use, weed resistance, and some economic issues. The case of the USA. *Sustainability* **2011**, *3*, 1302–1322. [CrossRef]

5. Myers, J.P.; Antoniou, M.N.; Blumberg, B.; Carroll, L.; Colborn, T.; Everett, L.G.; Hansen, M.; Landrigan, P.J.; Lanphear, B.P.; Mesnage, R.; et al. Concerns over use of glyphosate-based herbicides and risks associated with exposures: A consensus statement. *Environ. Health* **2016**, *15*, 19. [CrossRef] [PubMed]

6. Budzinski, H.; Couderchet, M. Environmental and human health issues related to pesticides: From usage and environmental fate to impact. *Environ. Sci. Pollut. Res.* **2018**, *25*, 14277–14279. [CrossRef] [PubMed]

7. Denholm, I.; Rowland, M.W. Tactics for managing pesticide resistance in arthropods: Theory and practice. *Annu. Rev. Entomol.* **1992**, *37*, 91–112. [CrossRef] [PubMed]

8. Jutsum, A.R.; Heaney, S.P.; Perrin, B.M.; Wege, P.J. Pesticide resistance: Assessment of risk and the development and implementation of effective management strategies. *Pest Manag. Sci.* **1998**, *54*, 435–446. [CrossRef]

9. Ferguson, J.C.; Chechetto, R.G.; O'Donnell, C.C.; Fritz, B.K.; Hoffmann, W.C.; Coleman, C.E.; Chauhan, B.S.; Adkins, S.W.; Kruger, G.R.; Hewitt, A.J. Assessing a novel smartphone application–SnapCard, compared to five imaging systems to quantify droplet deposition on artificial collectors. *Comput. Electron. Agric.* **2016**, *128*, 193–198. [CrossRef]

10. Mierzejewski, K. *Aerial Spray Technology: Possibilities and Limitations for Control of Pear Thrips in Towards Understanding Thysanoptera General Technical Report NE-147*; USDA Forest Service: New England, USA, 1991.

11. Hoffmann, W.C.; Hewitt, A.J. Comparison of three imaging systems for water sensitive papers. *Appl. Eng. Agric.* **2005**, *21*, 961–964. [CrossRef]

12. Wolf, R.E. Assessing the ability of Droplet Scan to analyze spray droplets from a ground operated sprayer. *Appl. Eng. Agric.* **2003**, *19*, 525.

13. Zhu, H.; Salyani, M.; Fox, R.D. A portable scanning system for evaluation of spray deposit distribution. *Comput. Electron. Agric.* **2011**, *76*, 38–43. [CrossRef]

14. Rasband, W.S. *Image J*; US National Institutes of Health: Bethesda, MD, USA, 2014.

15. Leading Edge Associates. DropVision®AG. 2015. Available online: https://www.leateam.com/product/129/1144/ (accessed on 25 December 2018).

16. Hewitt, A.J. Droplet size and agricultural spraying, Part I: Atomization, spray transport, deposition, drift, and droplet size measurement techniques. *At. Spray* **1997**, *7*, 235–244. [CrossRef]

17. Hewitt, A.J. Tracer and collector systems for field deposition research. *Aspects Appl. Biol.* **2010**, *99*, 283–289.

18. Salyani, M.; Serdynski, J. Development of a sensor for spray deposition assessment. *Trans. ASAE* **1990**, *33*, 1464–1468. [CrossRef]

19. Zhang, R.; Chen, L.; Lan, Y.; Xu, G.; Kan, J.; Zhang, D. Development of a deposit sensing system for aerial spraying application. *Trans. CSAM* **2014**, *45*, 123–127.

20. Da Costa, E.F.; De Oliveira, N.E.; Morais, F.J.O.; Carvalhaes-Dias, P.; Duarte, L.F.C.; Cabot, A.; Siqueira Dias, J.A. A self-powered and autonomous fringing field capacitive sensor integrated into a micro sprinkler spinner to measure soil water content. *Sensors* **2017**, *17*, 575. [CrossRef] [PubMed]

21. Lim, L.G.; Pao, W.K.; Hamid, N.H.; Tang, T.B. Design of helical capacitance sensor for holdup measurement in two-phase stratified flow: A sinusoidal function approach. *Sensors* **2016**, *16*, 1032. [CrossRef] [PubMed]

22. R Development Core Team. *R: A Language and Environment for Statistical Computing*; R Foundation for Statistical Computing: Vienna, Austria, 2013; ISBN 3-900051-07-0.

23. Föll, H. Chapter 3: Dielectrics. Available online: https://www.tf.uni-kiel.de/matwis/amat/elmat_en/kap_3/backbone/r3_3_5.html (accessed on 16 January 2019).

MDPI

St. Alban-Anlage 66

4052 Basel

Switzerland

Tel. +41 61 683 77 34

Fax +41 61 302 89 18

www.mdpi.com

Sensors Editorial Office

E-mail: sensors@mdpi.com

www.mdpi.com/journal/sensors